Automobility in Transition?

Routledge Studies in Sustainability Transitions

SERIES EDITORS: JOHAN SCHOT, JOHN GRIN AND JAN ROTMANS

1 Transitions to Sustainable
  Development
  New Directions in the Study
  of Long Term Transformative
  Change
  *John Grin, Jan Rotmans and
  Johan Schot
  In collaboration with Frank Geels
  and Derk Loorbach*

2 Automobility in Transition?
  A Socio-Technical Analysis of
  Sustainable Transport
  *Edited by Frank W. Geels,
  René Kemp, Geoff Dudley
  and Glenn Lyons*

Automobility in Transition?
A Socio-Technical Analysis of Sustainable Transport

Edited by Frank W. Geels, René Kemp, Geoff Dudley and Glenn Lyons

Routledge
Taylor & Francis Group
NEW YORK LONDON

First published 2012
by Routledge
711 Third Avenue, New York, NY 10017

Simultaneously published in the UK
by Routledge
2 Park Square, Milton Park, Abingdon, Oxon OX14 4RN

Routledge is an imprint of the Taylor & Francis Group, an informa business

© 2012 Taylor & Francis

The right of Frank W. Geels, René Kemp, Geoff Dudley and Glenn Lyons to be identified as the authors of the editorial material, and of the authors for their individual chapters, has been asserted in accordance with sections 77 and 78 of the Copyright, Designs and Patents Act 1988.

Typeset in Sabon by IBT Global.

All rights reserved. No part of this book may be reprinted or reproduced or utilised in any form or by any electronic, mechanical, or other means, now known or hereafter invented, including photocopying and recording, or in any information storage or retrieval system, without permission in writing from the publishers.

Trademark Notice: Product or corporate names may be trademarks or registered trademarks, and are used only for identification and explanation without intent to infringe.

Library of Congress Cataloging-in-Publication Data
  Automobility in transition? : a socio-technical analysis of sustainable transport / edited by Frank W. Geels . . . [et al.].
    p. cm. — (Routledge studies in sustainability transitions ; 2)
  Includes bibliographical references and index.
  1. Transportation, Automotive—Environmental
aspects. 2. Automobiles—Environmental aspects. 3. Automobiles—Social aspects. 4. Transportation and state. 5. Sustainable development. I. Geels, Frank W., 1971–
  HE5611.A983 2012
  388.3'21—dc23
  2011030241

ISBN13: 978-0-415-88505-8 (hbk)

Contents

List of Figures ix
List of Tables xi
Preface xiii
FRANK W. GEELS, RENÉ KEMP, GEOFF DUDLEY AND GLENN LYONS

PART I
The Transition Perspective and Problems Associated With Car Mobility

1 Introduction: Sustainability Transitions in the Automobility Regime and the Need for a New Perspective 3
   RENÉ KEMP, FRANK W. GEELS AND GEOFF DUDLEY

2 Visions for the Future and the Need for a Social Science Perspective in Transport Studies 29
   GLENN LYONS

3 The Multi-Level Perspective as a New Perspective for Studying Socio-Technical Transitions 49
   FRANK W. GEELS AND RENÉ KEMP

PART II
Stability and Regimes Pressures

4 The Dynamics of Regime Strength and Instability: Policy Challenges to the Dominance of the Private Car in the United Kingdom 83
   GEOFF DUDLEY AND KIRON CHATTERJEE

vi *Contents*

5 The Governance of Transport Policy 104
IAIN DOCHERTY AND JON SHAW

6 The Nature and Causes of Inertia in the Automotive Industry: Regime Stability and Non-Change 123
PETER WELLS, PAUL NIEUWENHUIS AND RENATO J. ORSATO

7 Providing Road Capacity for Automobility: The Continuing Transition 140
PHIL GOODWIN

8 A Socio-Spatial Perspective on the Car Regime 160
TOON ZIJLSTRA AND FLOR AVELINO

9 The Emergence of New Cultures of Mobility: Stability, Openings and Prospects 180
MIMI SHELLER

PART III
Dynamics of Change

10 The Electrification of Automobility: The Bumpy Ride of Electric Vehicles Toward Regime Transition 205
RENATO J. ORSATO, MARC DIJK, RENÉ KEMP AND MASARU YARIME

11 Introducing Hydrogen and Fuel Cell Vehicles in Germany 229
OLIVER EHRET AND MARLOES DIGNUM

12 Transition by Translation: The Dutch Traffic Intelligence Innovation Cascade 250
BONNO PEL, GEERT TEISMAN AND FRANK BOONS

13 The Emergent Role of User Innovation in Reshaping Traveler Information Services 268
GLENN LYONS, JULIET JAIN, VAL MITCHELL AND ANDREW MAY

14 Innovation in Public Transport 286
REG HARMAN, WIJNAND VEENEMAN AND PETER HARMAN

Contents   vii

15 Intermodal Personal Mobility: A Niche Caught Between Two
    Regimes   308
    GRAHAM PARKHURST, RENÉ KEMP, MARC DIJK AND HENRIETTA SHERWIN

16 Findings, Conclusions and Assessments of Sustainability
    Transitions in Automobility   335
    FRANK W. GEELS, GEOFF DUDLEY AND RENÉ KEMP

    Contributors   375
    Index   385

Figures

1.1	Passenger travel per capita by mode.	7
1.2	Congestion pressure on Dutch highways.	9
1.3	Total transport and road transport, European Union.	9
2.1	First white line in London.	32
2.2	The 'system of systems' embodied by the pursuit or application of intelligence.	43
3.1	Multiple levels as a nested hierarchy.	52
3.2	Emerging trajectory carried by local projects.	54
3.3	Meta-coordination through socio-technical regimes.	55
3.4	A dynamic multi-level perspective on system innovations.	57
3.5	From sailing ships to steamships through add-on and hybridization.	61
3.6	The Benz Velo horseless carriage.	64
3.7	Fit-stretch pattern in the co-evolution of form and function.	64
3.8	The hype-cycle.	66
3.9	Interacting hype-cycles in societal and policy debates on Dutch renewable energy.	66
3.10	Trajectory of niche-accumulation.	67
3.11	Trajectory of niche-accumulation for the diffusion of steamships.	68
4.1	Average distance (in miles) traveled per person per year in Great Britain.	84
4.2	Households with regular use of a car.	89
6.1	The automotive industry landscape, regime and niches.	124
7.1	Length of road in Great Britain, 1914–2009.	141
7.2	Length of motorway in Great Britain, 1950–2009.	141
7.3	The evolution of public support for a new policy idea.	149
7.4	Car traffic growth compared with 1989 and current official forecasts.	154

Figures

7.5   A turning point in traffic intensity showing decoupling since the early 1990s.   155
8.1   Multi-level analysis of automobility from socio-spatial perspective.   173
9.1   Dominant cultural structure for mobility.   182
10.1   Number of prototypes for hydrogen fuel cell vehicles and electric vehicles.   209
10.2   Influences on the electrification of automobility after 2005.   211
10.3   Announced national electric vehicle and plug-in hybrid-electric vehicle sales targets   213
10.4   Fit-stretch pattern for power train technologies.   222
11.1   National organization hydrogen and fuel cell technology.   233
13.1   CycleStreets.   274
13.2   Parkatmyhouse.com—http://www.parkatmyhouse.co.uk/.   275
13.3   PickupPal.   277
13.4   MyBikeLane.   278
14.1   The public transport regime.   302
15.1   Main mode in intermodal trips in the Netherlands.   316
15.2   Previous transport choice before use of OV-fiets.   323
15.3   Conceptual models of intermodal journey types.   325
15.4   Concept for park and ride with bus service as the main mode for serving multiple destinations   326
16.1   Hype-cycles for green car propulsion technologies.   345

Tables

2.1	Summaries of the Four Intelligent Infrastructure Systems Scenarios	39
2.2	Key Features That Have Shaped Images of the Future Role of New Technologies in Transport	45
6.1	Production Volume and Percentage Share of Total for All Motor Vehicles, 1998 and 2008	130
6.2	Production Volume and Percentage Share of Total for Cars, 1998 and 2008	130
6.3	Low Carbon Vehicles Currently Available (< 100 g/km CO^2 on NEDC)	135
9.1	Cultural Dimensions of the Mobility System	183
10.1	Key Differences Between PHEVs and EVs	223
14.1	Overall Proportions of Passenger Kilometers Traveled by Mode in Percentages	290
15.1	Percentages of Passenger Kilometers Modal Share of Principal Motorized Land Modes	315
15.2	Distances in Kilometers Traveled on Foot and by Bicycle per Person per Annum	315
15.3	Main Access Mode to UK Rail Stations for Certain Journey Purposes	317
16.1	Summary of Main Conclusions From the Book's Empirical Chapters	337
16.2	Factors That Are Important to Consumers in Deciding Which Car to Buy	354

Preface
Frank W. Geels, René Kemp, Geoff Dudley and Glenn Lyons

This book is part of a series on 'transitions to sustainable development', which aims to combine new theories of large-scale system change with empirical studies of transitions in the energy, transport, agri-food and health care domains. This book offers a socio-technical analysis of the transition to sustainable mobility. The title *Automobility in Transition? A Socio-Technical Analysis of Sustainable Transport* refers to three important characteristics of the book. First, the book focuses on the automobility system. Because car-based mobility is dominant in developed countries, transitions to sustainable transport cannot avoid taking automobility into account. But the book also pays attention to transport modes, such as bus, train, tram and taxi, which can interact with the automobility system in ways that reduce car-based mobility. We pay less attention, however, to transport modes that have limited interactions with automobility, such as walking, cycling, aviation and shipping.

Second, the title contains a question mark, because we do not take for granted that a transition to sustainable transport will happen. Transport and automobility may well be the 'hardest case', because there are many stabilizing mechanisms and secular trends that point in the direction of more, not less, mobility. So, it is an open question if and how fast a transition to sustainable mobility can happen on the ground. If a transition will take place, a further question is what kind of path will it follow? Will a future sustainable transport system be based on 'green' cars? Or will this system look very different from current transport systems, with intermodal linkages between various sub-systems and less prominence for cars?

Third, the book makes a socio-technical analysis, which aims to bridge the contemporary split in transport studies, which tends to focus either on technology fix or on behavior change. The book thus aims to go beyond disciplines such as engineering, planning, economics and psychology, which prevail in transport studies. While these disciplines remain important, this book also draws on insights from cultural studies, business studies, innovation studies, political science, critical theory and sociology. Most importantly, the book is interested in *interactions* between various dimensions and analyzes the activities, beliefs, motivations, strategies and resources of various social groups, the moves they make and how they respond to each other. The book therefore not only looks at the technical dimension of transport systems, but also at policies and their embedding in broader governance systems, cultural discourses and images around established and new mobility practices, innovations in traffic management, innovative user behavior in response to transport information devices, experiments with intermodal transport, urban planning and spatial innovations, mobility schemes and so forth.

These three characteristics make this book different from others on this topic. The book attempts to speak to two audiences simultaneously, those of transport studies and transition studies. The book offers a socio-technical 'transition theory', which indicates how transitions come about through co-evolutionary processes and multi-level alignments. Most chapters in the book use this theory, explicitly or as background, which gives the book a high degree of coherence in the sense that chapters talk to each other. For transition studies scholars, we hope that the deep and broad studies of the automobility system offer valuable empirical insights that can be used to challenge and further develop transition theory. In the concluding chapter, we make an effort to highlight several characteristics of the transport system that challenge transition theory and offer opportunities for further theorization.

To realize the book's ambitions, we have chosen book contributors from both communities. Glenn Lyons, Geoff Dudley, Kiron Chatterjee, Iain Docherty, Jon Shaw, Peter Wells, Paul Nieuwenhuis, Renato Orsato, Phil Goodwin, Mimi Sheller, Juliet Jain, Val Mitchell, Andrew May, Reg Harman, Peter Harman, Wijnand Veeneman, Graham Parkhurst, Toon Zijlstra, Oliver Ehret and Henrietta Sherwin are part of the transport studies community. Frank Geels, René Kemp, Flor Avelino, Marc Dijk, Masaru Yarime, Marloes Dignum, Bonno Pel, Geert Teisman and Frank Boons are associated with the transition studies field and the related field of innovation studies.

We want to thank the book contributors for their extraordinary willingness to engage with the transition topic in this book. M. Gell-Mann once said that "*a scientist would rather use someone else's toothbrush than another scientist's terminology.*" However, in our case, the spirit of cooperation and engagement has been exceptional and an important reason for the book's success. Most contributors participated in two 2-day workshops, where we discussed the book's outline, message, contribution and drafts of chapters. These discussions were lively, respectful, intellectually stimulating and very useful in crafting a coherent book. People were willing to consider other viewpoints and contributions on the same issue. Beyond that, the contributors have been willing to revise their chapters, sometimes three times, to enhance the fit with, and contribution to, the book's topics and questions. We want to sincerely thank them for their dedication and for the insights they brought to the book, which would not have come to fruition without them.

Preface xv

We also want to thank the Dutch Knowledge Network on System Innovation and Transitions (KSI) for their financial and intellectual support to this book. Many of the transition scholars contributing to this book have been working in the KSI program, which ran between 2005 and 2010 and was funded by the Dutch government. The KSI program has been a fertile breeding ground for many of the theoretical ideas that we have used in this book. In particular, we want to thank the KSI directors, Jan Rotmans, Johan Schot, John Grin and Marjan Minnesma, for their support for this book and for organizing several workshops with book editors from all volumes in the Routledge series. These workshops formed excellent space for sharing experiences, exchanging ideas and comparing findings across the different empirical domains. We therefore want to extend our thanks to the editors of other volumes: Geert Verbong, Derk Loorbach, Gert Spaargaren, Anne Loeber, Peter Oosterveer, Jacqueline Broerse and Harry Lintsen. We also thank Mieke Rossou-Rompen, Julie Triggle, Annet Grol and Anja Bogaerts for their help on practical matters and Hans Horsten, Pepik Henneman, Joris Lohman and Diederik van der Hoeven for book publicity services (the fruits of which can be found at http://sustainabilitytransitions.com/). We especially want to thank Johan Schot for his efforts in realizing the book series and for his intellectual and personal encouragements. We also thank Arjan van Binsbergen and three anonymous book reviewers for their feedback on the book proposal, which was encouraging and helpful, prompting us to deal with identified weaknesses. In the final stage of the book project we received comments from the series editors, Matthias Weber and an anonymous reviewer on the book manuscript, whose constructive comments led us to revise the introduction, chapter 3 and the concluding chapter. We also thank Tim Schwanen for his comments on chapter 16, which were very helpful. We want to express out deep gratitude to all authors and commenters.

Last, but not least, we want to thank Max Novick and Jennifer Morrow from Routledge for their support and productive cooperation in bringing the book project to a successful completion.

We hope the book has established a fruitful intellectual marriage between transport studies and transition studies and that academics, policy makers, professionals and other stakeholders in the transport domain find the book's insights not only interesting, but also relevant as a guide for their activities in the 'real world'.

Part I
The Transition Perspective and Problems Associated With Car Mobility

1 Introduction
Sustainability Transitions in the Automobility Regime and the Need for a New Perspective

René Kemp, Frank W. Geels and Geoff Dudley

1.1. MISSION STATEMENT

> ... while we flatter ourselves that things remain the same, they are changing under our very eyes from year to year, from day to day—
> Charlotte Perkins Gilman

> *The more things change, the more they stay the same*—French saying

We begin the book with these two statements because they refer to the fundamental issue of change and stability. On the one hand, automobility faces a need for change to address persistent problems such as increasing traffic congestion and atmospheric pollution (including emissions that contribute to climate change). On the other hand, automobility is deeply embedded in western lifestyles and stabilized through sunk investments, interests vested in its continuation and taken-for-granted beliefs and practices. While the last two decades saw many attempts to introduce radical innovations with higher sustainability performance, the wider automobility regime still seems relatively stable. But under the surface, cracks may be appearing that create opportunities for wider system change and transitions to sustainability. The fundamental goal of this book is to examine these and other dynamic tensions between stability and change in and around automobility and the interactions between different types of change. These tensions have worldwide salience in the case of motorized transport. While the explosion in vehicle ownership and use, initially in the developed nations but now spreading to the emerging economies in Asia, Africa and the Americas, has brought about a revolution in personal mobility with many positive consequences, it increasingly threatens both the quality of life for the individual and the wider global environment. The complexity of these problems relates to the fact that solutions in one area may aggravate problems in another. For example, while alternatively fueled vehicles (battery, fuel cell, biofuels) may offer solutions to pollution problems caused by the internal combustion engine, they may also encourage a new wave of vehicle growth that, in turn, can aggravate problems of congestion and the quality of the spatial environment.

The way forward therefore involves not only technological solutions, but also the development of fresh perspectives that offer novel ways of understanding how society as a whole can make transport transitions that encompass more radical change in mobility behavior, spatial planning, traffic management and infrastructure. At the same time, it must be recognized that change processes work alongside powerful forces of system stability. Consequently, a greater understanding is required of stability and change and of interactions between different types of change. The book aims to achieve this by developing a "transport in society" perspective grounded in transition studies. To that end, the book consists of contributions from scholars with different types of expertise that relate to culture, governance, traffic management and behavior, infrastructural and spatial planning, the car industry, emerging technologies and transitions. Knowledge about transitions helps to put emerging technologies into a socio-technical perspective, and the involvement of transport experts helps to anticipate system-wide effects. Empirically, the book draws on primary research, but most importantly on the expertise from renowned specialists from transport studies and transition studies, which is combined and integrated in this book. With support from the Dutch Knowledge Network on System Innovation (KSI), the editors have organized two workshops to facilitate interactions and discussions between book contributors, aimed at fostering mutual understandings and creating coherence in the book.

The book investigates whether the current regime of automobility is in transition or not. Transitions are long-term processes (40–50 years), which are the outcome of alignments between multiple developments; they are not caused by a single factor such as a high oil price, a transport innovation or a government intervention. Transitions are a special research topic, because they are large-scale and relatively rare, only occurring now and then. During the 20th century developed countries experienced a transition from existing regimes of public transport to a regime of automobility, with the privately owned and driven car as the main means of personal transport. In some of those countries the regime of automobility may have stopped expanding (in the United Kingdom automobility is no longer growing, as Chapter 4 shows), but it is not at all clear how personal mobility will develop in the face of current pressures.

In examining the mechanisms of stability and change, the book offers a novel perspective by differentiating between incremental evolutionary change, technological discontinuities and more comprehensive systemic change. An important question that guides the book is the following: Will we see a greening of cars, based on technological innovations that sustain the existing car-based system? Or is something more radical possible and likely, for example, the development of travel regimes in which car use is less dominant and in which the logic of travel is based on *combining* different forms of transport, leading to a more sustainable transport system? Other questions for investigation are whether crossovers between private transport and public transport are occurring and gathering pace and whether we are moving to a greater diversity and variety in transport modes and travel behavior.

Over and above these questions, the book will address questions such as why automobility has remained dominant, despite its association with growing societal problems; why motor manufacturers have only recently (re)promoted the use of electric vehicles; why public transport has failed to benefit from problems of car-based mobility; why intermodal travel might hold greater promise than modal shift; and which developments, innovations and policy measures *jointly* could break the dominance of cars and promote opportunities for broader change in transport systems (or not).

Every expert holds implicit or explicit views on the evolution of mobility, based on disciplinary backgrounds and specialist knowledge. Such views are based on assumptions of what people want, technological expectations, views on what government can usefully do and beliefs about future oil prices. The transition perspective used in this book helps to scrutinize these assumptions and knowledge from experts by taking the authors outside their traditional field of expertise and by taking a longer (historical) view. An important aspect of the transition perspective is that our beliefs are historically bounded and part and parcel of the process of change. This has important implications for the *study* of transition processes: We have to be mindful that our own viewpoints as well as those of real-world actors (government, companies, consumers, social movements, engineers and traffic planners) are evolving in connection with events, circumstances and possibilities. By bringing together insider perspectives on the car industry, with those on (transport) governance, planning, traffic management, innovation and car culture, we are able to reveal the multitude of factors at play and thus obtain a richer and deeper understanding of processes of change and forces of stability.

The transition perspective plays an important role in integrating various kinds of knowledge: It helps to put specific dynamics into a broader context which pays attention to lateral and unexpected developments, hype-disillusionment cycles, innovation cascades, redefinition of goals and interests and knock-on effects, as well as inertia. By examining developments within and outside the transport sector, the book aims not only to analyze more closely the mechanisms behind stability, evolutionary change and more discontinuous systemic change, but also to gain a greater understanding of how these dynamics may shape the interactions between transport systems and society in the decades to come. The transition perspective has developed a specific way of looking at these dynamics that recognizes recurring patterns, for example, of regimes resisting change, the role of special (local) niches for the exploration of transformative change and conditions under which these changes can spread to regimes and societal landscapes. The perspective thus helps to understand what is currently happening in the transport system, as well as to anticipate possible outcomes of new developments, and to identify useful intervention strategies for working towards more sustainable systems of mobility.

The case studies in the book focus primarily on developments in and around automobility in the Netherlands and the United Kingdom. These countries strongly experience the problem of stability and change. On the one hand, they have mature and relatively stable systems of automobility; on the other hand, they have undertaken distinctive attempts of radical change, for instance with new propulsion technologies, intermodal transport, traffic information systems, congestion charging and mobility behavior. The case studies in these countries therefore contribute to developing the principal themes of the book. We recognize that transport systems and mobility cultures in other parts of the world (United States, China, India, South Africa) differ substantially from those in the United Kingdom and the Netherlands. We therefore do not claim simple geographical generalizability of the findings in this book. Instead, we aim for analytical generalization via theoretical patterns and underlying mechanisms. These findings can be applied for transitions in other countries, although this requires in-depth knowledge of the transport systems, actors and contexts in these countries. The book does not address all transport modes. Because we want to investigate if automobility remains dominant or not, the book focuses on the car and on transport modes that may affect the car system (either via replacement or reconfiguration into hybrid systems). We therefore do not address slow modes (walking, cycling), nor air and water travel. Having set the scene with this mission statement, the subsequent sections further elaborate transport achievements and problems, transitions to sustainability, research topics, book aims, the transition perspective and the contributions from the various chapters.

1.2. ACHIEVEMENTS AND PROBLEMS IN PERSONAL MOBILITY

During the last half century, personal mobility has rapidly expanded with many positive consequences in terms of convenience, speed, comfort and freedom. In particular, the use of private cars has increased enormously, compared to other transport modalities such as train, bus/metro/tram, and bicycles. Figure 1.1 describes the evolution of passenger kilometers per capita in eight industrialized countries—the United States, Canada, Sweden, France, Germany, the United Kingdom, Japan and Australia. In each country, personal mobility increased enormously between 1973 and 2007–2008. The greatest increase in passenger kilometers is for cars, the dominant mode of transport. The mode share of bus and rail has remained relatively constant or declined slightly, except for Japan where it fell significantly (Millard-Ball & Schipper, 2011, p. 366).

Introduction 7

The relative prominence of other transport modalities may differ somewhat between countries, depending on public policies, public support and geographical circumstances. The Netherlands, for instance, is characterized by relatively high bicycle use, which was more or less on a par with car use around 1960 in terms of passenger kilometers. The share of bicycle kilometers fell gradually owing to the rise of car-based transport, but today total passenger kilometers by bike is almost as high as that of train (14.1 million against 15.4 in 2007).[1] Nevertheless, the overall pattern is that other transport modalities are currently relatively small compared to car use. Terrestrial passenger transport is thus clearly dominated by cars, which is the reason this book focuses on automobility. The book addresses other transport modalities mainly with regard to how they relate to car transport, possibly via new intermodal mobility services such as bicycle train schemes and park and ride schemes.

The long-term change in transport modalities coincided with the increasing adoption of cars by consumers. In the Netherlands, car ownership

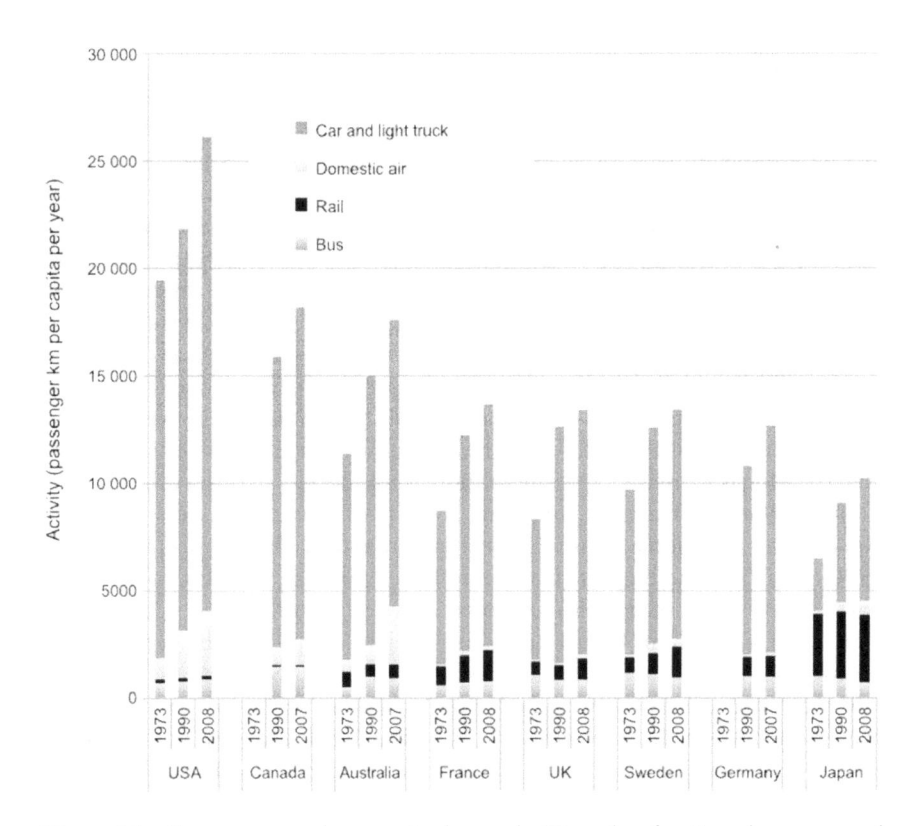

Figure 1.1 Passenger travel per capita by mode. Note that for Canada, metro and other local rail services are included in the 'bus' category (Millard-Ball & Schipper, 2011, p. 367).

increased from 165 per 1,000 households in 1960 to 1,005 in 2007. In the United Kingdom, 80% of the households possessed a car in 2005,[2] against 41% in 1965 (see Dudley and Chatterjee in Chapter 4).

The expansion of car-based transport has given rise to a range of persistent social problems such as congestion, deaths and accidents, climate change, local air pollution, social exclusion, land fragmentation, noise pollution, end-of-life disposal, oil dependence and energy security (Cohen, 2006; see also Chapter 4). Car use is also contributing to obesity and dehumanization of public space (Bassett, Pucher, Buehler, Thompson & Crouter, 2008). Some of these problems have been substantially reduced in the past few decades. Traffic safety, for instance, has generally improved. The fatalities per million inhabitants in the European Union has fallen steadily, thanks to enhanced driver education, better vehicle design, safety technologies (seatbelts, air bags) and better road design. Still, about 40,000 Europeans die each year in fatal traffic accidents. In 2006, 42,950 persons lost their lives in road accidents: car drivers and passengers, occupants of buses and coaches, riders and passengers of powered two-wheelers, cyclists, pedestrians and commercial vehicle drivers (Eurostat, 2009, 139). This remains a staggering number, amounting to the crash of an average size aircraft each day.

In many places, problems of local air pollution have also diminished, especially due to catalytic converters, improved engine design and changes in fuel composition. The development is different from South East Asia where cities suffer from very bad air quality as a result of motorized transport. Despite improvements, air quality is not very good in developed world cities. A particular cause of concern is small particulate matter, because scientific research has shown that these small particles can diffuse deeply into the lungs, where they cause more damage than previously thought.

Other problems show only modest signs of improvement, such as CO_2 emissions per kilometer, fuel economy of new cars, or in some cases, such as congestion pressure, are even getting worse. Figure 1.2 shows that in the Netherlands the congestion pressure (defined as the length of traffic jams times the period they lasted) has increased at a rate of more than 6% per year since 1990.

In absolute terms, CO_2 emissions from transport increased, with over 90% of the emissions coming from road transport (Figure 1.3). Between 1990 and 2007 the biggest increase in CO_2 emissions was for road transport (+200.7%; European Environment Agency, 2009).

Today, more than 1 billion motor vehicles populate the world, and we are moving toward 2 billion vehicles in 2020 (Sperling & Gordon, 2009, p. 2). Behind this projected growth is demand for personal motorization in emerging economies such as China and India with a population of 2.4 billion people, whose markets are targeted by automakers, and which are creating their own motor car industries (Sperling & Gordon, 2009, p. 4). Because CO_2 emissions from motor cars cannot be captured and stored, total CO_2 emissions from motor cars are likely to rise if the majority of new cars are internal combustion engine cars, as is widely

Introduction 9

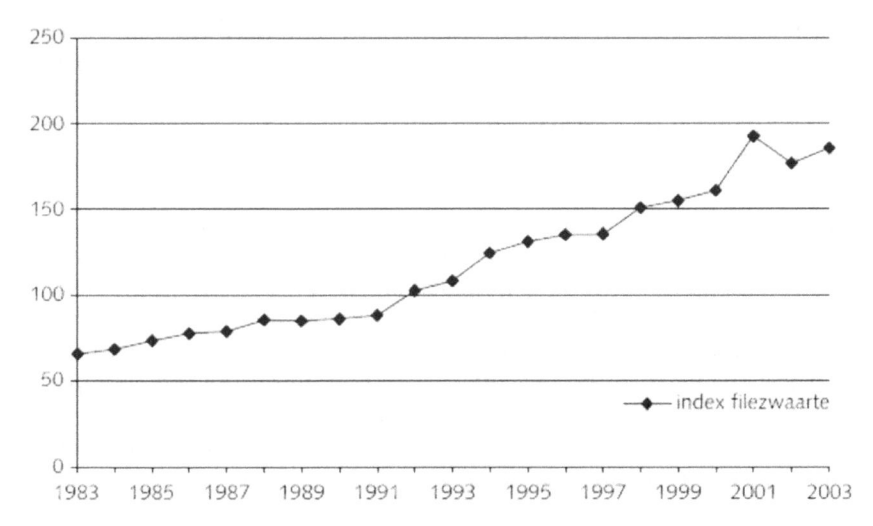

Figure 1.2 Congestion pressure on Dutch highways (Rijkswaterstaat, 2004, p. 12).

expected. Improvements in fuel economy are not expected to keep up with the increase in motor cars. This increase in the coming decade will change urban life and the landscape of countries. The growth in cars will come at a big cost for society.

Figure 1.3 Total transport and road transport, European Union: Greenhouse gas emissions and share of Road in Transport emissions, 1990 to 2006 (million tonnes CO2-equivalent and percent; Eurostat, 2009, p. 171).

Economically, socially and environmentally, motorized transport based on fossil fuels is not sustainable. This raises the question: What is sustainable mobility? A useful attempt to define sustainable mobility has been provided by David Banister in a prize-winning paper published in *Transport Policy*. Elements of the sustainable mobility paradigm are reasonable travel time rather than travel time minimization, reducing the need to travel (through distance reduction and home working), seeing transport as a valued activity rather than derived demand, achieving a modal shift (especially to walking and cycling), lower levels of pollution and noise from transport, greater energy efficiency, more efficient use of infrastructures (through higher vehicle occupancy and demand management) and increasing the quality of places and spaces (Banister, 2008). As authors, we think this is the best attempt to define sustainable mobility, but it is not one that is universally agreed and acted upon. Cleaner vehicles are supported chiefly for making a contribution to air quality, not to achieve sustainable mobility. Intermodal travel is being promoted by train companies to attract travelers to trains, not for sustainable mobility reasons. Traffic information is supported by transport authorities for increasing the efficiency of roads and reducing congestions, but for the providers of traffic information products, it is just a product. For transport authorities, sustainable mobility is not something they aspire to achieve on an everyday basis. Many of the things they do are not consistent with it.

In the book we could have assessed the contribution made to sustainable mobility of each innovation and development process studied (such as intermodal travel, battery electric vehicles). Although attention is given to various sustainability aspects, we have not sought to quantify this or to draw conclusions about it. Instead, our primary interest is with socio-technical dynamics of greener cars, traffic information systems, intermodal travel and sustainable mobility planning. In particular, we are interested in the following aspects: what *motivates* different actors to engage in those activities, *how* did they come about (through what actions, developments and special circumstances), their success and whether the developments and innovations are within the regime of automobility or an element of alternative mobility.

Accepting that sustainability is something normative, subjective and contested (Jordan, 2008; Kemp & Martens, 2007), we employ a more sociological and action-oriented analysis that focuses on how sustainability is understood by different social groups and what it is "doing": how sustainability claims and appeals are used by social actors to legitimize actions and to attribute blame. The book is foremost an analysis of processes of change, relevant to problems of sustainability expressed in society.

1.3. TRANSITIONS TO SUSTAINABILITY

The persistent or 'wicked' problems discussed previously may be difficult to address within the existing transport system. There is therefore increasing interest in transitions to new transport systems with higher sustainability performance. There is no agreement, however, about the specifics of these transitions, nor about what constitutes 'sustainability'. Some people advocate technological changes, for example, new car engines and fuels, which promise to reduce CO_2 emissions. The President of the European Commission Barroso, for instance, champions a transition towards fuel cell vehicles and hydrogen. Rifkin (2002) also advocates the hydrogen economy, but his vision is broader and advocates not just changes in cars, but also suggests that citizens may use car-based fuel cells to generate electricity for their own houses, thus creating new linkages between transport and energy systems. But the hydrogen economy has become increasingly contested, as scholars draw attention to the many barriers and problems for realizing this transition (see, e.g., Romm, 2005).

Other people criticize the exclusive focus on climate change, or environmental problems more generally, because it neglects other persistent problems (Lanberg, 2001). Alternatively fueled vehicles may be more environmentally friendly, but their introduction is unlikely to have significant effects on the numbers of vehicles on the road. Thus they do not significantly address issues such as traffic congestion, road accidents and casualties, nor geographical and spatial problems, such as the role and place of the car in the built and natural environment. Broader visions of 'sustainable transport' therefore exist, with people advocating transitions towards multi-modal transport, car sharing, automated people movers, or even suggesting that future systems may be characterized as 'after the car' (Dennis & Urry, 2009). Spatial planners and geographers further suggest that changes in mobility will require transitions in spatial structures, for example, a move towards more concentrated cities with smaller distances between work, home, leisure, school and so forth (Henderson, 2009; Newman & Kenworthy, 1999). Transport planners and traffic managers, in turn, suggest that the integration of new information and communications technologies into highway systems may lead to intelligent highways, dynamic traffic management or even automated vehicle guidance that allow cars to drive at similar speeds on very short distances from each other, thus improving the efficiency of road use (Mitchell, Borroni-Bird & Burns, 2010).

Such grand technological schemes have been criticized, however, by sociologists of innovation (Geels & Smit, 2000), who found that many historical and contemporary grand schemes failed because of lack of social support or because car drivers and consumers developed alternative, unexpected behaviors. Congestion charging, for instance, remains a politically and socially contested issue, despite ongoing claims by transport planners about the efficiency of pricing mechanisms. While the London Congestion Charge, introduced in 2003, has been relatively successful in reducing inner city traffic and congestion, proposals to introduce congestion charging in Edinburgh and Manchester have recently been heavily defeated in referendums. The social and political obstacles inherent in the introduction of urban congestion charging illustrate the difficulties in bringing about (partial) system innovation and the importance of public acceptance. Indeed, technological advances may themselves create new problems of public acceptance. For example, road pricing that works through satellite controlled guided positional systems is able to track, and then illustrate to the user, when and where the individual has driven. This can arouse public concern about a 'big brother' system that challenges established norms of privacy and confidentiality. Sociologists of mobility therefore suggest that visions, policies and discussions of transport transitions should pay more attention to 'darker' possibilities and to cultural, motivational and behavioral dimensions (see Sheller in Chapter 9).

This variety in views and visions, which has changed over time, has resulted in a flurry of activities, ranging from local projects such as improved light-rail, transformations of city centers into pedestrian areas, promotion schemes of bicycle use, park and ride schemes, urban congestion charging, car sharing projects, automated vehicle projects to transfer commuters to business parks, to large-scale programs such as the American FreedomCAR and Vehicle Technologies (FCVT) program, the Fuel Cells and Hydrogen Joint Undertaking (FCH JU) of the European Union and international intelligent highway systems programs.

But while these activities contain seeds for substantial change, they do not yet appear to have substantially influenced the automobility system, which still accommodates the majority of passenger miles (around 90%, with some differences between countries). Despite two decades of work on alternatives, the internal combustion engine still reigns supreme, although hybrid-electric vehicles have gained a market foothold, and battery electric vehicles are much discussed again. Use of public transport and bicycles is still small, compared to cars. Cars are also deeply embedded in lifestyles (e.g., bringing children to school or sport, shopping, family visits, holidays), supported by cultural discourses (around freedom, individuality, adventure) and stabilized by positive feelings and emotions. Sheller (2004) therefore concludes that "cars will not easily be given up just (!) because they are dangerous to health and life, environmentally destructive, based on unsustainable energy consumption, and damaging to public life and civic space. Too many people find them too comfortable, enjoyable, exciting, even enthralling. They are deeply embedded in ways of life, networks of friends and sociality, and moral commitments to family and care for others" (p. 236).

1.4. FURTHER ELABORATION OF RESEARCH TOPICS

Against this background, the book addresses two important research topics that relate to sustainability transitions: (a) the forces of stability and change and (b) the prospects of a transformation in personal mobility. The first topic concerns stability and change in automobility. On the one hand, there are many visions of sustainable transport systems and many (local) change activities. On the other hand, the existing automobility system is characterized by stability and lock-in. Following the literature on path dependence and lock-in (Arthur, 1989; Unruh, 2000; Walker, 2000), we can distinguish several factors that contribute to stability, for example, (a) low costs of existing technologies due to economies of scale and learning-by-doing; (b) sunk investments in infrastructure, machines and people; (c) people's life styles and behavioral patterns; (d) legislation, institutions and subsidy schemes that favor existing technical systems and hinder new ones; (e) mental maps and cognitive schemes that blind incumbent actors to alternatives that fall outside their scope of attention; and (f) resistance from powerful actors who aim to protect their vested interests.

With regard to this first problem, the book will empirically investigate (a) the degree of stability of the existing automobility system and the possible presence of certain 'cracks' that may create 'windows of opportunity' for sustainability transitions and (b) the degree to which several change initiatives are ready to take advantage of these windows of opportunity.

By addressing both processes, the book will further address the puzzle and prospect of 'tipping points', raised by Sheller (2004): "Despite incremental change and experimentation in new transportation policies (regulation, taxation, road pricing, congestion charging) there has not been a radical transformation of the car and the road system itself, nor of the patterns of habituation and feeling that underlie existing car cultures. However, there are signs that suggest we may be approaching a 'tipping point' in the demise of current configurations of the dominant culture of automobility" (p. 236).

The attention given to climate change, which has increased strongly in the last 5 years, and the current economic problems in the car industry, are just two developments that may push the automobility system firmly in a new direction. The bankruptcy of General Motors (GM; the second largest automobile company in the world) in June 2009, which was the third largest US bankruptcy ever, may have provided a shock that will stimulate car manufacturers to search more actively for new technologies and business models. The GM bankruptcy entailed the creation of a new GM company and was supported by $50 billion in US Treasury loans, which gave the US government a 60.8% stake in the new company. In addition, the Canadian government invested $10 billion for a 12% stake. These controversial public investments indicated both the fact that GM would cease to exist without official support and also that the US and Canadian governments could not afford to pay the political cost of seeing such a leviathan fall. During 2009–2010, a major restructuring of GM took place that included a reduction of more than 65,000 jobs in the United States. In November 2010, GM returned to the stock exchange and raised $20.1 billion dollars through a share offering. This reduced the shareholding of the US government from 60.8% to around 26% (*BBC News*, 18 November 2010). The gradually improving financial position of the company was also reflected in GM making a profit in 2010 for the first time since 2004. Nevertheless, the

14 *René Kemp, Frank W. Geels and Geoff Dudley*

continuing major public investment in GM is likely to mean it maintains a commitment to environmental and energy efficiency. For example, GM is planning to produce a small car at one of its formerly closed factories. This will be the first mini type vehicle produced by a major manufacturer in the United States. The company is also prominent in the development of hydrogen and electric powered vehicles.

Public concerns about climate change, and 'sustainability' more generally, provide another window of opportunity, although there are still substantial uncertainties about the degree to which public concerns translate into real consumer demand for green cars and the 'willingness to pay' for sustainable transport options. Citizens are unlikely to vote for car restraining policies, but certain cities such as the German town of Freiburg have moved in that direction. In many cities car-free zones have been introduced, and more cities are creating special lanes for cyclists and 30 kph home zones. Another trend that could create windows of opportunity is the increasing integration of information and communications technologies (ICT) in cars and transport systems. ICT may strengthen the car transport system in the form of electronic navigation devices or dynamic traffic management systems. But ICT can also act as a linking pin in, and catalyst for, intermodal transport systems, allowing people to determine en route what their transport options are, how to transfer from one mode to another and buy tickets in advance.

If we would move towards a transition to sustainability, the book's second research question is if a green technology pathway is more likely than a broad transformation of the mobility system. The green technology pathway would consist of green cars and car-facilitating measures. The transformation pathway would consist of the development of travel regimes in which car use is less dominant, in which bicycles are used for small trips and high speed rail for longer trips and in which the logic of travel is increasingly based on *combining* different forms of transport.

One important difference between both transition paths is the degree of change in mobility patterns and travel behavior on the demand side. In the first path, user preferences and mobility patterns remain more or less unchanged. People buy a 'greener' car but do not really change their travel behavior (although high penetration of ICT in cars and infrastructures may change car-based travel experience). The second transition assumes more change in mobility behavior, especially more active travel planning, mixed use of multiple transport modes, perhaps less private car ownership and so forth. This second path also assumes technological change (e.g., new ICT devices), investments in modal transfer and parking spaces that allow the linking of transport modes and policy change (e.g., new taxes, subsidies, visions and experimentation programs), but the main change concerns consumer behavior.

The book will also give attention to policy and governance. We will examine the various roles of transport policy and the tensions between policies that aim to sustain or change the present automobility system. We will examine

Introduction 15

the beliefs, expertise and motivations behind those policies, the outcomes of such policies and why they did (or did not) have their intended effects. Special attention will be given to how sustainability is framed and translated in policy acts, the support for (better) public transport, bicycle infrastructure, traffic management and/or the promotion of green cars. The way in which sustainability is defined and translated in policy is an interesting transition issue. Sustainable mobility may be defined in terms of access rather than mobility, the use of cars may or may not be assumed, and demand management may be viewed as an essential element (which is currently not the case). The book will look at the various ways in which policy is involved in fostering change, and also at the ways in which it seeks to protect non-sustainable practices and businesses. Over many decades, the motor industry has developed as a hugely powerful institution that encompasses mutually reinforcing interests including oil, raw materials, engineering and component industries, as well as vehicle dealerships and consumers. For many people, the car remains a powerful status symbol that helps to define personal identity and shapes social behavior. This probably means that policymakers are inhibited from acting too much against car-based modes of transport.

With this dual focus on the elements that give the system of automobility stability, together with the elements of change, the book hopes to avoid two potential mistakes:

- Wishful thinking about certain solutions 'solving' transport problems. The book will not only show that solutions often have unanticipated effects (e.g., the effects of creating more sustainable transport systems typically fall short of what is expected), but also that 'innovation journeys' (Van de Ven et al., 1999) often experience ups and downs and twists and turns. Innovative solutions may also face a mis-match with other dimensions of transport systems (e.g., lack of infrastructure, market demand or regulations), which hinders wider diffusion.
- The assumption that private car use will simply continue as we know it, because car mobility reflects people's true preferences. Because the future of transport systems depends on choices and interactions between various social groups, the book will show that people's travel modes and mobility choices reflect situational characteristics, cultural values and cultural ways of thinking that help define and frame individual preferences. The success of public bike systems in France and other countries shows that there is a dormant interest in cycling.

1.5. THE NEED FOR A NEW ANALYTICAL PERSPECTIVE

With regard to transport studies, the book argues that a new theoretical perspective is needed to analyze systemic transitions. This perspective has the following characteristics, which make it suitable to address the topics discussed previously:

(a) *Co-evolutionary and 'systemic' view on transport*: We view the transport system as consisting of a semi-coherent configuration of mutually aligned elements, which include technology, industry, markets, consumer behavior, policy, infrastructure, spatial arrangements and cultural meaning (Geels, 2004). Although the configuration is semi-coherent, tensions and mis-alignments may (temporarily) exist between elements, which create windows of opportunity for wider change. This means that system change is rarely driven by single factors such as prices or technological change, but usually involves co-evolution between multiple developments.
(b) *Actor-based approach*: The book differs from mainstream transport approaches that tend to focus on technology (e.g., engineering assessments and comparisons of various technologies) or economic transport modeling (where cost, performance, prices and incentives are the main variables). The book instead takes an actor-based approach, which focuses on framing, strategies, perceptions, actions and interactions between car drivers, transport planners, car manufacturing firms and public opinion.
(c) *Stability and change*: The perspective should encompass dynamic stability and incremental change on the one hand and radical innovations and system change on the other. We are especially interested in the co-existence, synergies and competition between various developments and emergence of mixed forms (tram-trains, shared taxis, hybrid electric cars).
(d) *Complex dynamics*: The book adopts a particular view on dynamics, which deviates from simple drivers and linear cause-and-effect relationships. Instead, the emphasis is on mutually reinforcing developments and (sometimes unexpected) alignments, co-evolution, mixed forms, circular causality, innovation cascades, knock-on effects and hype-disappointment cycles.

Because, to our knowledge, such a dynamics and actor-centered perspective does not exist in the transport and mobility domain, the book opens up new ground. With regard to policy the book aspires to bring out the various ways in which policy is involved in transport and how such policies are entangled in multilevel processes of change. For example, it is valuable to look at problem–solution sequences, the use of visions, and non-intended effects of policies. Policy support for radical change will be investigated, alongside policy efforts that support automobility but mitigate negative effects. The evaluation of past policies may be used to say something about the need for policy to be more concerned with transformative change and what policymakers could usefully do.

Introduction 17

The book will use a socio-technical perspective on transitions, which we think meets the four criteria formulated previously. The socio-technical transition perspective comes from an 'evolutionary' system approach of innovation which does not prioritize social and technical elements but sees these as inexorably linked (Geels, 2002, 2005, 2010; Geels & Schot, 2007; Hoogma, Kemp, Schot & Truffer, 2002; Rip & Kemp, 1998).

The multi-level perspective (MLP), which is further discussed in Chapter 3 can deal with both stability and change. To explain *change*, it uses concepts such as 'niches', which are protected spaces where radical innovations emerge, and 'socio-technical landscape', which are external developments that create pressure on existing systems (or better 'regimes'). To explain *stability*, the notion of socio-technical regime plays an important role, which says that we are locked into car-based modes of transport because societies have adapted themselves to their use in terms of car ownership, infrastructure, training and knowledge, communities of practice, regulations, social practices and cultural acceptance.

The interactions between niche, regime and landscape developments are enacted by social groups (firms, policymakers, customers and car drivers, social movements, transport planners, engineers), who have their own perceptions, interests and resources but are also linked together to maintain and reproduce the regime. These various social groups navigate a transition, finding their way through searching and learning, while also engaging in power struggles, controversies and debates. The dynamics are not mechanical but socially constructed and enacted. Because perceptions and strategies of actors change over time, transition dynamics are not linear, as the chapters in this book will show.

1.6. THE ANALYSIS OF SOCIO-TECHNICAL TRANSITIONS

The transitions perspective brings an innovative approach to the analysis of change processes, particularly through the manner in which it allows us to not only trace the often complex dynamics of the interactions between technological and social change but also how these relate to the framing of ideas and the associated shifting perspectives with regard to problems and their solutions. As an example of this type of analysis, we offer a brief discussion how the motor industry responded to the emergence of ideas concerning sustainable development and the derived transport concept of sustainable mobility.

1.6.1 Competing Solutions

Over the past two decades, one of the primary challenges to the motor industry is the rise of environmental ideas and concerns, often expressed in concepts such as sustainable development and sustainable mobility. From the 1980s, these challenges were expressed in terms of the harmful effects of vehicle emissions on public health, such as the impact of lead in petrol on children's learning processes and the contribution made by a variety of emission gases to the rising incidence of respiratory diseases such as asthma. Pressed by regulations, car manufacturers sought ways to reduce emissions. The lean burn option lost against the catalytic converter, which was effectively prescribed by regulation, first in the United States and later in other countries (Nill & Tiessen, 2005). For lean burn engines a window of opportunity existed only for a small period.

With the rise of the climate change issue over the past 10 years, research efforts shifted toward emissions of carbon dioxide. In the absence of a control device, car manufacturers had to look for alternative propulsion and energy sources. The emerging long-term response of the motor vehicle is to develop effective alternative technologies to the internal combustion engine that have less harmful emissions, such as electric, hybrid electric-petrol or hydrogen fueled vehicles.

Market introduction of alternative propulsion cars has been slow. However, the hybrid electric-petrol Toyota Prius is the first alternatively fueled vehicle to make significant inroads into the mass car market, with 2 million sold worldwide by September 2010 (*Financial Times*, 9 October 2010). Two interesting issues here are why Toyota devoted considerable resources to the Prius in a time (the 1990s) when climate policies were absent and why the Prius did so well. The introduction of the Prius car was based on perceived first-mover benefits and the expectation that there is a market for cars equipped with electric propulsion. Their thought was that the Prius was the leader for the cars to come, which is reflected in the car's name: Prius, which in Latin means [to go] before.[3] The success of the Prius led other manufacturers to follow suit. The success owed a great deal to the Prius being redesigned for the American customer and being singled out for public attention, not so much because of its fuel efficiency but because of its fit with cultural values of greenness and sensations of electric drive. It became a fashionable consumer status symbol. Government was enrolled and provided support through subsidy schemes. The response of the car manufacturers to the success of the Prius car is interesting. Many companies started to develop hybrid-electric vehicles but they also continued to invest large sums of money in the improvement of internal combustion engine vehicles. They may shift their resources more firmly into electric cars if the market develops. What this shows is that cultural change interacts with car manufacturing product choices and technical change.

1.6.2. Changing Configurations and Co-evolution

In terms of a taxonomy of change used in this book, the development of electric vehicles is interesting in that it may contribute to a transformation of the electricity systems, when vehicles start to deliver back electricity to the grid, and may facilitate intermodal travel, when cars are used in combination with other modes of transport. The latter necessitates shifting perceptions on the part of users, planners, politicians and companies, so that issues are framed in a new way that challenges existing dominant ideas and interests.

The stimulus for a rethink may come from discussions about sustainable mobility, but it may also come from somewhere else. For example, with regard to demand management policies, it is significant that early examples of urban road pricing, such as those in Norway, were designed to raise revenues for infrastructure projects, rather than to reduce the number of vehicles on the road. In general, successful new developments in mobility behavior can be expected to spread, but there is no guarantee that this will happen. The London Congestion Charge proved to be a success in terms of achieving its principal aim: Compared to the previous year, 60,000 fewer cars and delivery vehicles per day entered the zone in the first 6 months after implementation. This experience raised expectations that it might become a template for the introduction of schemes of this type worldwide. However, progress here has so far proved elusive, with only Stockholm implementing a similar type of road pricing. In the United Kingdom itself, proposals to introduce congestion charging in Edinburgh and Manchester have both been heavily defeated in referendums. The social and political obstacles inherent in the introduction of urban congestion charging illustrates well the difficulties in bringing about system innovation through framing a problem in a new way. Even if the technology exists, public acceptance can present formidable barriers. The socio-technical perspective helps to appreciate this and leads one to perceive issues of change more as *co-evolution* rather than diffusion. It also points to a general dynamic, which is that technology and society adapt to each other.

All the issues just mentioned can be given a proper place in a multi-level transition analysis, which looks beyond immediate transport issues.

1.7. THE STRUCTURE OF THE BOOK

The book's structure reflects the transition perspective. The chapters in Part I introduce the "transport in society" perspective of which the transition perspective is a special case. The chapters in Part II focus on the regime level, analyzing various dimensions of automobility. These chapters will address which factors and trends contribute to the lock-in and stability of the existing automobility regime and which ones create tensions and 'cracks' that may provide windows of opportunity for wider change. Several chapters in this part include a historical analysis of developments in recent decades, because this is the best way to investigate path-dependencies and lock-in, and possibly aspects of 'unlocking'. The chapters in Part III focus on developments that may benefit from regime problems and pressures: niche innovations and modes of public transport. Part III contains chapters on electric mobility (battery electric and fuel cell electric) cars, public transport, user innovation, new ICT devices and forms of information provision, spatial innovations and intermodal transport. The chapters talk to each other and are especially written towards this aim, giving the book a coherence which is unusual for an edited book. We also made a great effort to synthesize the findings in a final chapter.

The following is a short outline of the individual chapters.

1.7.1. Part I: The Transition Perspective and Problems Associated With Car Mobility

Chapter 2, "Visions for the Future and the Need for a Social Science Perspective in Transport Studies" by Glenn Lyons, discusses the shift in transport studies to incorporate social and behavioral factors into the analysis, as exemplified by foresight studies in the past 16 years. The transition perspective of this book is believed to take the analysis one step further, in giving attention to social and technical interaction, tensions between stability and change and addressing all relevant actors and subsystems.

Chapter 3, "The Multi-Level Perspective as a New Perspective for Studying Socio-Technical Transitions" by transition experts Frank Geels and René Kemp, describes the transition perspective as a theoretical framework for the study of stability and change. It explains the multi-level framework and presents transition patterns and actor-related dynamics identified in innovation and transition studies, using examples from transport. One such pattern is the fit-stretch pattern, which is illustrated with early cars moving from a 'fit' with the horse-based regime, toward 'stretch' in terms of articulating their own technical and use principles. The seeds of a new auto regime emerged with the T-Ford (1908), as a new technical form with more practical, utilitarian types of use (initially by farmers, doctors and taxi drivers). Technology and use thus co-evolve both materially and socially in terms of user groups and beliefs. These co-evolution processes mean that innovations may gradually evolve to have more radical and transformational implications. Two contemporary examples are road pricing and batteries, two regime-preserving innovations which may be game changing in the longer term. Other patterns in transitions are the hype-disillusionment cycle, innovation races, domestication, societal embedding, cultural framing and transport issues becoming part of wider power struggles.

1.7.2. Part II: Stability and Regime Pressures

Chapter 4, "The Dynamics of Regime Strength and Instability" by Geoff Dudley and Kiron Chatterjee, offers a historical analysis of UK government policies toward cars, with special attention to policies for sustainable mobility. Automobility developed rather late, with the Minister of Transport rejecting the concept of a motorway network in the 1940s. The first motorway was build no earlier than 1959 in an attempt to please car drivers. It also describes how the motor car transcended social groups and assumed a place at the forefront of the consumer revolution. In the 1990s the credibility of the roads strategy was undermined by sustainable development, becoming a new point of orientation. Optimistic plans and high ambitions for sustainable mobility were formulated between 1997 and 2000, but material achievements were low, as car-restraining policies proved unpopular.

Chapter 5, "The Governance of Transport Policy" by transport governance experts Iain Docherty and Jon Shaw, explores the impacts of changing forms of governance on the formulation of transport studies in two UK jurisdictions: Scotland and London. Particular attention is given to how the ideology of marketization and devolution of responsibilities are affecting transport policies. It is found that changes in governance have profound effects on transport policies but not in a deterministic way, as transport authorities in Scotland and London made quite different choices. Whereas London curtailed car mobility and invested in public transport, Scotland invested in new roads and public transport (more of everything). Their chapter shows that it is not easy to repeat the success of congestion charging in other jurisdictions and that the transition to sustainable mobility is affected by national and decentralized systems of governance, in which many considerations come into play.

Chapter 6, "The Nature and Causes of Inertia in the Automotive Industry" by Peter Wells, Paul Nieuwenhuis and Renato Orsato, offers an insider discussion of why a technologically sophisticated industry since the 1920s kept relying on the internal combustion engine and steel body. The reasons for this obdurance have to do with scale economies and associated capital costs, safety regulations that are based on the steel body, exit barriers as well as entry barriers, consumers becoming habituated to certain performance attributes, car-accident risks (which are lower for drivers of heavy vehicles) and apparently non-trivial matters such as paint finish. The trend has been towards achieving ever-greater economies of scale and scope. The car industry is prepared to meet the challenge of greening of cars but at their own preferred pace, with a great desire to stay within the internal combustion engine trajectory. They are supported in this by consumers preferring low-cost, heavy vehicles (with shiny paint finish) over light-weight, plastic-body electric vehicles. The prospects for newcomers are that they either fail or are being bought up. Big companies are not allowed to fail by governments because of their economic importance and the perceived national interest. Radical change is possible and being tried, but as this chapter shows, there are powerful forces working against it, with companies focusing the bulk of their research and investment on internal combustion engine vehicles.

Chapter 7, "Providing Road Capacity for Automobility" by Phil Goodwin, takes up the theme of road building and transport policy principles. Goodwin, a transport policy expert and first author of the influential report

Transport: The New Realism, describes how the policy principle of *predict-and-provide* came under increased criticism. He views 1989 as a pivotal year in this respect, in terms of the new policy principle of predict-and-prevent becoming established. Road building continued, but at a slower pace. As also noted in Chapters 4 and 5, the new planning regime, based on reducing car mobility through traffic management, suffered from problems of implementation. Ever since the 1960s the use of road pricing for traffic management has been postponed, essentially for political reasons. In Goodwin's words, "everybody can be kept happy while the discussion is still about principles but not when the devil of the detail emerges." He offers an interesting discussion about whether we are in a transition away from automobility or not. He proposes that a transition in thinking already occurred 20 years ago and that traffic intensity (vehicle kms/GDP) has fallen since 1994. The share of car traffic, however, hardly fell, leading him to the conclusion that a transition to sustainable mobility requires many years of persistent effort, creative imagination, political courage and consistency. Sustainable mobility is, and will remain, a contested issue when it comes to the introduction of real measures to reduce automobility.

Chapter 8, "A Socio-Spatial Perspective on the Car Regime" by Toon Zijlstra and Flor Avelino, examines the link between car mobility and spatial planning. The authors start their chapter by questioning the rationality of car-based choices, even in a world in which mobility is organized around cars. In a similar manner to Chapters 4 and 7, they note that, despite the strong support for automobility, there has always been an undercurrent of radical criticism. This is elaborated in actor terms by describing the social movements that protested against car-based pollution, the decline of the urban and rural landscape, atomized individualism, forced car ownership, illusion of speed and freedom, oil dependence and road safety problems. The social criticisms set the stage for a discussion of four radical socio-spatial mobility niches, in which car mobility is less dominant: modal-split based planning regimes, low-speed and shared space areas, sustainable urban planning regimes based on reducing the need for mobility (as in compact cities) and self-reliant communities (with *Transition Towns* and *Slow Cities* as examples). Some of the niches have a strong modernist element, others are decidedly non-modernist and take a critical stance against consumerism and globalization. The niches undermine conventional assumptions that car-based mobility is always desirable and rational. At the end of their chapter the authors offer a criticism of the transition approach for being too functionalist. They also make a plea for a spatial planning-based approach towards sustainable mobility.

Chapter 9, "The Emergence of New Cultures of Mobility" by Mimi Sheller, adopts a cultural perspective in analyzing the factors contributing to stability of the car-based mobility regime in the United States, potential openings and prospects for a future transition. Culture is viewed as a crucial performative part of transitions: It is part and parcel of transition processes.

Culture is present in practices, actor networks and discourses. Automotive emotions are viewed as an essential element of car-based cultures. In her chapter, Sheller analyzes emergent cultural openings in the national and urban transport regimes in the United States, where she discusses the Transit Oriented Development regime of the city of Philadelphia. The creation of bike lanes, better public transit and electric vehicles is viewed as not enough for a transition. According to Sheller, new cultural articulations are needed for a transition. An interesting conclusion from this chapter is that a transition to alternative mobility may not be driven by sustainability concerns but by pervasive market forces assembled around personal entertainment and surveillance as well as by diffuse cultural forces.

1.7.3. Part III: Dynamics of Change

Chapter 10, "The Electrification of Automobility" by Renato Orsato, Marc Dijk, René Kemp and Masaru Yarime, examines what they call the "bumpy road" for pure battery electric vehicles (BEVs). They show that the history of BEVs has been one of twists and turns. For a short period, BEVs dominated the world of motor vehicles, losing their dominant position against noisy and polluting internal combustion engine vehicles around 1915. In the 1970s and 1990s there were brief revivals of battery electric cars, but the revival was short-lived and highly localized. Climate change concerns, high oil prices and the success of the Prius (a hybrid electric car) together with advances in battery technology helped to generate new interest in BEVs. Today almost all major car manufacturers are working on prototypes to be commercialized soon, with some companies such as Renault-Nissan, Mitsubishi Motors and the Chinese company BYD already offering battery electric vehicles to customers. Better Place, a Californian start-up, has been working with governments, businesses and energy producers to provide electric mobility services. Their business model is based on customers paying for electric mobility on a use-basis, with revenues being used to pay for charging and battery swapping infrastructure. A trajectory for electric mobility is underway, involving different types of vehicles: (plug-in) hybrid electric vehicles, battery electric vehicles and hydrogen fuel cell vehicles. The chapter considers fit-stretch patterns for BEVs in terms of technical forms and type of use (e.g., short-term rental). The authors do not expect that cars with electric drive will change patterns of car-mobility fundamentally. It may nevertheless fundamentally change the car industry, and there are interesting spillovers to the power sector and to bicycles and scooters. A significant difference from the 1970s and 1990s is that electric drive systems are now accepted culturally.

Chapter 11, "Introducing Hydrogen and Fuel Cell Vehicles in Germany" by Oliver Ehret and Marloes Dignum, describes the activities for hydrogen and fuel cell vehicles in Germany in the 1998–2010 period. The German government and energy companies became enrolled in research and demonstration projects for hydrogen fuel cell vehicles (FCV). With hydrogen produced from low-carbon energy sources, CO_2 emissions from cars and light duty vehicles can be reduced to 20g CO_2/km (tank to wheel) by 2050, compared to some 160g CO_2/km in 2010 (car fleet average). The chapter describes how hydrogen became the favored transport fuel in Germany and the programs and widening network of actors involved in hydrogen fuel cell initiatives for road vehicles (cars and buses). The strategic selection and introduction of FCV in the transport sector is an example of strategic niche management, a model propagated by innovation scholars to escape lock-in (Hoogma et al., 2002). Interestingly, ideas from the model, which was originally developed to help challengers in their fight against incumbents, seem to fit well with activities of regime actors. Another interesting finding is that FCV are regime-preserving as well as regime-changing. Within the regime of automobility they are regime-preserving as they fit with current driver preferences, but for the energy sector they are a disruptive innovation, as the hydrogen is expected to be produced from wind power and biomass. The chapter also considers the relationship with battery electric vehicles, which is characterized as competitive *and* synergetic. They both compete for government support but the markets and technology are to some degree complementary. FCV incorporate advanced batteries and electric engines (just as electric vehicles do), but its users are believed to be different, making it sensible for car manufacturers to invest in both vehicle types.

Chapter 12, "Transition by Translation" by Bonno Pel, Geert Teisman and Frank Boons, examines the innovation journey for travel information systems as a journey driven by cascading events. Drawing on complexity theory, and the concept of translation from actor network theory, it describes the creation of traffic information products. These include the different moments at which innovation processes got stuck and the subsequent cascading events that broke the stalemate. The problem of creating a business case for traffic information was resolved by TomTom's decision to integrate dynamic traffic data in their car navigation products without charging users specially for this. Public transport information provision met with opposition from transport operators to disclose dynamic data, revealing deficiencies in punctuality. The government stepped in to resolve this tension. The (ongoing) innovation journey for traffic information problems required translations and interventions. It also shows how regime-preserving change may cascade into regime-shifting change, an issue of great significance for the transition to sustainable mobility and management thereof. The traffic intelligence cascades developed an infrastructure that enables the introduction of road pricing, a vital measure for traffic control. Regime-preserving may thus lay seeds for regime-shifting change.

Chapter 13, "The Emergent Role of User Innovation in Reshaping Traveler Information Services" by Glenn Lyons, Juliet Jain, Val Mitchell and Andrew May, examines traveler information services from a transition perspective in which 'Intelligent Transport Systems' (ITS) is identified as

Introduction 25

the (sub)regime and 'user innovation' as an emergent niche development. There is a growing interest in people's travel choices and how these can be supported and influenced; information is seen to be a key aspect of this. This chapter looks at Web-based user innovations in the area of transport: information services that are being created by users themselves to tackle the mobility challenges they and others like them face. The chapter presents six examples of user innovation: CycleStreets, ParkatmyHouse.com, PickupPal, TrainDelays, MyBikeLane and Slugging. After highlighting the sorts of factors which characterize these niche developments, the chapter goes on to look at the prospect for a regime transition. It questions the apparent (as yet) limited uptake of user innovations, suggesting that these innovations may, as with mainstream information systems developments in the ITS regime, not be taking sufficient account of human behavior and limitations to behavior change where travel and mobility are concerned. The chapter finally explores what is considered an important question: Will the emerging user innovations reconfigure the ITS regime, or will they just be filling in the gaps?

Chapter 14, "Innovation in Public Transport" by Reg Harman, Wijnand Veeneman and Peter Harman, offers a study of innovation in public transport. Despite the reputation of public transport systems for being inflexible and non-innovative, several innovations have occurred in them. Rail innovations include high speed trains and local tram networks serving the hinterland of cities. Innovations in busing include special bus lanes, demand-dependent services, information provision about arrival times and short distance radio systems allowing buses to get priority at traffic lights. In urban transit, innovations include rapid personal transit and tram-trains. Green and energy-saving propulsion also have been introduced into public transport, with public sector services acting as a lead market for hydrogen and natural gas. In the countries studied, Great Britain and the Netherlands, shares of public transport (in passenger kilometers) went up slightly between 1997 and 2007. But goals of a better modal split for public transport have not been achieved. Better public transport does not automatically attract car drivers. Apart from the usual explanation of drivers being wedded to their car, offering better transport services presents a dilemma for public transport companies because commuters want different services than regular public transport users do. Commuters want rapid services with few stops, whereas people dependent on public transport want many local stops. Logistically both things cannot be easily achieved. The chapter furthermore learns that the Netherlands has been more successful in creating an integrated transport system and public transport-based spatial planning than Great Britain for geographical and institutional reasons. Of the various innovations, the tram-train concept is believed to hold great prospects of attracting commuters and widening access to cities. Providing free buses is not a good idea from the viewpoint of sustainable development, as they attract few car drivers, give rise to additional travel and increase public spending. When this was trialed in Hasselt (Belgium), more than half of the travelers were new travelers making low-value trips. Former cyclists started to take the bus.

Chapter 15, "Intermodal Personal Mobility" by Graham Parkhurst, René Kemp, Marc Dijk and Henrietta Sherwin, examines the development of intermodal travel in Europe, with case studies from the United Kingdom and the Netherlands. After defining intermodality and the reasons for its existence, the authors examine three cases of intermodality: access to trunk rail (by bike, taxi and public transport); short-range park and ride (to buses in the United Kingdom and urban rail in the Netherlands); and services to facilitate intermodal travel (mobility cards, integrated ticketing, intermodal information). The experiences and developments in both countries show that intermodal travel is a niche phenomenon both in terms of use and in terms of there not being a well-developed coalition behind it in terms of providers, policy makers, spokespersons and a community of experts. It is a niche caught between the regimes of car mobility and public transport, in which intermodal activities are not core. Both jurisdictions show that policy interest is unstable, often implicit, and over-dependent on local factors. In those cases where successful schemes of intermodality have been introduced, the impetus in a number of cases came from regime 'outsiders' joining up with traditional transport companies. An example is the OV-fiets (a public bike available from railway stations), invented by a bicycle organization and Prorail, which was brought 'inside' the regime through direct NS control (the national railway operator). Given the context-specificity, intermodal travel cannot be planned from the top, but must be developed from below, with carrots and sticks emanating from the top. Intermodal travel is interesting from a transition point of view, as it may constitute a potential mechanism of resilience, enabling society to 'divest' from car dependence once this regime is no longer sustained, with cars increasingly used in combination with other modes of transport, alongside the use of feet and bicycles for short-range trips and public transport for longer trips. This hoped for effect does not happen easily. Car-restraining policies are needed to complement investment in intermodal interchanges, both for reasons of promoting intermodal travel and for making sure that there are sustainability benefits. One effect of better intermodality is that it promotes more travel. For example, in some places park and ride has encouraged car use.

The final chapter of the book (Chapter 16) brings the various contributions together. It draws conclusions about the sources of stability, mechanisms of change, the potential of various developments (i.e., which developments can be expected to break out) and the long-term effects of multiple developments. The reader will see that a major turning point has been reached on various issues, such as the trend towards hybrid forms of mobility, both in personal transport and public transport. At present, more activities, beliefs and resources are oriented towards a 'greening of cars', which sustains the existing car-based system, than towards more comprehensive system innovations based on combinations of different forms of transport. For the latter to become more likely, we conclude that stronger landscape pressures, such as related to climate change and Peak Oil, and their translation into stronger policies are needed. The chapter also uses the empirical findings to draw theoretical conclusions that are relevant for transition studies (e.g., the importance of interactions between *multiple* regimes, the continuing importance of regime actors in sustainability transitions and the fact that some landscape developments work to *stabilize* existing regimes). The chapter also evaluates current transport policies, concludes that these are too weak to bring about transitions and develops suggestions for a better sustainable transitions policy. Eight generizable policy lessons are derived from findings of the various chapters.

Together, the various chapters offer a wide and comprehensive analysis of stability and change in the automobility system.

NOTES

1. See http://www.swov.nl/nl/research/kennisbank/inhoud/00_trend/05_mobiliteit/reizigerskilometers.htm.
2. In the United Kingdom, 26% of households possessed two cars and 5% three cars or more in 2005.
3. See http://www.allaboutprius.com/blog/1014178_toyota-prius-a-brief-history-in-time.

REFERENCES

Arthur, W.B. (1989) 'Competing technologies, increasing returns, and lock-in by historical events', *Economic Journal*, 99(394): 116–131.
Banister, D. (2008) 'The sustainable mobility paradigm', *Transport Policy*, 15(2): 72–80.
Bassett, Jr., D.R., Pucher, J., Buehler, R., Thompson, D.L. & Crouter, S.C. (2008) Walking, cycling, and obesity rates in Europe, North America, and Australia, *Journal of Physical Activity and Health*, 2008(5): 795–814.
Dennis, K. & Urry, J. (2009) *After the Car*, Cambridge, England: Polity Press.
Department for Transport (2010) Transport Trends, Department of Transport, United Kingdom.
European Environment Agency (2009) *Greenhouse Gas Emission Trends and Projections in Europe 2009. Tracking Progress Towards Kyoto Targets*. EEA Rep. No.9/2009, Copenhagen: European Environment Agency.
Eurostat (2009) *Panorama of Transport, 1990–2006*, Luxemburg: Eurostat, European Commission.
Geels, F.W. (2002) 'Technological transitions as evolutionary reconfiguration processes: A multi-level perspective and a case-study', *Research Policy* 31(8/9): 1257–1274.
Geels, F.W. (2004) 'From sectoral systems of innovation to socio-technical systems: Insights about dynamics and change from sociology and institutional theory', *Research Policy*, 33(6–7): 897–920.
Geels, F.W. (2005) 'The dynamics of transitions in socio-technical systems: A multi-level analysis of the transition pathway from horse-drawn carriages to automobiles (1860–1930)', *Technology Analysis & Strategic Management*, 17(4): 445–476.

Geels, F.W. (2010) 'Ontologies, socio-technical transitions (to sustainability), and the multi-level perspective', *Research Policy*, 39(4): 495–510.
Geels, F.W. & Schot, J.W. (2007) 'Typology of sociotechnical transition pathways', *Research Policy*, 36(3): 399–417.
Geels, F.W. & Smit, W.A. (2000) 'Failed technology futures: Pitfalls and lessons from a historical survey', *Futures*, 32(9/10): 867–885.
Henderson, J. (2009) The politics of mobility: De-essentializing Automobility and Contesting Urban Space, in J. Conley & A. Tigar McLaren (eds.) *Car Troubles: Critical Studies of Automobility and Auto-Mobility*.
Hoogma, R., Kemp, R., Schot, J. & Truffer, B. (2002) *Experimenting for Sustainable Transport. The Approach of Strategic Niche Management*, London: EF&N Spon.
Jordan, A. (2008) 'The governance of sustainable development: Taking stock and looking forwards', *Environment and Planning C: Government and Policy* 26(1): 17–33.
Kemp, R. & Martens, P. (2007) 'Sustainable Development: How to manage something that is subjective and that never can be reached?' *Sustainability: Science, Practice & Policy*, Fall 2007, 3(2): 1–10, Available at http://ejournal.nbii.org/archives/vol3iss2/0703-007.kemp.html.
Lanberg, B. (2001) *The Skeptical Environmentalist Measuring the Real State of the World*, Cambridge University Press, New York.
Mitchell, M., Borroni-Bird, C. & Burns, L. (2010) *Reinventing the Automobile: Personal Urban Mobility for the 21st Century*, Cambridge, MA: MIT Press.
Newman, P. & Kenworthy, J. (1999) *Sustainability and Cities: Overcoming Automobile Dependence*, Washington, DC: Island Press.
Nill, J. & J. Tiessen (2005) Policy, Time and Technological Competition: Leanburn versus catalytic converter in Japan and Europe, in C. Sartorius & S. Zundel (eds) *Time Strategies, Innovation and Environmental Policy*, Cheltenham, England: Edward Elgar.
Rifkin, J. (2002) *The Hydrogen Economy: The Creation of the Worldwide Energy Web and the Redistribution of Power on Earth*, New York: Tarcher/Putnam.
Rijkswaterstaat (2004) *Ontwikkelingen Verkeer en Vervoer, 1990–2020: Probleemverkenning voor de Nota Mobiliteit*, Rotterdam.
Rip, A. & Kemp, R. (1998) Technological Change, in S. Rayner & L. Malone (eds) *Human Choice and Climate Change, Vol 2 Resources and Technology*, Washington, DC: Batelle Press.
Romm, J.J. (2005) *The Hype About Hydrogen: Fact and Fiction in the Race to Save the Climate*, Washington, DC: Island Press.
Sheller, M. (2004) 'Automotive emotions: Feeling the car', *Theory, Culture & Society*, 21: 221–242.
Sperling, D. & Gordon, D. (2009) *Two Billion Cars. Driving Toward Sustainability*, Oxford, England: Oxford University Press.
Unruh, G.C. (2000) 'Understanding carbon lock-in', *Energy Policy*, 28(12): 817–830.
Walker, W. (2000) 'Entrapment in large technology systems: Institutional commitments and power relations', *Research Policy*, 29(7–8): 833–846.

2 Visions for the Future and the Need for a Social Science Perspective in Transport Studies

Glenn Lyons

2.1. INTRODUCTION

This chapter sets the scene for the book from the perspective of transport studies. It describes how transport studies has evolved in the last 40 years, in tandem with real-world developments in the transport sector, from a primary focus on technology fix to a broader orientation that also includes behavior change. This shift has been accompanied by a greater role for the social sciences in transport studies, as this chapter will demonstrate with a case study of the UK Foresight Programme, in particular the Intelligent Infrastructure Systems (IIS) project in which the author participated. This was a major exercise that afforded the transport sector the opportunity to look beyond its day-to-day priorities and consider prospects on the far horizon. The study of this Foresight project shows how perspectives on the future of transport, and by implication our interpretations of the development of our present transport system or regime, are evolving. What the study reveals is a growing recognition of the need, as pointed to in Chapter 1, for a 'transport in society' perspective among the actors responsible for both the delivery and use of our transport systems. With regard to the new challenge of sustainability transitions, it can be argued that transport studies may need to enter a new phase, where the dynamic interaction between technology and behavior is investigated from a socio-technical perspective. While this chapter highlights the importance of such a 'way of thinking' about the automobility regime, later chapters of the book will have to substantiate this suggestion.

At the outset of the chapter, it is appropriate to clarify its specific reference to social science. A recent commission into the state of social sciences in Britain (Commission on the Social Sciences, 2003) acknowledged that there is "no simple or unambiguous specification of the social sciences." (31) It viewed social sciences as "'disciplined curiosity about societies in which we all live' leading to the creation and sharing of social knowledge." (32) The commission referred to "the old-style view—still present in some thinking—that sees social sciences as a 'back-end fix' to the problems arising from new scientific developments." In this chapter, the intention is to underline a (continuing need to) move away from the back-end fix mentality in transport. In order to do so, a rigorous and prescriptive definition of the heterogeneous array of disciplinary elements deemed to constitute social science is not necessary. Instead the following is offered as a guide to the reader: *In contrast to technology and data-driven advances leveled at the transport system associated with the disciplines of engineering and physical science, social science in this chapter is taken to refer to a discipline of thinking that aims to place transport in the broader system of society and social practices and to enforce a view that advance is derived from human behavior—facilitated by rather than achieved through scientific development.*

2.1.1. Technology Fix and Behavior Change

Two important strands in the real-world transport domain and in transport studies can be described in simple terms as a 'technology fix' and 'behavior change' approach (Lyons, 2011). Technology fix concerns itself with seeking to improve or preserve the regime of automobility as we know it- using technological advances to improve the environmental and economic efficiency of transport systems' operation. Behavior change points to prospects of adaptation if not transition in the automobility regime. Change can encompass the spatial and temporal configurations of travel. It also concerns the means of travel used ranging from motorized mobility through walking and cycling to the prospect of virtual travel where information rather than people or goods moves to meet an individual's need for access and social or economic participation.

Both strands have their supporters, and tensions between them are apparent—reflective of the tensions between stability and change. The chapter illustrates the growing recognition that when placing transport in society there is a legitimate place for both strands. However, while perspectives on transport are evolving and have come to recognize transport as more than only a system of mobility, it seems we have yet to embrace a sociotechnical transition perspective that seeks to understand how technology and the behavior change of all actors (individuals, firms, policymakers, etc.) dynamically interact and co-evolve.

From a historical perspective, transport studies can be said to have three phases to its development with a respective focus on the following:

(a) *Infrastructure and vehicles*—creating and maintaining the asset
(b) *Traffic management*—pursuing efficient operation of the asset in light of the demands placed upon it
(c) *Demand management*—influencing how and how much the asset is used in an effort to bring supply and demand into balance

Technology fix is concerned with (a) and (b), and behavior change is concerned with (c). While the three phases may appear to be sequential, they in fact co-exist. An enduring problem is that we struggle to appreciate how the different phases influence one another and act together; and we also struggle to situate this treatment of the transport system in the wider regime of automobility and in turn struggle to fully appreciate how the regime is, if at all, changing or how it could be changed. The next two sub-sections describe real-world transport developments and the evolution of transport studies. The chapter then turns to the case study of the Foresight Programme and the IIS project, and ends with conclusions and suggestions about a possible next step towards a new transition perspective on transport.

2.1.2. Real-world Transport Developments

Transport as a sector or as a discipline of academic study might be perceived as being principally concerned with the movement of people and goods from one place to another. This in turn involves investment in and the study of the infrastructure, vehicles and users of the transport system. The goals of such endeavor have been to ensure the efficient and effective operation of the system in ways that seek to offer fast(er), (more) reliable and safe(r) journeys. Journey speed has, for decades if not centuries, been an imperative that has driven progress which has seen motorization take us from a horse-drawn world to one of trains, planes and automobiles. A strong theme in transport policy of the past if not the present has come to be known as 'predict and provide'. This has been particularly associated with road traffic whereby predictions of future traffic growth have informed investment decisions to expand the capacity of the road network to accommodate that growth.

A brief selection of statistics highlights how transport has 'evolved' (placed in inverted commas to question the positive connotation of the term) in recent times (Department for Transport [DfT], 2009; Office for National Statistics, 2006). In Great Britain, the distance we travel annually has increased massively in the past half a century. In 1961 the figure was 295 billion passenger kilometers. By 2007 this had increased to 817 billion passenger kilometers. Domestic air travel, by distance, has increased by nearly ten times over the same period. The car has come to be and remains by far the dominant mode, accounting, in 2007, for 84% of all recorded distance traveled. In 1961, 69% of households did not have regular use of a car—by 2007 this figure had reduced to 24% (with 32% of households with two or more cars). This inexorable growth in motorized mobility has come to represent, in many people's view, a losing battle in the sense that providing and managing transport system capacity cannot keep pace with the demand to use the system. The familiar adverse effects of congestion and pollution have tended to be seen as the most evident and troublesome consequences, and indeed 'solving congestion' (Goodwin, 1997) has become a holy grail of the transport profession (though perhaps more recently overshadowed by the challenge of 'solving carbon emissions').

Looking back a little over 80 years, at the early stages of the motor age, the presence of congestion was already being faced. The experimental solution of the time to address this was to paint the first white line in a London street (see Figure 2.1).

If only our contemporary endeavors could be a simple as this! Existing as we do in the early stages of the information age and faced with transport challenges much greater in scale than in 1924, we have looked increasingly to new technologies and telecommunications in a bid to help address the ongoing development of our transport systems. Many national and international programs of research and implementation initiatives have looked to science and technology in a bid to further improve the efficient movement of vehicles (and people) through our transport systems and to gather data and transform that data into information and knowledge that can positively assist the decisions of policymakers, transport system operators and system users.

However, the information age is also, it seems, bringing with it an age of social science in relation to transport. At the highest level, Government now realizes that it can no longer hope (or afford) to build its way out of congestion (DfT, 2004). Accordingly, increasing attention has been given to demand management—addressing and influencing where, when and how we travel in an effort to better match transport supply and demand. In essence, demand management is about carrots (e.g., better public transport) and sticks (e.g., road pricing) which seek changes in individual behavior. To change behavior requires an *understanding* of behavior and an understanding of *why* we travel. Social and environmental policy has also exerted greater influence on transport policy objectives with climate change and social exclusion as key concerns. Transport more so perhaps than ever

Figure 2.1 First white line in London (image reproduced from Morton, 1934).

before is expected to play its part in creating a more sustainable and inclusive society. The need to understand how people think and behave, the attitudes and beliefs they hold, the social practices in which they engage and the physical organization of society is receiving growing attention.

2.1.3. The Development of Transport Studies

In line with changing 'real-world' priorities, the academic field of transport studies has also changed over the last couple of decades. Defining the scope of 'transport studies' is as difficult as to do so for social science. However, for over 40 years an umbrella organization for UK academic research in transport has existed called the Universities Transport Study Group (UTSG).[1] The activity of this community offers an instinctive sense of scope and its members have, over 40 years, often provided the intellectual voices inputting to the ongoing transport policy debate. As one of the founding figures in this research community, Professor Richard Allsop recently undertook a review of four decades of papers presented at the annual UTSG conference. This offered a crude but valuable and effective means of charting how the content and focus of transport studies have changed over this period. He observed that there "has been a substantial shift of emphasis towards the behavioral, social and environmental aspects of transport studies alongside continuing activity concerned with engineering, technology and operational aspects of transport systems" (Allsop, 2006, 5).

The UTSG community has offered the most substantial UK academic voice on transport issues. However, it is not the only voice—there are, for example, overlapping communities of transport economists and transport geographers. There has also been a recent 'awakening' of sociology and other dimensions of social science to the merits of transport becoming the subject of greater study. However, here the term 'mobility' dominates and is seen to represent much more than the physical travel from origin to destination with which the term 'transport' is associated. The recently formed Centre for Mobilities Research (CeMoRe) at Lancaster University in the United Kingdom and more recently founded Mobilities Research and Policy Center at Drexel University in the United States are significant focal points of an evolving 'new mobilities paradigm' (Sheller & Urry, 2006).[2] CeMoRe's Web site suggests the following. "The concept of 'mobilities' encompasses both the large-scale movements of people, objects, capital, and information across the world, as well as the more local processes of daily transportation, movement through public space, and the travel of material things within everyday life."[3]

It would seem that the nature of transport, or now transport and mobilities, study is changing or expanding in a way which is bringing to the fore a greater diversity of thinking and perspectives. This is not to suggest that the importance of science and technology within such study is diminished but rather that a greater richness of understanding is evolving as contemporary practice and debate concerns itself with a new interpretation of transport. It may once have been sufficient to believe that transport was here to *serve* society in an era when constructing new infrastructure and managing the flows of people and vehicles on it was the business of the transport profession. However, as the problems of a mobility dependent society are faced, a more sophisticated understanding of transport has been unfolding. We must now recognize that far from merely serving society and social practices, transport *shapes* them as in turn they shape transport. With such recognition comes a need for new endeavors in policy and practice that seek to ensure transport *supports* society, social patterns and practices in an appropriate manner (Lyons, 2004).

One such new endeavor is the application of Foresight to the transport sector. The case study in the next section will show that the Foresight Programme experienced a similar 'socialization of transport' as was discussed previously for the field of transport studies. While the first of three rounds of Foresight considering transport mainly focused on promising new technologies, later rounds increasingly focused on the social dimension of future transport. This will be elaborated in detail for the IIS project. The aims of the case study are as follows:

- To provide an insight into the 'Foresight' approach to facilitating evidence-based policy and strategy as a case study of how perspectives on transport are changing
- To critically reflect upon the Foresight project's social science dimension and the progression towards a point where a socio-technical transition approach could now be beneficial

The next section of the chapter first describes the Foresight Programme and then the IIS project, indicating the methodological process followed and summarizing the materials and messages that emerged. The chapter's concluding discussion then focuses upon a selection of issues that relate to the social science 'discipline of thinking' and which have featured in discussions that shaped the Foresight project. The chapter argues that this discipline of thinking while needing to continue to be encouraged needs to go further towards a transition perspective in order to assist the formulation of policy and strategy.

2.2. THE FORESIGHT PROGRAMME

The UK Government's Foresight Programme was initiated by the then UK Government's Office of Science and Technology, along with the Department for Transport as the sponsor. The study looked at the future of transport in order to inform and influence key public and private sector decision makers. An important feature of the ensuing IIS project under the Foresight Programme was its early recognition of the need to place transport in its social context.

The Foresight Programme introduces itself as follows: "Foresight, and its associated horizon scanning centre aims to provide challenging visions of the future, to ensure effective strategies now. It does this by providing a core of skills in science-based futures projects and unequalled access to leaders in government, business and science."[4] With its aim of improving "the relative performance of UK science and engineering and its use by government and society" the focus is upon identifying opportunities for science and technology to address challenges facing society. The Foresight Programme "brings together key people, knowledge and ideas to look beyond normal planning horizons to identify potential opportunities from new science and technologies and actions to help realise those opportunities."

Foresight is currently in its third round. The Technology Foresight Programme was initiated in 1994 following the Government's White Paper 'Realising Our Potential: A Strategy for Science, Engineering and Technology' (1993). The first round was comprised of 16 sector-facing panels consisting of experts from academia, industry and government. The remit was to look 20 years ahead examining emerging market and technological opportunities and associated research priorities. Following published visions and recommendation for action in 1995, 4 years of development and implementation followed. Reporting for the Transport Panel noted that "the growth of personal mobility in particular is a story of remarkable success for transport and expanding opportunity for suppliers and operators"[5] but went on to acknowledge it had come at a price—congestion delays, accidents and environmental impact. The opportunities for technology were highlighted including the prospect of the information age reducing the need to travel and assisting in the undertaking of travel. It was suggested that the greatest challenge was to develop more eco-friendly vehicles to be able to meet demand for mobility and meet environmental targets. Many of the Transport Panel's priorities focused upon three "innovative ideas": The Informed Traveler (providing integrated multi-modal travel information to the traveler), Foresight Vehicle (stimulating the UK car industry to produce more environmentally friendly vehicles suitable for the mass market) and Clear Zones (creating more livable urban centers).

The second round of the Foresight Programme began in 1999. This round again centered upon sector (and thematic) panels looking 20 years ahead. The Built Environment and Transport sector Panel ran until 2001 and involved three task forces including the Social and Motivational Behavior Task Force. The chairmen of this task force noted that "This Task Force was established to examine the social dimension of future transport technologies. Simply looking at the technological advances in the transport sector without considering the social and in particular, the behavioral and motivational dimensions is incredibly wasteful" (see endnote 5). It looked at three groups in society: people aged over 65, one-person households and those on low income. By examining literature, technology projects and considering two case studies (involving smart cards, travel information and online shopping), the task force's concluding recommendations were as follows: Improvements to public transport, greater transparency for the public about the transport system, more research into the impacts on travel behavior of demographic change, greater flexibility of smart card payments, and greater consultation with users in the design of new technologies.

The Commission on the Social Sciences (2003, 29) observed that "the government's two large Foresight exercises differed dramatically: the first was very largely technologically-driven whilst the latest one incorporated somewhat more social science in the setting of the starting point and in various scenarios." According to the available summarizing text concerning the treatment of transport in the two rounds, the first round is firmly centered upon the technological possibility of facilitating mobility while ameliorating the adverse environmental effects (technology fix). The second round is somewhat more attentive to behavior, attitude and variation across social groups though in its recommendations appears still to concentrate quite narrowly on the enabling technologies themselves and seems not to place transport in its wider social context.

The third (current) round of the Foresight Programme follows a review of the Programme in which it was determined that greater focus was needed to direct attention to where it was most needed. The Programme has thus moved away from its panel structure to instead concentrate upon a rolling program of projects which began in 2002. The IIS project began in September 2004. Within the Foresight Programme, consideration of transport has thus spanned over 10 years. The third round has underlined a further and quite pronounced increase in the significance of social science input to envisioning the future of transport within the Programme.

The author would suggest that of the many transport visioning exercises that have taken place, the IIS project has proved to be one of the most progressive in its engagement with social science. Social science has been embedded in the project from its early conception. This is important since the "social sciences contribute best to central concerns of society . . . by being involved in 'big questions' from the very outset, rather than as a 'back-end fix'" (Commission on the Social Sciences, 2003, 30).

The chapter now turns to examine the IIS project itself in terms of its approach and considerations.

2.3. INTELLIGENT INFRASTRUCTURE SYSTEMS

The stated aim of this project has been to "explore how science and technology may be applied over the next 50 years to the design and implementation of IIS that are robust, sustainable and safe."[6] The project's title and indeed the aim itself do not evidently point towards an exercise that will look at the future of transport and accommodate a social science dimension. The project initially defined 'infrastructure' as any platform used in the delivery of shared services to people. The transport system connects people to opportunities, social networks, goods and services (Kenyon, Lyons, & Rafferty, 2002) and was taken to constitute the infrastructure system in question. The title of the project might then have been 'Intelligent Transport Systems' (ITS). However, this phrase has been the long-adopted label of a specific facet of transport that sees technology at its heart (see Chapter 13). A recent Government framework document has set out the role of ITS in supporting the delivery of transport objectives (DfT, 2005, 44). The following answers the question 'what are Intelligent Transport Systems?': "Combinations of information processing, maps, databases, communications and real-time data from a range of sensors, to produce solutions that enable:

- infrastructure owners and operators to improve the quality, safety and management of transport networks;
- individual travelers, drivers, hauliers, transport operators and authorities to make better informed, more 'intelligent' journey decisions;
- network operators and 'third party' service providers to supply advanced information services, increasingly on a multi-modal basis, to all types of traveler; and road users to drive safer, 'smarter' vehicles."

Such issues are important, but they align more readily with a 'transport is here to serve' perspective than with one of 'transport is here to support society' in relation to how new advances in science and technology can be put to good effect. The IIS project certainly embraced ITS but only as part of a broader outlook.

2.3.1. Methodology

The research process within the project involved a number of strands and nearly 300 people in different capacities. There is not the space within this chapter to provide a comprehensive description of the project structure and process. Such information is, however, available on the project website.[7] The author's focus here is on two of the core activities within the project, namely the development of a set of state-of-research reviews and the production of a set of scenarios. These have been core in the sense that they constitute a substantial proportion of the published material from the project in their own right and also because it is from the act of creating the reviews and scenarios (alongside also the production of a technology forward look report[8]) that the thinking of key figures within the project has evolved and enabled in turn the creation of an overview report which seeks to capture the essence of valued considerations and messages that have existed and evolved within the project as a whole.

38   *Glenn Lyons*

At this point it is important to clarify the purpose of this and other Foresight projects. The intention is not to develop policy or strategy or to attempt to foretell what the future will have in store.[9] For some this has perhaps appeared a surprise or a frustration, and it might seem that the project in not so doing is evading confronting the pressing challenge of mapping an effective way forward toward a better (transport) future. However, the very strength of the project approach is to underline that the future is not predetermined and waiting to happen—it is ours to shape. The task of shaping must rightly fall to decision makers who face the challenge of reconciling often competing objectives in making choices. The purpose of the project therefore is to provide a resource of both state-of-the-art knowledge and a way of thinking and assessing that knowledge that can then be used to engage key decision makers to enable them to make better informed choices.

A total of 18 science reviews were commissioned by the project. A team of commissioning editors was responsible for overseeing this process, including this chapter's author. The reviews were grouped at the launch of the project into 'society', 'environment', 'technology', 'information' and 'policy and economics'. There were 5 reviews in the first grouping:[10]

- *Social factors in travel* (Axhausen, 2006)—This examined the consequences of a world in which costs of travel and communication are reducing, highlighting not only the changes brought to where we live, work and shop but also the importance and our limited understanding of how social networks impact upon and are impacted upon by personal travel.
- *The social impacts of intelligent infrastructure on transport* (Little, 2006)—This paper considered the role of social science in studying and understanding both the possibilities that technologies can deliver and the needs that we would, knowingly or otherwise, like them to fulfill ('user pulls').
- *The psychology of travel* (Stradling, 2006)—This review examined the question of why people travel and challenged the simple notion of time/cost minimization in travel choices by highlighting the (un)welcome expenditure of physical, mental and emotional effort associated with making journeys and how this can differently affect people's aspirations and decisions.
- *The role of information in decision making in transport* (Lyons, 2006)—This considered the importance of individuals' strategic and tactical decisions in determining patterns of travel and the place of information provision in supporting or influencing decisions, but it also highlighted the significant barriers to technological possibility in this regard of satisficing behavior and habit.
- *Public perceptions of risk* (Eiser, 2006)—This review considered how the public respond to uncertainty in their daily lives and in the face

of change and innovation and examined the notions of trust, caution and cognitive heuristics as well as social amplification and social attenuation as determinants of evolving public behavior and thus the nature of science and technology's impacts on society.

The reviews' preparation fed into the scenario planning exercise within the project. Scenario planning exposes uncertainty about the future, highlighting

Table 2.1 Summaries of the Four Intelligent Infrastructure Systems Scenarios

Good Intentions	Perpetual Motion
• The need to reduce carbon emissions constrains personal mobility. • Traffic volumes have fallen and mass transportation is used more widely. • Businesses have adopted energy-efficient practices: they use wireless identification and tracking systems to optimize logistics and distribution. • Some rural areas pool community carbon credits for local transport provision, but many are struggling. • Airlines continue to exploit loopholes in the carbon en-forcement framework.	• Society is driven by constant information, consumption and competition. In this world, instant communication and continuing globalization has fueled growth: Demand for travel remains strong. • New, cleaner, fuel technologies are increasingly popular. Road use is causing less environmental damage, although the volume and speed of traffic remains high. Aviation still relies on carbon fuels—it remains expensive and is increasingly replaced by 'telepresencing' for business, and rapid trains for travel.
Tribal Trading	**Urban Colonies**
• The world has been through a sharp and savage energy shock. The global economic system is severely damaged and infrastructure is falling into disrepair. • Long-distance travel is a luxury that few can afford, and for most people, the world has shrunk to their own community. • Cities have declined and local food production and services have increased. • There are still some cars, but local transport is typically by bike and by horse. • There are local conflicts over resources: Lawlessness and mistrust are high.	• Investment in technology primarily focuses on mini-mizing environmental impact. • Good environmental practice is at the heart of the United Kingdom's economic and social policies: Sustainable buildings, distributed power generation and new urban planning policies have created com-pact, dense cities. • Transport is permitted only if green and clean—car use is energy-expensive and restricted. • Public transport—electric and low energy—is efficient and widely used.

Note: Reproduced from the main project report; Office of Science and Technology (2006).

the multiplicity of different futures that could unfold dependent upon the nature and interaction of a large number of drivers for change. The development of scenarios must not be confused with the preparation of a vision. The latter is an end-state toward which one is seeking or aspiring to work; the former are outcomes which together can illustrate the diversity of future possibility and assist decision makers in considering policy formulation that is likely to be either most effective or most resilient in the face of the different possibilities. Looking ahead to the year 2056 and considering varying extents of people's acceptance of intelligent infrastructure and extents of transport's environmental impact, four scenarios were developed and given the following names: 'perpetual motion', 'urban colonies', 'tribal trading' and 'good intentions'. Table 2.1 summarizes the four scenarios[11] or 'sociologies of the future'.

During the course of the project, a high level of liaison with different government departments and key decision makers was forged with the intention of both informing the project development but also of engaging key individuals and organizations in the process of Foresight. The launch event for the project in January 2006 then marked the end of one process—the generation and documentation of thinking and knowledge—and the beginning of another—the encouragement of public and private sector decision makers to engage with and use this thinking and knowledge.

2.3.2. Findings

Perhaps the very essence of the project has resided with the term 'intelligence'. There is a danger when the term is used in association with technological advance—it can bestow undeserved anthropomorphic qualities on the technology itself. Intelligence can often implicitly refer to technological sophistication rather than the efficacy of the technology in achieving a greater good. It risks raised expectations of effect. Technology itself is not intelligent, it is enabling. Intelligence comes from how the technology is used. What follows is that the pathway of adaptation will be governed less by the technology than by the policies, control mechanisms and choices it enables. This concern about the treatment of 'intelligence' was acknowledged at a relatively early stage in the project. Project reporting (Office of Science and Technology, 2006, 2) then chose very fittingly to return to the matter of intelligence as a key focal point. It identified a need to invest in intelligence on four levels. "We need:

- *intelligent design*, minimising the need to move, through urban design, efficient integration and management of public transport and local production
- a system that can *provide intelligence*, with sensors and data mining providing information to support the decisions of individuals and service providers

- *infrastructure that is intelligent,* processing the mass of information we collect and adapting in real time to provide the most effective services
- *intelligent use* of the system where people modify their behaviors to use infrastructure in a sustainable way"

These are important principles to abide by and are strongly associated with what science and technology can *enable* rather than (only) associated with technological endeavor itself. However, espousing such principles and putting them into practice presents a need to confront the many more detailed considerations that have been summarized in the main project report.

The project has of course given substantial consideration to the technological capabilities of the future and what they can enable in pursuing intelligence. This includes the following: vast networks of tiny and inexpensive sensors to monitor the use of the transport system, data mining capabilities to yield nuggets of knowledge and understanding from gigantic sets of data, software agents capable of investigating travel and other options on behalf of their 'owners', complex modeling and simulation; major advances in the speed of transfer of information (e.g., allowing feature-film downloads in seconds), speech interfacing with computing, and complex information systems able to self-monitor for signs of instability. Seen in isolation, such a list might suggest an overshadowing of social science input.

However, the reporting notes, in relation to such technological advance the following. "Historically, when we have improved the transport system and reduced costs, people have travelled more ... A key issue is how to use the technologies to ensure that we not only improve efficiency, but also deliver sustainable and robust solutions" (ibid, 7). A number of 'softer' issues are highlighted accordingly and interposed with technological capabilities, including:

- a recognition of fundamental human needs to travel, an understanding of the positive utility of traveling itself;
- the need to consider lifestyle decisions (such as where we choose to live) that impact upon travel rather than only travel decisions themselves;
- the unexpected uses of technology compared to those originally intended (Goodwin, 2006);
- an overriding importance of using spatial planning to prevent excesses of motorized mobility dependence (Banister & Hickman, 2006), the greater achievement of access through virtual communications as an alternative to physical travel and the need for increased travel costs as well as increased telepresence capability for such potential to be realized;
- the realization that sustainability now deserves as much attention as economic growth (Köhler, 2006);
- a need to consider charging for the full cost of travel in conjunction with education and public awareness raising to achieve fundamental changes in behavior;

- a realization that communications are no longer restricted to connecting from place to place but are often now from person to person in a mobile world;
- an underlining that travel opportunity transforms from luxury to necessity;
- an understanding that travel enables social change and expanding and sustained social networks; and
- recognizing that human resistance to change that can sometimes subvert optimization and rationality.

This 'rebalancing' of the place of science and technology and of social science in examining the future of transport is evident in the project report's examination of delivering intelligence: "Tempting as it may be to see the delivery of IIS in terms of technology, and technology will certainly be important, we have to remember that it is people who travel, not their cars, for example. So, delivering IIS will be as much a matter of understanding the psychology of travel, the social circumstances of the travelling public and of influencing their decisions as it is about technological development." (Office of Science and Technology, 2006, 34)

2.4. CONCLUDING DISCUSSION

The Foresight exercise has sought to broaden the thinking on transport, and it provides a reminder of the increased complexity that is introduced when one moves beyond a 'transport is here to serve' way of thinking.

2.4.1. A System of Systems Automobility Regime

Figure 2.2 is offered as an attempt to (partially) reflect this complexity. It is portrayed as a 'system of systems' and in a sense is reflective of what is being referred to as the automobility regime—much more than just a technical and functional operation of a transport system.

The diagram cannot pretend to be comprehensive. Notably, it is not possible to easily or fully portray the complex and poorly understood patterns and flows of cause and effect—certainly these will seldom be linear or unidirectional.

What the diagram is intended to illustrate is the array of sometimes conflicting inter-dependencies around which cause and effect are occurring and across which choices are being considered and decisions made. The outer layer of the diagram depicts society and its social practices as a whole and three of the main aspirations deemed to signify societal well-being. Meeting all three of these aspirations simultaneously is far from easy with many trade-offs occurring and seemingly conflicting priorities. The next layer points to the different forms of societal governance that interact, with their own (sometimes conflicting) objectives. Governance plays a significant part in determining whether and

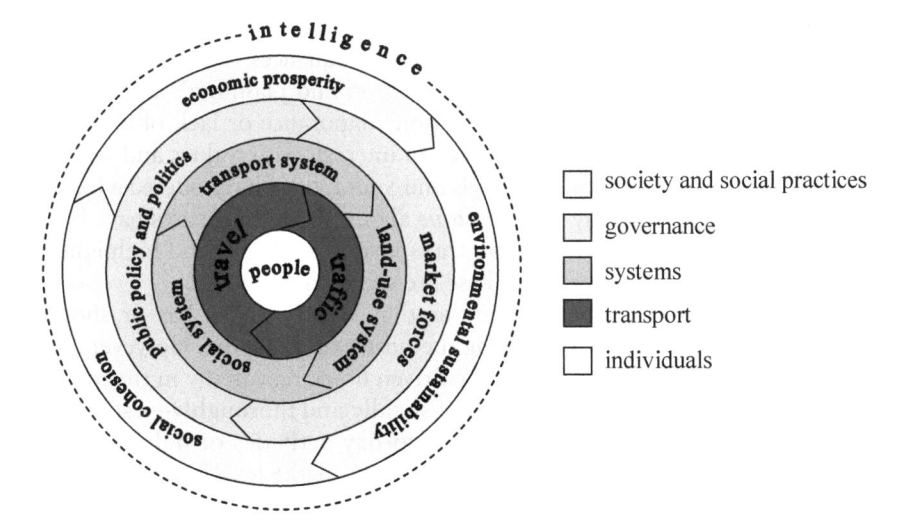

Figure 2.2 The 'system of systems' embodied by the pursuit or application of intelligence.

how societal aspirations and society itself evolves. It also significantly defines how three of the core systems within society operate and evolve. Importantly, the three systems themselves are strongly inter-dependent in their evolution, and this is something which has been well recognized in the IIS project. It is in the context of such inter-related and interacting systems that transport itself operates with the movement of people and goods (and information) and the associated movement of vehicles. At the center point of this system of systems is the largest set of actors within the system—the general public. Importantly, while they are impacted upon by the decisions and developments associated with the outer layers, the general public emanates a significant outward set of impacts through the other layers as people make not only transport choices but a wider array of lifestyle choices. Thus, each layer impacts upon every other layer. A key deduction is that effective decision making in any one layer needs to have a sufficient understanding of the whole system or regime if there is to be a reasonable understanding of the consequences of actions.

2.4.2. Challenges of Interpretation and Effecting Change

It is of course relatively easy to conceptualize such an overview interpretation. Joined-up-thinking in practice is well recognized as a much more challenging goal to pursue. However, it is a goal that must be pursued if we are to avoid deflecting from integrated decision making towards disintegration and disarray.

Unintended consequences are ever waiting in the wings if policy and strategy are not sufficiently well thought out and informed. Such a warning emanated from social science itself 70 years ago. A paper by the sociologist

Merton (1936) is seen by many as a seminal article which first framed the notion of unintended (or 'unanticipated') consequences. Merton identified five limitations to "successful social prediction and planning" and "a correct anticipation of consequences of action": ignorance or lack of adequate knowledge, error, imperious immediacy of interest, basic values and self-defeating prediction. Allied to this, Geels and Smit (2000) have offered a highly instructive insight into why many *visions* about transport futures have been wrong. They focused specifically on transport technologies and highlighted a number of pitfalls and lessons. These are shown in Table 2.2.

The previous paragraph reminds us that social science has or should have a central part to play in shaping the future of transport. However, this may be a message that will need to be driven home repeatedly in the face of the very real difficulties of thinking holistically and thoroughly—especially in the context of the more focused day-to-day activities of individuals in their professional lives. Bringing about change at the level of key decision makers takes time and effort, and one can never assume that once a point is made or principle agreed upon that it will endure in the course of future decision making and balancing of priorities.

For example, it took a period of many years to fully acknowledge the flaws in the predict and provide approach to transport policy. While we now have policy statements appearing to endorse a move away from this, there are also some stark indications that it remains an embedded orthodoxy. For instance, the Highways Agency in the United Kingdom is pursuing the use of the hard-shoulder on motorways to create an extra running lane at times of congestion[12] (Chase, 2006) which to some appears to be an alarming encouragement of predict and provide. Likewise, once we acknowledge the social consequences of cheaper and more readily available air travel—sustained international social networks, long distance business and leisure travel and so forth—then it becomes clear that air travel is moving from a luxury to a necessity such that plane dependence may be the new alternative to car dependence. Accordingly it could be seen as alarming that predict and provide appears to remain an accepted practice for the aviation industry.

Perhaps one of the challenges for social science is that it aligns less readily than physical science and technology with industry and the private sector. Science and technology can thus 'enjoy' strong support in the governance and systems rings of Figure 2.2 such that technology push rather than user pull can be at risk of prevailing. This said, one must be careful not to leave a sense from this chapter that it is believed that science and technology are not important. Far from it. They provide incredible opportunities. The information age in which we now live is substantially permeated by a use of and reliance upon a myriad of technological advances. The enduring concern is that science and social science must *both* feature significantly in our examination of the future and the formulation of policies and strategies. Foresight has contributed greatly to addressing and vocalizing this concern.

Visions for the Future 45

Table 2.2 Key Features That Have Shaped Images of the Future Role of New Technologies in Transport

Contemporary concerns and hopes	Perceptions of the future are shaped and colored by current problems and aspirations resulting in optimistic rather than plausible scenarios.
New technological trajectories	The pathway of technological innovation and product development may significantly change introducing new possibilities and expectations concerning the role in, and impacts on society of the technology.
New for old substitution	The role of a new technology is often phrased in terms of replacing or substituting the old technology while in reality old and new technologies often coexist, serving different markets, circumstances or purposes.
Social practices neutral	It is often wrongly assumed that the pool of social practices and needs remains unchanged thereby implying that new technology will (only) substitute certain social practices. In reality the pool of social practices can increase.
Narrow functional thinking	Through only functional thinking, new technologies can be judged capable of enabling the purpose of an activity to be fulfilled. This neglects to consider other social and psychological aspects of an activity that may not be addressed.
Societal embedding	The process of societal embedding of new technologies can be viewed as unproblematic when in practice many social and institutional adjustment processes have to take place which may not be straightforward and can take some time to achieve.
Promise games by product champions	Promoters in particular of an emerging technology can voice unrealistically high expectations. This may be to serve the purpose of creating a 'breathing space' for investment and development to continue. It may also be a consequence of neglecting the coevolution of technology and society, and underestimating the practical difficulties and resulting slowness of processes of societal embedding of technology.

Note: Adapted from Geels and Smit (2000).

2.4.3. The Next Step—A Transition Perspective

This chapter, it is hoped, has revealed how different perspectives prevail and evolve in addressing transport. The Foresight project has been progressive in its elevation of social science thinking and in its four-pronged interpretation of 'intelligence'. This must not be undervalued. However, in the context of this book it can be suggested that such progress still falls short in our attempts to robustly and effectively understand and influence how our transport system and the wider automobility regime are evolving. The following challenges among others remain, reflective of the potential value in a transition perspective being embraced in future strategy and policy formulation:

- *Social and technical interaction*—It is significant that Foresight has brought the social dimension forward from its prior position as an afterthought. However, the four levels of intelligence synthesized by the project tend to highlight a co-existence of social and technical considerations. What is needed is more attention embedded into the thinking regarding how, in practice, the social and technical elements *interact* dynamically in a co-evolution (Geels, Hekkert & Jacobsson, 2008).
- *Tensions between stability and change*—Foresight in its employment of scenario planning recognizes and exposes uncertainty about how the automobility regime will evolve. However, it has tended to do this with greater attention to the transport system than to the regime. Attention has been given to different sorts of change, but what is yet to be thoroughly addressed is an acknowledgement, understanding and interpretation of how different actors, developments and decisions may create tensions between (implicit) moves to stabilize (parts of) the regime and change it. Addressing these tensions holds the prospect of enhanced policy and strategy formulation.
- *Addressing all the actors*—The Foresight project has engaged with one set of actors (transport and social experts) to assess the behavior of another set (transport system users). However, to better understand and address prospects for transition requires that a socio-technical perspective also embrace how the behavior of organizations and policy makers is impacting upon the automobility regime.

It can be argued that these challenges not only apply to the Foresight Programme but also to transport studies. To deal with the new sustainability problems, the next phase of transport may consist of addressing the interactions between technology and behavior, stability and change and multi-actor dynamics. The following chapter fully introduces the transition perspective, expanding upon how such challenges as those discussed previously can be embraced.

NOTES

1. See http://www.utsg.net/
2. See also the international academic journal *Mobilities* available at http://www.tandf.co.uk/journals/titles/17450101.asp.
3. See http://www.lancs.ac.uk/fass/centres/cemore/
4. See http://www.foresight.gov.uk
5. Quote taken from online archival information on Foresight now no longer available—for historical overview see http://www.bis.gov.uk/foresight/about-us/history

Visions for the Future 47

6. See http://www.bis.gov.uk/foresight/our-work/projects/published-projects/intelligent-infrastructure-systems
7. See http://www.bis.gov.uk/foresight/our-work/projects/published-projects/intelligent-infrastructure-systems
8. See http://www.bis.gov.uk/assets/bispartners/foresight/docs/intelligent-infrastructure-systems/technology-forward-look.pdf
9. This is comparable to what is noted in Chapter 3 concerning the multi-level perspective: not being a 'truth machine' that automatically solves the puzzle when the right data are entered.
10. In fact at the stage of the reviews being commissioned, papers by Goodwin and by Banister and Hickman were also under the 'social science' umbrella which is reflected in their content—these papers are referenced later in the chapter.
11. For the full report "The Scenarios—Towards 2055" go to http://www.bis.gov.uk/assets/bispartners/foresight/docs/intelligent-infrastructure-systems/the-scenarios-2055.pdf
12. See http://www.highways.gov.uk/knowledge/1353.aspx.

REFERENCES

Allsop, R. (2006, January) 'Some reflections on forty years evolution of transport studies', Proceedings of the 38th UTSG Conference, Dublin, Ireland.

Axhausen, K. (2006) Social factors in future travel: A qualitative assessment, Review commissioned for the Foresight 'Intelligent Infrastructure Systems' project, Office of Science and Technology, Department for Trade and Industry, London.

Banister, D. & Hickman, R. (2006) How to design a sustainable and fairer built environment, Review commissioned for the Foresight 'Intelligent Infrastructure Systems' project, Office of Science and Technology, Department for Trade and Industry, London.

Chase, P. (2006, January) 'Maximising motorway capacity through hard shoulder running: UK perspective', Proceedings of the 38th Universities Transport Study Group Conference, Dublin.

Commission on the Social Sciences (2003) Great expectations: The social sciences in Britain, Academy of Learned Societies for the Social Sciences, March.

Department for Transport (2004) The future of transport, White Paper, Department for Transport, July, London: The Stationary Office.

Department for Transport (2005) Intelligent transport systems (ITS): The policy framework for the roads sector, Department for Transport, November, London: The Stationary Office.

Department for Transport (2009) Transport trends 2009 edition, Department for Transport, November, London: The Stationary Office.

Eiser, J.R. (2006) Public perception of risk, Review commissioned for the Foresight 'Intelligent Infrastructure Systems' project, Office of Science and Technology, Department for Trade and Industry, London.

Geels, F.W., Hekkert, M. & Jacobsson, S. (2008) 'The micro-dynamics of sustainable innovation journeys: Editorial', *Technology Analysis & Strategic Management*, 20(5): 521–536.

Geels, F.W. & Smit, W.A. (2000) 'Failed technology futures: Pitfalls and lessons from a historical survey', *Futures*, 32: 867–885.

Goodwin, P.B. (1997) Solving congestion (when we must not build roads, increase spending, lose votes, damage the economy or harm the environment, and will never find equilibrium), Inaugural Lecture for the Professorship of Transport Policy, University College London, 23 October.

Goodwin, P. (2006) The Economic factors that will affect future transport modes—Conjectures on the dynamic functional transformation of intelligent infrastructure, Review commissioned for the Foresight 'Intelligent Infrastructure Systems' project, Office of Science and Technology, Department for Trade and Industry, London.

HMSO (1993) *Realising Our Potential: A Strategy for Science, Engineering and Technology*, Cmnd. 2250, London: HMSO.

Kenyon, S., Lyons, G. & Rafferty, J. (2002) 'Transport and social exclusion: Investigating the possibility of promoting inclusion through virtual mobility', *Journal of Transport Geography*, 10(3): 207–219.

Köhler, J. (2006) Transport and the environment: Policy and Economic Consideration, Review commissioned for the Foresight 'Intelligent Infrastructure Systems' project, Office of Science and Technology, Department for Trade and Industry, London.

Little, S. (2006) The intersection of technology and society: Evaluating the impact of intelligent infrastructure on transport and travel, Review commissioned for the Foresight 'Intelligent Infrastructure Systems' project, Office of Science and Technology, Department for Trade and Industry, London.

Lyons, G. (2004) 'Transport and society', *Transport Reviews*, 24(4): 485–509.

Lyons, G. (2006) The role of information in decision-making with regard to travel, Review commissioned for the Foresight 'Intelligent Infrastructure Systems' project, Office of Science and Technology, Department for Trade and Industry, London.

Lyons, G. (2011) Technology fix versus behavior change, in M. Grieco & J. Urry (eds) *Mobilities: New Perspectives on Transport and Society*, Aldershot, England: Ashgate.

Merton, R.K. (1936) 'The unanticipated consequences of purposive social action', *American Sociological Review*, 1(6): 894–904.

Morton, H.V. (ed) (1934) *The Pageant of the Century*, London: Odhams.

Office for National Statistics (2006) *Social trends*. No. 36, London: Office for National Statistics/TSO.

Office of Science and Technology (2006) Intelligent infrastructure futures: Project overview, Foresight Programme, Office of Science and Technology, Department of Trade and Industry, London.

Sheller, M. & Urry, J. (2006) 'New mobilities paradigm', *Environment and Planning A*, 38(2): 207–226.

Stradling, S. (2006) Moving around: Some aspects of the psychology of transport. Review commissioned for the Foresight 'Intelligent Infrastructure Systems' project, Office of Science and Technology, Department for Trade and Industry, London.

# 3 The Multi-Level Perspective as a New Perspective for Studying Socio-Technical Transitions

Frank W. Geels and René Kemp

3.1. INTRODUCTION

The starting point for transitions research is a recognition that many environmental problems, such as climate change, loss of biodiversity and resource depletion (clean water, oil, forests, fish stocks), are formidable societal challenges. Because solutions that stay within present systems are insufficient to address these problems, policy makers, academics and civil society organizations increasingly argue that more radical change and transitions to new systems are needed, e.g., to achieve an 80% reduction in CO_2 emissions by 2050. Examples of prospective sustainability transitions are the transition from a fossil-fuel based electricity system to a low-carbon electricity systems, the transition from a throughput economy to one based on closed material loops and the transition from car-based mobility to alternative mobility systems which are less car-centered.[1]

The socio-technical approach to transitions conceptualizes transport, energy and agri-food systems as a configuration of elements that include technology, policy, markets, consumer practices, infrastructure, cultural meaning and scientific knowledge (Elzen, Geels & Green, 2004; Geels, 2004; Geels & Schot, 2010a; Kemp, Schot & Hoogma, 1998; Smith, Stirling & Berkhout, 2005). As a shorthand these systems are labeled 'socio-technical' systems, and major shifts are indicated as socio-technical transitions. The elements in socio-technical systems are linked to various actor groups (e.g., firms and industries, policy makers and politicians, consumers, civil society, engineers and researchers), who maintain, reproduce and change system elements. Transitions are seen as co-evolutionary processes, which take decades to unfold and involve many actors and social groups.

The socio-technical approach to transitions is broader than other approaches to sustainable development. To bring out the differences, we give a stylized characterization of the most prominent alternative approaches.

1. Neo-classical economists view environmental problems as negative externalities resulting from market failures. The government can help internalize external costs by changing incentives and frame conditions

(e.g., taxes, emissions trading) but should then let private initiative do the real work. If the prices are right, private actors (firms and consumers) will then find individual optimal (profit or utility maximizing or cost-effective) solutions, which are supposed to lead to a socially more desirable outcome. This framing often dominates policy discussions (e.g., Stern, 2006).
2. Psychologists focus on individuals and their attitude, behavior, and choice (Kaiser, Ranney, Hartig & Bowler, 1999). Assuming that behavior change is caused by changes in attitudes, they make policy recommendations that highlight information provision and education campaigns.
3. Deep ecologists relate environmental problems to the failures of modernism, capitalism and anthropocentrism (Næss, 1973). They therefore advocate eco-centrist approaches, which include 'green values' and ideological change (e.g., localism, self-sufficiency).
4. Engineers and industrial ecologists see environmental problems arising from inefficient and polluting production processes. Building on a strong belief that science and technology can deliver solutions, they advocate clean technology, eco-efficiency, dematerialization and the closing of material loops (Huber, 2000).
5. Political scientists study the development and struggles over formal goals and targets as embedded international treaties (e.g., Kyoto, Millennium Development Goals). These goals are subsequently translated into regulations and policy programs, which are then implemented and controlled by bureaucrats and backed up by sanctions. Environmental management standards (e.g., ISO 14001), performance reporting and environmental impact assessments may be part of this process.

Each of these approaches focuses on a limited set of dimensions of (un)sustainability. The socio-technical approach to transitions instead highlights co-evolution and multi-dimensional *interactions* between industry, technology, markets, policy, culture and civil society. Understanding large-scale transitions to new transport, energy, agri-food and other systems requires analytical frameworks that encompass multiple approaches in ways that addresses interactions between them. The multi-level perspective (MLP), which will be discussed in this chapter, is one such framework. While elements of other approaches, such as markets, technologies, political institutions, behavior and culture, remain relevant, the MLP articulates a longitudinal process approach that addresses the co-evolution of these elements (Geels & Schot, 2010a).

The MLP also provides a way of addressing the core analytical puzzle of transitions, namely stability and change. On the one hand, existing systems are characterized by stability, lock-in and path dependence, which give rise to incremental change along predictable trajectories. The following are examples of lock-in mechanisms: shared beliefs that make actors are blind for developments outside their scope; consumer lifestyles, regulations and laws that create market entry barriers; sunk investments in machines, people and infrastructure; resistance from vested interests; low costs because of economies of scale (Unruh, 2000; Walker, 2000). For sectors that are economically important (such as the car industry) there also exist exit barriers (see Chapter 6 for a discussion of entry and exit barriers for automobile companies). On the other hand, radical alternatives are being proposed, developed and tried by pioneers, entrepreneurs, social movements and other relative outsiders (to the existing regime). These alternatives typically face an uphill struggle against stable systems, because they are more expensive (since they have not yet benefited from economies of scale and learning curves), may require changes in user practices, face a mis-match with existing regulations or may lack an appropriate infrastructure. The core puzzle in transitions thus centers around (dynamic) stability and (radical) change, in particular their interactions which are played out on multiple dimensions. The MLP provides a way of investigating these issues.

This chapter offers an introduction into the MLP of niches, regimes and landscape as a socio-technical approach to study transitions in automobility. It offers a discussion of key processes, patterns and mechanisms in transitions (e.g., novelties growing in niches, regime actors resisting disruptive change, societal groups supporting alternatives, hype-disappointment cycles, regime actors becoming involved in transformation, bandwagon effects, issues of naming and shaming, lock-in) and how government policy is part and parcel of processes of change it seeks to manage. The transition perspective is discussed on the basis of examples from passenger road transport.

The chapter proceeds as follows. Section 2 delineates the transition phenomenon and explains its background. Section 3 describes the MLP which is central to the study of transitions. Section 4 describes typical transition patterns identified in empirical studies. Actor-related mechanisms in these patterns are discussed in Section 5. Section 6 deals with limitations for government to foster change, and Section 7 offers concluding remarks.

3.2. THE FIELD OF TRANSITION STUDIES

In the last decade three traditions to study transitions have emerged, each with their own specific focus. The book *Transitions to Sustainable Development* (Grin, Rotmans, Schot, Geels & Loorbach, 2010) provides a further description of these traditions. The first and perhaps best established tradition is that of socio-technical transitions, consisting of retrospective and future studies of socio-technical change. Important publications in this tradition are Geels (2002, 2004, 2005c), Elzen et al. (2004), Geels and Kemp (2007), Geels and Raven (2006), Geels and Schot (2007), Smith (2007), Verbong and Geels (2007) and Van Bree, Verbong and Kramer (2010). The second tradition is that of transition management (TM), developed by Jan Rotmans, René Kemp, and Derk Loorbach, drawing on a complex systems and (societal) governance perspective (Kemp, Rotmans & Loorbach, 2007; Loorbach, 2007; Rotmans, Kemp, & Van Asselt, 2001). The third tradition is that of reflexive modernization associated with John Grin and co-workers (Hendriks & Grin, 2007). A cross-cutting tradition is that of transition criticism, where scholars are asking critical questions about the MLP, the transformative power of niche developments and (mostly constructive) criticisms of the model of TM as a model for reflexively changing societal systems. Important publications in that tradition are Smith et al. (2005), Shove and Walker (2007, 2008), Hendriks (2008), Genus and Coles (2008), Voß, Smith and Grin (2009), Markard and Truffer (2008), Smith and Kern (2009), and Meadowcroft (2009).

3.3. THE MULTILEVEL PERSPECTIVE ON TRANSITIONS

The basic premise of the MLP is that transitions are non-linear processes that result from the interplay of multiple developments at three analytical levels: niches (the locus for radical innovations), socio-technical regimes (the locus of established practices and associated rules) and an exogenous sociotechnical landscape (Geels, 2002, 2005c; Rip & Kemp, 1998). These 'levels' refer to heterogeneous configurations of increasing stability, which can be seen as a nested hierarchy with regimes being embedded within landscapes, and niches existing inside or outside regimes (see Figure 3.1). Niches are 'protected spaces' such as research and development laboratories, subsidized demonstration projects or small market niches where users have special demands

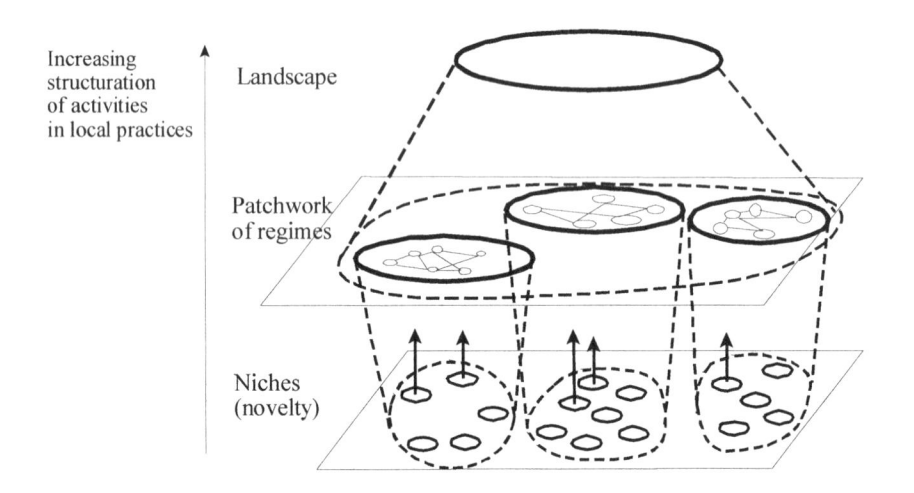

Figure 3.1 Multiple levels as a nested hierarchy (Geels, 2002, p. 1261). Reprinted by permission of the publisher (Elsevier).

and are willing to support emerging innovations (e.g., the military). Niche actors work on radical innovations that deviate from existing regimes. Niche actors hope that their promising novelties are eventually used in the regime or even replace it. This is not easy, however, because the existing regime is stabilized by many lock-in mechanisms. Nevertheless, niches are crucial for transitions because they provide the seeds for systemic change.

The MLP helps explain why there may simultaneously be a flurry of change activities (at the niche level) and relative stability of existing regimes. We briefly describe the three analytical levels next and operationalize them for the transport domain.

3.3.1. Niches

Within the MLP, novelties emerge in niches, that is, particular domains of use, actor constellations and geographical areas with special characteristics. The novelty may be a new practice, a new technology or special government intervention. What happens in the niche is shaped by external developments. For example, the use of bicycles or electric cars is shaped by the road infrastructure, priority rules, fiscal measures, climate change concerns and the economics of using other means of transport. These developments not only shape the willingness of individuals to engage in the use of a bicycle or electric car but also shape the expectations and strategies of companies and government.

The literature on niche-innovation (Kemp et al., 1998; Hoogma, Kemp, Schot & Truffer, 2002; Schot & Geels, 2008) distinguishes three social processes within niches:

- The articulation (and adjustment) of *expectations* or *visions*, which on the one hand provide guidance and direction to the internal innovation activities and on the other hand aim to attract attention and funding from external actors
- The building of social *networks* and the enrollment of more actors, which expand the social and resource base of niche-innovations
- *Learning processes* on various dimensions, about imperfections of technology and how they may be overcome, issues of organization, market demand, user behavior, infrastructure requirements, policy instruments and symbolic meanings

Niches are often carried by experimental or demonstration projects, which allow niche actors to learn about innovations in real-life circumstances. There is a 'local' element (of use and experimentation in concrete projects) and a 'global' element, which refers to the community of niche actors who share certain rules such as problem agendas, search heuristics, expectations and understandings (see Figure 3.2). Niches gain momentum if visions (and expectations) become more precise and more broadly accepted, if the

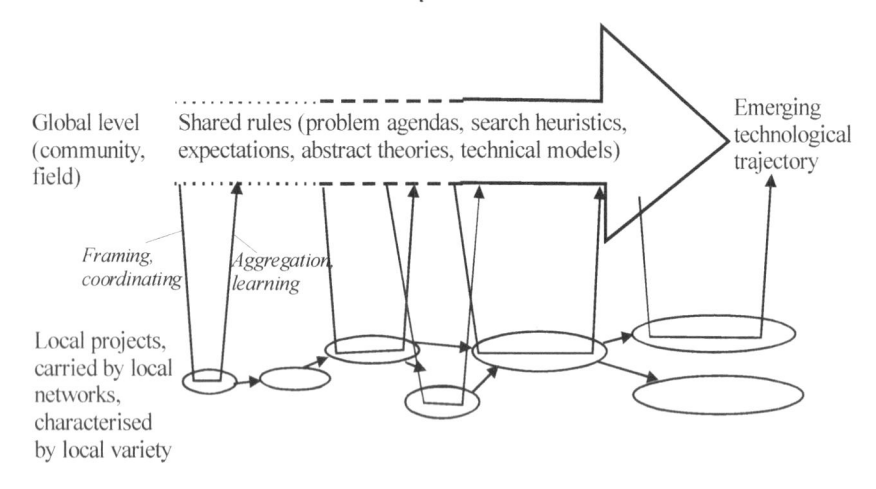

```
Global level       Shared rules (problem agendas, search heuristics,      Emerging
(community,        expectations, abstract theories, technical models)     technological
field)                                                                    trajectory

        Framing,      Aggregation
        coordinating  learning

Local projects,
carried by local
networks,
characterised
by local variety
```

Figure 3.2 Emerging trajectory carried by local projects (Geels & Raven, 2006, p. 379). Reprinted by permission of the publisher (Taylor & Francis Group, http://www.informaworld.com).

alignment of various learning processes results in a stable configuration ('dominant design') and if networks become bigger (especially the participation of powerful actors may add legitimacy and bring more resources into niches). This latter mechanism has occurred with regard to alternative engine/motor technologies: While fuel cells and battery electric drive were initially pioneered by outsiders and start-ups, the big car manufacturers have moved into these areas, often creating strategic alliances with these small firms (or taking them over; Dyerson & Pilkington, 2005, and Chapters 6 and 11).

3.3.2. Socio-technical Regime

Novelties must compete with technologies that benefit from well-developed systems around them. The alignment of existing technologies, regulations, user patterns, infrastructures and cultural discourses results in socio-technical systems (Geels, 2004). The system elements are reproduced, maintained and changed by various social groups and actors. These actors do no act in a vacuum but are instead embedded in socio-technical regimes, which are the deep-structural rules that coordinate and guide actors' perceptions and actions (Giddens, 1984). In existing regimes, innovation is mostly incremental because of lock-in mechanisms and path dependence. Change still occurs but proceeds in predictable directions, giving rise to stable trajectories.

Rip and Kemp (1998) defined a technological regime as "the rule-set or grammar embedded in a complex of engineering practices, production process technologies, product characteristics, skills and procedures, ways of handling relevant artifacts and persons, ways of defining problems—all of them embedded in institutions and infrastructures" (p. 388). Technological regimes are outcomes of earlier changes, and they structure subsequent change. Geels (2004) proposed the broader notion of *socio*-technical regimes to encompass not only firms and the activities of engineers but also other social groups, such as users, policy makers, special-interest groups and civil society actors. This broader concept thus helps overcome the tendency, which is prominent in innovation studies, to view manufacturers, such as the car industry, as the pivotal actors in regimes (such as automobility). Although car manufacturers are undoubtedly an important actor (who exert much influence through their product offerings, marketing strategies and political lobbying), automobility regimes are also sustained by habits of use, prevailing normality and mindsets and established practices of professionals, such as transport planners, whose logic and choices help to reproduce a regime. The automobility regime is also sustained by everyday conversation and politics and by cultural associations with freedom, modernity and individual identity.

Geels (2004) suggests that a socio-technical regime is made up of alignments between regimes that refer to specific populations (e.g., policy makers, academics, industries, civil society, users/consumers). Actors within these groups share a set of rules or regime. Because the different groups share different rules, we may distinguish different regimes, for example, technological or industrial regimes, policy regimes, science regimes, financial regimes and societal or user regimes (see Figure 3.3.). Actors in these different communities tend to read particular professional journals, meet at specialized conferences, have professional associations and lobby clubs, share aims, values and problem agendas and so forth. The alignment of different groups results in linkages between regimes. The concept of socio-technical regimes aims to capture this meta-coordination.

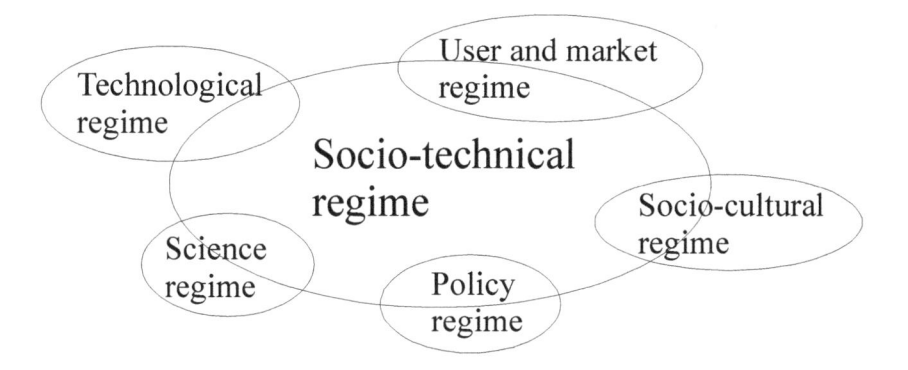

Figure 3.3 Meta-coordination through socio-technical regimes.

The auto-mobility regime can thus be understood as consisting of more specific regimes. Car manufacturers are part of a car manufacturing regime, which is characterized by international competition, entry barriers and exit barriers. Transport planners are part of a planning and policy regime. We can also distinguish regimes that relate to use and mobility behavior. In the present use/mobility regime, people use cars for commuting, shopping, social visits, holidays and so on. They use cars to cover short distances, where walking and bicycles are (potentially) valid alternatives, and for long-distance trips, where high-speed trains and airplanes are (potentially) valid alternatives. Although cars can be rented on a per ride basis and people may share rides, the (currently) mainstream practice is that people individually own their car. Of course, this does not hold true for everyone and for every trip, but these use practices are common (the mode around which variations exist). It is possible thus to define specific regimes empirically and conceptually, as various chapters in this book do.

Another important qualification for this book is that there is not just one regime (auto-mobility) but also other regimes (e.g., train, tram, bus, cycling). These transport modes have been around for many decades and are carried by specific communities of actors that have developed institutionalized practices, beliefs, capabilities and so forth. It makes no sense to call these transport modes 'niches' in the sense of being radically new and precarious innovations. But these transport modes capture only a small percentage of total mobility (in terms of passenger kilometers) and in that sense occupy small market niches. We propose to call these transport modes *subaltern* regimes in contrast to the *dominant* auto-mobility regime.

While the notion of socio-technical regime is useful to understand comprehensive wholes such as automobility, it has also been criticized for not being precise and fully operationalized (Genus & Coles, 2008; Markard & Truffer, 2008). A useful attempt to determine whether something is a regime (or not) is offered by Holz, Brugnach and Pahl-Wostl (2008). They distinguish five characteristics of a regime: (a) there is a clear "goal" in the sense of functionality for users and society that is being fulfilled (such as the function of mobility), (b) many things are being combined and aligned with each other, (c) a high degree of stability (actors do not suddenly orient themselves to something completely different), (d) the absence of a central actor in charge of everything and (e) autonomy of regime internal dynamics. According to these criteria the system of automobility is a regime.

The notion of a regime introduces a structuralist element in the analysis, by assuming that actor behavior is constrained by rules located at the collective level of a regime, which cannot easily be changed at the micro-level of individual action (Rip & Kemp, 1998). It is important to emphasize that 'regime' is an interpretive analytical concept, which invites the analyst to investigate the 'deep structure' behind activities, for example, shared beliefs, norms, standardized ways of doing things, heuristics and rules of thumb. While the notion of socio-technical 'system' refers to tangible and measurable elements (such as artifacts, market shares, infrastructure, regulations, consumption patterns, public opinion), the notion of 'regimes' refer to more intangible rules on which actors draw in concrete actions. In this book we are interested in how such regime rules shape behavior and how these rules change as a result of new practices, new ideas and processes of socio-technical evolution. The book's general challenge is to investigate if the various regime elements are still well-aligned or if tensions and cracks are appearing. If so, which niche-innovations have sufficient momentum to take advantage of these windows of opportunity?

3.3.3. Socio-technical Landscape

The socio-technical landscape is the wider context, which influences niche and regime dynamics. The *socio-technical landscape* is a landscape in the literal sense, something around us that we can travel through, and in a metaphorical sense, something that we are part of, that sustains us (Rip & Kemp, 1998, p. 334). It includes infrastructure and other physical aspects (such as houses and cities), political ideologies, societal values, beliefs,

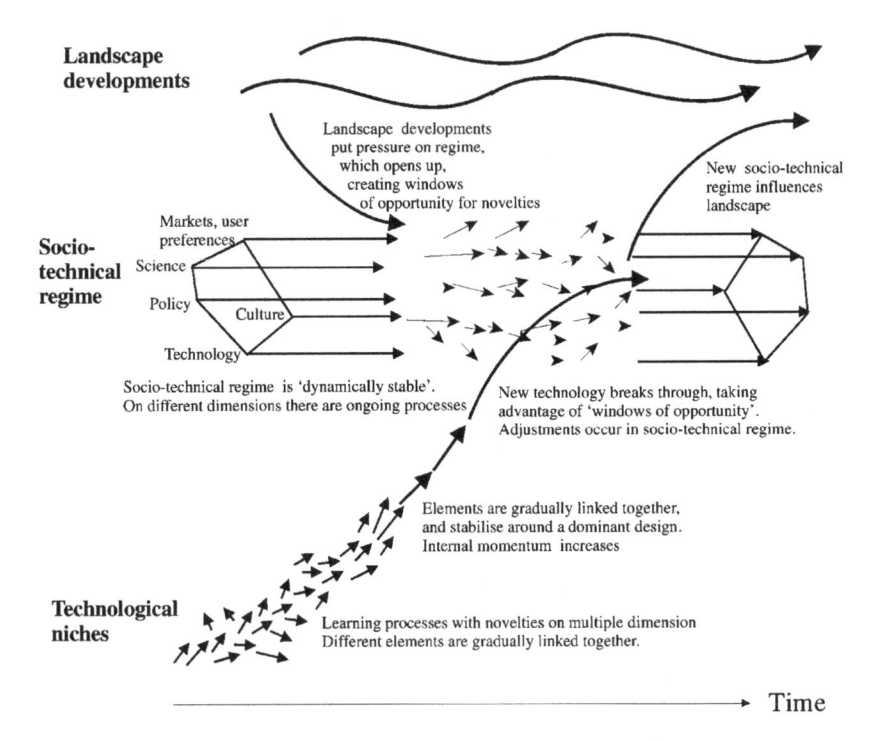

Figure 3.4 A dynamic multi-level perspective on system innovations (Geels, 2002, p. 1263). Reprinted by permission of the publisher (Elsevier).

concerns, the media landscape and macro-economic trends. The landscape will differ from place to place and perhaps we should be talking about multiple landscapes. Regimes are nested in landscapes, which represent the greatest degree of structuration in the sense of being beyond the influence of individual actors.

Figure 3.4 provides an ideal-typical representation of how the three levels interact dynamically in the unfolding of socio-technical transitions. Although each transition is unique, the general dynamic is that transitions come about through the interaction between processes at different levels: (a) Niche-innovations build up internal momentum, (b) changes at the landscape level create pressure on the regime and (c) destabilization of the regime creates windows of opportunity for niche-innovations. An important implication is that the MLP does away with simple causality in transitions. There is no single 'cause' or driver. Instead, there are processes on multiple dimensions and at different levels which link up and reinforce each other ('circular causality'). Transition studies and the chapters in this book therefore emphasize lateral alignments, unexpected linkages, thresholds and tipping points.

The MLP encompasses both stability and change. To explain *stability*, the notion of socio-technical regime emphasizes lock-in mechanisms that stabilize car-based modes of transport, for example, sunk investments (in road infrastructures, plants, skills), user patterns and lifestyles oriented around the car, favorable regulations, cultural values (such as speed, freedom, individuality, identity) and resistance from vested interests (industry, car lobby, road-building lobby). *Incremental change* occurs within regimes, while *radical change* often emerges outside regimes in relatively protected and isolated niches.

To illustrate some of these abstract concepts, Box 3.1 provides a brief and incomplete illustration of the American transition from horse-and-carriages to automobiles (1880–1930).

Textbox 3.1 The Shift from Horse-drawn Carriages to Cars as a Multilevel Process.

A historical transition that shows the multitude of developments at play is the shift from horse-drawn carriages to electric trams and later to automobiles in America from 1880 to 1930 (Geels, 2005a). This transition was not driven by a single factor but the outcome of a cumulative, multi-layered process in which problems of congestion, horse droppings, lack of safety and high costs associated with the use of horses for urban transport created space for transport alternatives. These problems coincided with various developments in technology, products and society (immigration, middle-class people moving out to suburbs, etc.). In the

Continued

1880s electricity was the new technological wonder, first used for lighting but gradually finding other applications such as battery electric vehicles (BEVs). In the same period, bicycle use emerged as a new phenomenon: People started to make trips on bicycles, especially in the country side, not for reasons of travel but for fun. Both electric trams and bicycles diffused rapidly. First used commercially in 1888, the electric trams obtained a market share of 97% of American street railways in 1902. The reason for this very fast diffusion was that it was supported by horse-tram companies, real estate promoters, electric light companies and local authorities. The electric tram and bicycle paved the way for the automobile. The bicycle contributed to two new phenomena: touring and racing. It also gave rise to traffic regulations and bicycle clubs and magazines. It was a form of individual transport, which was new for many people. The tram contributed to suburbanization and altered the function of streets: Streets became arteries for transport instead of places where one encounters and meets others. The middle classes moved out of crowded and filthy cities and into suburbs and engaged themselves in travel for fun and recreation. In the 1890s cars emerged in special niches: Electric cars were used as taxi and as luxury vehicles, and gasoline vehicles were used for racing and touring, just as steam cars. Around 1905 the market was tipping toward internal combustion cars. Doctors, rich farmers, salesmen and insurance agents started to use cars for utilitarian functions. Following the introduction of the T-Ford car in 1908 and use of an electric starter invented in 1911, gasoline cars started to compete with electric trams. A car culture developed, in tandem with the practice of living in the suburbs. Highways were built, shopping molls appeared on the edge of cities and people went on vacation with their own car. From the 1950s onwards, one could even watch a move at a drive-in cinema and eat a meal from a drive-in restaurant in the United States. Cars were gradually improved and became affordable for a wide range of people. Cars became the desired form of transport, used for the majority of trips including short trips and long-distance trips.

3.4. TRANSITION PATTERNS

While the MLP provides an overarching analytical framework for entire transitions (like a helicopter view), this section and the next one discuss specific patterns and mechanisms that are played out on shorter time periods or on specific dimensions. These patterns and mechanisms help to put more flesh on the bones of the MLP. In this section, we discuss six patterns and give examples of how these patterns can be related to sustainability transitions. The patterns are (a) transition pathways, (b) add-on and hybridization pattern, (c) knock-on effects and innovation cascades, (d) fit-stretch patterns, (e) hype-disappointment cycles and (f) niche-accumulation patterns.[2]

3.4.1. Transition Pathways

Having studied a large number of transitions, Geels and Schot (2007) identified four transition pathways that result from differences in the *timing* of multi-level interactions (whether or not niche-innovations are relatively well-developed when landscape pressures occur) and the *nature* of multi-level interactions (whether niche-innovations have a competitive relationship with the existing regime or a symbiotic relationship). The transition pathways are as follows:

- *Transformation*: This pathway occurs when there is moderate landscape pressure at a moment when niche-innovations have *not* yet been sufficiently developed, leading regime actors to respond by modifying the direction of development paths and innovation activities.
- *De-alignment and re-alignment path*: Major landscape pressures first cause big internal problems for regimes leading to their disintegration (de-alignment of elements); this erosion then creates space for various niche-innovations that co-exist for prolonged periods. The variety of niches creates uncertainty and may delay important actors to make full-scale commitments for fear of betting on the wrong horse. Eventually, processes of re-alignment occur around one of the innovations, leading to a new regime.
- *Technological substitution*: This pathway occurs when there is much landscape pressure at a moment when niche-innovations have developed sufficiently, causing the latter to break through and replace the existing regime.
- *Reconfiguration pathway:* Niche-innovations are initially adopted in the regime to solve local problems but subsequently trigger adjustments in the basic architecture of the regime.

In the current automobility transition 'in the making' we can distinguish the onset of various pathways. There are characteristics of *de-alignment and re-alignment*, in the sense that *multiple* niche-innovations are present and challenging the regime. In contrast to the ideal-typical path, however, the existing automobility regime is not yet falling apart, although tensions and 'cracks' may be getting larger (see Chapter 4 and 16). The automobility regime also contains aspects of *transformation*, in the sense that many car companies make large investments in incremental engine improvements to increase the fuel economy of internal combustion engine (ICE) vehicles (e.g., in response to European CO_2 emission regulations). And there are also aspects of *reconfiguration*, for example, the incorporation of fuel cells or batteries require reconfigurations in the car architecture, and the addition of Information and

Communication Technologies (ICT) to highways lead to dynamic traffic management and intelligent highways (Geels, 2007). And if we look at engines/motors and fuel infrastructures there are also aspects of *substitution* (with petrol, biofuels, electricity and hydrogen competing with each other). The future of automobility is thus intrinsically open. Which transition path will emerge depends on the interactions between multiple levels, which depend on the strategies, beliefs, interests and actions of various actors.

3.4.2. Add-on and Hybridization Pattern

Niche-innovations do not necessarily *compete* with existing (regime) technologies but can also enter into 'new combinations' with them (as in the reconfiguration pathway). Initially, new technologies may just form an 'add-on' to solve particular problems (Geels, 2002). Early steam engines, for instance, entered sailing ships as add-ons that were used when there was little wind or in canals with limited maneuverability. As new technologies

Figure 3.5 From sailing ships to steamships through add-on and hybridization (clockwise from top left: van Oosten, 1972; Fletcher, 1910; public domain photos from www.wikipedia.org). © F.C. van Oosten, *Schepen onder stoom: De geboorte van het stoomschip*, 1972, Unieboek BV, Bussum/Holland: 16; Fletcher, R.A. (1910), *Steamships: The story of their development to the present day*, Philadelphia: J.B. Lippincott.

are improved and actors gain more experience, the new technology may gain in prominence leading to a more hybrid form. Eventually, the new technology may take over and become dominant, leaving the old technology to wither away. The pictures in Figure 3.5 show the resulting transformation from sailing to steamships: The Comet (1812), The Rising Star (1822), the Great Britain (1843), the Great Eastern (1858).

In existing cars, the battery is an add-on, used for starting the ICE. In hybrid-electric vehicles (HEVs), the drive train consists of a more equal (parallel or sequential) combination of batteries and ICE. The issue of hybridization extends beyond technical aspects: It may also apply to borrowing other elements that are new to the domain of passenger transport such as green finance and pay per use schemes. Something like this is happening in the case of electric vehicles, where the Californian startup company Better Place is mimicking the cell phone business model and entering strategic alliances with electricity generating and distribution companies (see Chapter 10). Better Place is setting up recharging systems and investing in battery swapping stations. This example thus shows innovation through heterogeneous recombination.

3.4.3. Knock-on Effects and Innovation Cascades

The introduction of innovations in existing technologies or socio-technical systems may lead to knock-on effects and innovation cascades. As an example, the transition from propeller-piston engine aircraft to turbojets began with a component substitution: The piston engine and propeller were replaced by a jet engine (Geels, 2006). This substitution subsequently triggered further changes in the aircraft, for example, new wing designs (swept-back wings), new skin materials and larger size. Because jet engines are relatively inefficient at low speeds, longer runways were needed for takeoff. As planes got heavier, runways also needed to be stronger, and as jetliners entered commercial service, the speed difference with propeller aircraft required extra coordination around airports, which led to new air traffic control technologies. Knock-on effects were not only technological. As aviation markets grew in the 1940s and 1950s, there was demand for new aircraft that were safer, stronger, bigger, faster and could fly longer distances and at high altitudes ('above the weather'). The resulting jetliners (Boeing 707 in 1958), in turn, opened up new routes and new user groups ('flying for the masses'), which created new demands for aircraft (Boeing 747 in 1969). But they also led to problems in societal embedding, particularly noise, which led to heated debates and stricter regulations.

These kinds of socio-technical cascades are the bread and butter of transitions and are also to be expected with new engine technologies, new ICT devices in cars and highways, changing user preferences, regulations and so forth.

Innovation cascades do not need to start with *one* technology that has knock-on effects. They can also arise from multiple innovations coming together. For innovators as well as for transition analysts the challenge is to creatively think up "novel combinations" leading to "configurations that work" (Rip & Kemp, 1998). Transitions in automobility may arise from linkages with developments in neighboring fields: ICT, nanotechnology, electricity companies, cooperation between non-governmental organization (NGOs) and business, car-free streets, transition towns, carbon prices and carbon budgets, etcetera.

Some of the developments can be anticipated, others are more speculative. Technically, we can anticipate better batteries, because the agenda of battery development is shifting from laptops toward batteries for automobiles. Most likely, battery development will benefit from advances in neighboring technology fields such as nanotechnology. Something like this is already happening. A nanotechnology-based method of recharging lithium ion batteries has been developed by two MIT researchers. With this invention, batteries may be charged and discharged in a matter of seconds rather than hours.[3] Rental electric vehicles may also benefit from information and communication technology for making reservations, for leaving cars at the point of destination and for payment.

3.4.4. Fit-stretch Pattern

The thinking about the shape and functionality of niche-innovations initially tends to be close to the established regime, that is, has a *fit* with the established concepts and categories (Geels, 2005b). Historians of technology have commonly found that "When drastically new technologies make their appearance, thinking about their eventual impact is severely handicapped by the tendency to think about them in terms of the old technology. It is difficult even to visualize the complete displacement of an old, long-dominant technology, let alone apprehend a new technology as an entire system." (Rosenberg, 1986, p. 24).[4] Early cars, for instance, were compared with horse-and-carriages. They were therefore called the 'horseless carriage' and also looked very similar to them (see Figure 3.6). These early cars were usually driven by a chauffeur and used for entertainment and leisure purposes (e.g., promenading or driving in parks).

Views and perceptions about the 'horseless carriage' gradually changed when people gained more technical and user experience. As people began to use cars for racing and touring in the countryside, the technical form changed (lower and wider cars with a hood). From a 'fit' with the horse-based regime, cars were moving toward 'stretch', that is, articulating their own technical and use principles. The seeds of a new auto-regime emerged with the T-Ford (1908) as a new technical form and more utlitarian types of use, initially by farmers, doctors and taxi drivers, followed by suburban residents who began using cars for commuting (which also led to the

64   Frank W. Geels and René Kemp

Figure 3.6   The Benz Velo horseless carriage (1894; this Wikimedia Commons image is from user Chris 73 and is freely available at http://commons.wikimedia.org/wikiFile: Benz-Velo-1894.jpg under the creative commons cc-by-sa 2.5 license).

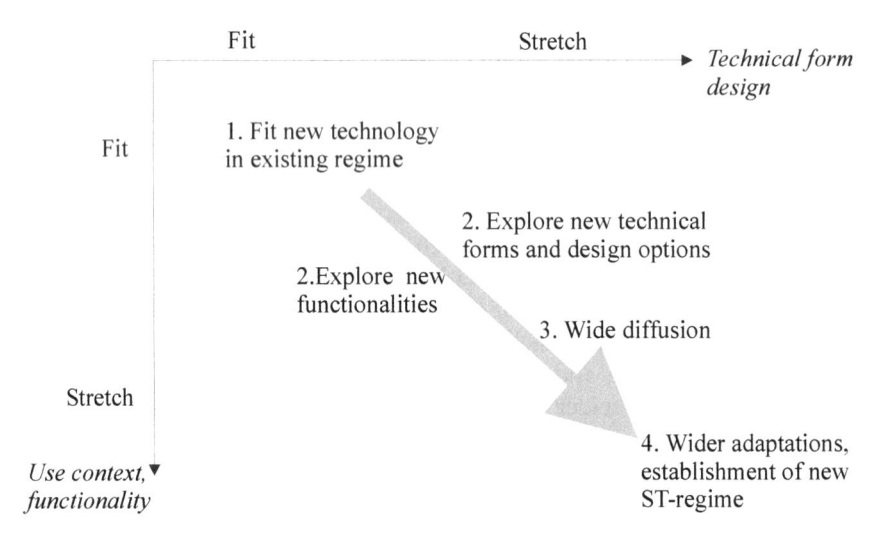

```
                    Fit                  Stretch
                                                          ► Technical form
                                                            design
              
              1. Fit new technology
         Fit  in existing regime
              
                                  2. Explore new technical
                                  forms and design options
                        2. Explore new
                        functionalities
                                          3. Wide diffusion
              
         Stretch
                                          4. Wider adaptations,
   Use context, ▼                         establishment of new
   functionality                          ST-regime
```

Figure 3.7   Fit-stretch pattern in the co-evolution of form and function.

The Multi-Level Perspective 65

introduction of car locks and closed bodies). Technology and use thus typically co-evolve, both materially and in terms of beliefs and perceptions. Figure 3.7 provides an analytical generalization of this pattern.

The fit-stretch pattern complicates the dichotomy of system-improving change and system-altering change (and associated notions of sustaining innovations and disruptive innovations). Innovations that may initially appear as incremental refinements of existing regimes may gradually evolve to have more radical and transformative implications. Road pricing, for instance, is an innovation of traffic management that is regime-preserving in the short term but may be a game changer in the longer term. At present it facilitates car use by using the roads more efficiently, offering benefits to car users; in the longer term it could contribute to intermodal travel and multi-modal travel through stimulating more selective use of cars and a modal split.

A fit-stretch pattern may also happen to batteries. At present, HEVs and BEVs are emerging, following the success of the Toyota Prius. At least three scenarios are possible for (hybrid) BEVs: (a) The vehicles could be used in the same way as normal cars, that is, for long-distance travel and for the majority of trips; (b) they could be used as fleet cars and self-service rental cars; and (c) they may be used not only for driving but also for storing *electricity for non-mobility use* in a vehicle-to-grid system configuration. While the first two scenarios have 'fit' characteristics with the existing regime, the third one moves towards 'stretch'. Additionally, a shift to electric vehicles may have unintended side-effects besides helping to achieve air benefits. For instance, in the Netherlands, bicycle users are shifting to electric bikes which are less environmentally sustainable than traditional bikes and may encourage people to make additional trips. It is difficult to predict how new technologies will be used and what this will mean in terms of overall mobility. Transition analysis aims to draw attention to such knock-on effects and fit-stretch patterns.

3.4.5. Hype-disappointment cycles

Niche-innovations often experience a hype-disappointment pattern, which relates to (cognitive) changes in visions and expectations. Gartner consultancy distinguishes several phases in this pattern (see Figure 3.8): (a) *Technology trigger*: Presentations and positive reports about a new technology generate press and industry interest; (b) *Peak of inflated expectations*: Technology leaders and product champions raise high expectations about performance and functionality of the technology, leading to over-enthusiasm; (c) *Trough of disillusionment*: The technology does not live up to its inflated expectations, and it rapidly becomes unfashionable; (d) *Slope of gradual improvements*: Ongoing work by technology developers leads to improved models, and real-world benefits are gradually demonstrated, leading to gradual market penetration and renewed enthusiasm.

66 *Frank W. Geels and René Kemp*

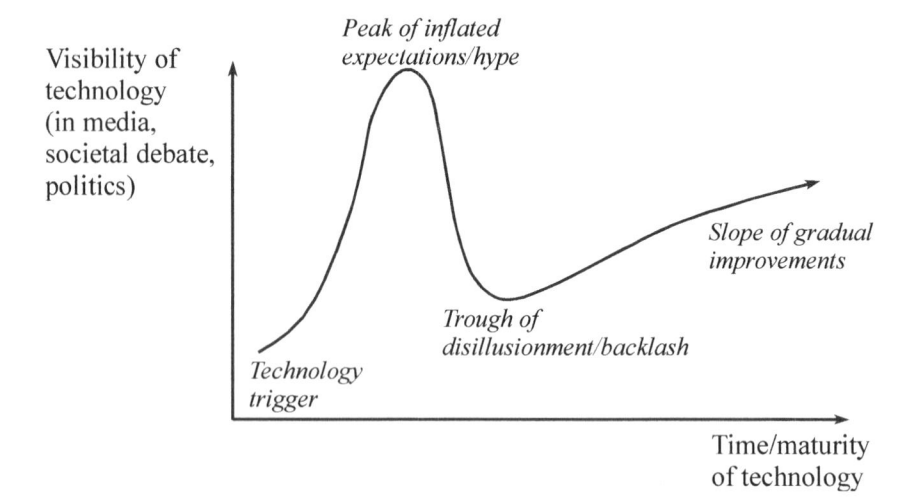

Figure 3.8 The hype-cycle (redrawn and adapted from Gartner.com, accessed 15 March 2005).

Recent examples are the dot.com bubble and Universal Mobile Telecommunications System (a third-generation mobile telephony). The revival of technologies in the fourth phase does, of course, not always happen. Sometimes, technologies die out in the 'trough of disillusionment' as actors withdraw their investments and support. An additional complication is that

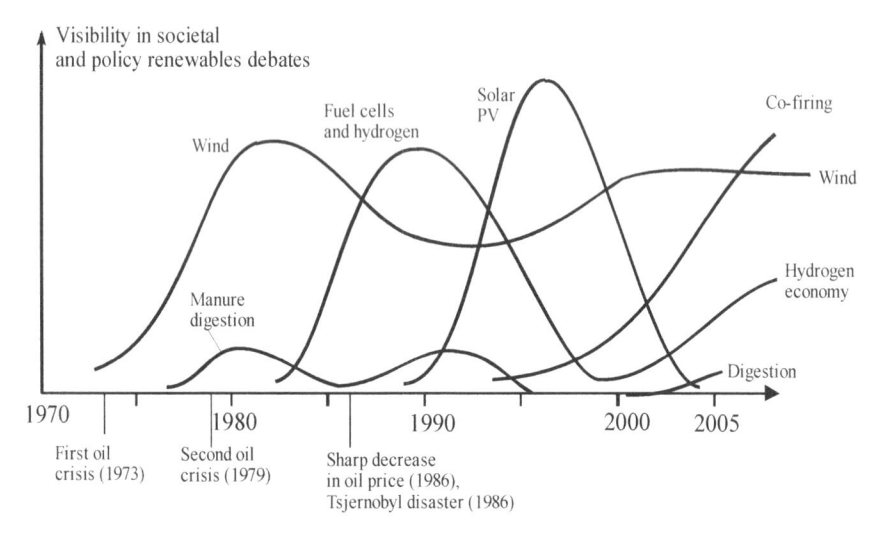

Figure 3.9 Interacting hype-cycles in societal and policy debates on Dutch renewable energy (Verbong, Geels and Raven, 2008, p. 568). Reprinted by permission of the publisher (Taylor & Francis Group, http://www.informaworld.com).

multiple hype-cycles may interact. Positive outcomes and learning processes in one niche trajectory may negatively influence the expectations of another trajectory. For Dutch renewable energy, Verbong, Geels and Raven (2008) found a pattern of interacting hype-cycles (see Figure 3.9).

Verbong et al. (2008) found that government policy contributed to this pattern. Because policy makers face societal credibility pressures because of sustainability problems, they are eager to show visions and successes. Because they are looking for positive stories, policy makers may jump on new technology bandwagons, thus reinforcing hype cycles. But when setbacks occur a few years later, they may become disappointed and withdraw support and move on to a new promising option. In the conclusion chapter of this book, we will return to hype cycles in 'green' car propulsion technologies.

3.4.6. Niche-accumulation pattern

Radical innovations can move out of initially small niches through a pattern of niche-accumulation (Geels, 2002). The initial (technological) niche often depends on the efforts of innovation actors, such as dedicated entrepreneurs. Innovations become somewhat more self-sustaining, if they can subsequently gain a foothold in a small market niche of users with special preferences. Broader diffusion tends to follow a pattern of niche-accumulation, with the technology entering increasingly larger market niches (see Figure 3.10).

Niche-accumulation may be a prolonged and complex process giving rise to various lineages. This is illustrated in Figure 3.11 with an example of niche-accumulation in steamship diffusion from the 1780s to the 1860s. The breakthrough in Atlantic passenger transport was related to landscape developments in the 1840s such as European revolutions, potato famines and migration to the United States. The breakthrough in freight transport, especially to China and India, was boosted by the Suez Canal (1869), which sailing ships could not use.

For the diffusion of fuel cells or BEVs it is probably also fruitful to think how various application niches can build on each other, for example, fleet vehicles (such as buses, taxis, delivery vans, postal vehicles), various types of passenger cars (saloon, sports car, second family car for shorter distances, station car) or even (electric) scooters and bicycles.

Technology A → Technological niche X → Market niche Y → Market niche Z

Figure 3.10 Trajectory of niche-accumulation.

68 *Frank W. Geels and René Kemp*

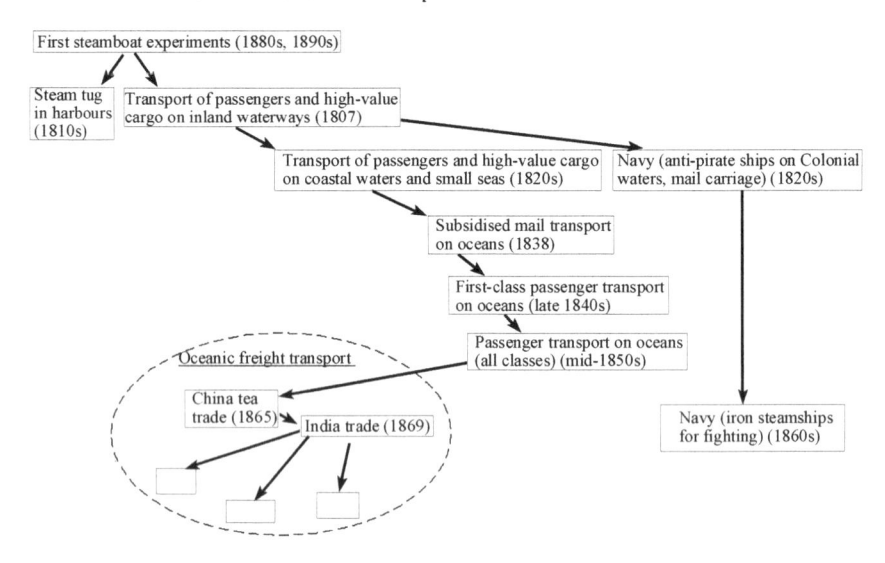

```
First steamboat experiments (1880s, 1890s)
  ↓    ↘
Steam tug    Transport of passengers and high-value
in harbours  cargo on inland waterways (1807)
(1810s)
                    ↓
          Transport of passengers and high-value cargo    Navy (anti-pirate ships on Colonial
          on coastal waters and small seas (1820s)        waters, mail carriage) (1820s)
                    ↓
          Subsidised mail transport
          on oceans (1838)
                    ↓
          First-class passenger transport
          on oceans (late 1840s)
                    ↓
          Passenger transport on oceans
Oceanic freight transport    (all classes) (mid-1850s)
   China tea                                              Navy (iron steamships
   trade (1865)   India trade (1869)                      for fighting) (1860s)
```

Figure 3.11 Trajectories of niche-accumulation for the diffusion of steamships (Geels, 2002: 1271). Reprinted by permission of the publisher (Elsevier).

3.5. TRANSITION MECHANISMS (ACTOR-RELATED)

Further dynamics can be added to the transition perspective by distinguishing recurring mechanisms in the activities and strategies of regime actors and niche players. Drawing on various empirical transitions studies Geels (2005c) identified the following mechanisms (to which we add examples from passenger transport):

- A cartel of fear may slow down the development or diffusion of new technologies. For instance, American airline companies long hesitated to buy jetliners, because there were many uncertainties about their performance (in particular fuel costs). No airline company wanted to take the first step because of the risks involved. So long as no airline company introduced jets in the American market, no one needed to buy it. The shared fear led to inertia. Firms watched each other, but no one acted.
- A cartel of fear can turn into rapid acceleration and an innovation race when one of the actors makes a move and breaks the deadlock. When Pan Am ordered jets in 1954, other airline companies quickly followed for fear of being left behind. This led to a domino effect and rapidly increasing orders for jetliners. Innovation races can also occur on the supply side. For instance, Boeing and Douglas were involved in a strategic game with regard to civil jetliners in the 1950s. Douglas chose a wait-and-see strategy, letting Boeing carry the first-mover risk in developing jetliners. When Boeing's efforts seemed to become

productive, Douglas followed quickly, leading to an acceleration in the development path. A recent example is the Prius car. When Toyota launched the car in 1997, other car manufacturers ridiculed it, pointing to technical complexity and higher costs. But when sales began to take off around 2003, other companies quickly followed, and they are now all developing hybrid electric cars. Maybe the same will happen with BEVs. Although governments were initially fearful to promote BEVs, many have developed goals and plans for the introduction of BEVs. As described in Chapter 10, the Nissan UK plant for the Leaf BEV is supported by a £20.7m grant from the UK government and a proposed finance package from the European Investment Bank of up to £197.3m.[5] The United Kingdom, Denmark and the Netherlands are following suit to Israel and California in promoting electric vehicles.

- Saturation of existing markets (or a sales crisis) can form a stimulus for firms to diversify to other markets and technologies. For instance, the bicycle crisis of 1898 stimulated bicycle producers to diversify into automobiles, leading to new entries in the emerging automobile sector and an acceleration of development. An interesting example of a firm diversifying into the transport sector is the JC Deveaux advertising company which in return for control of a substantial portion of the city's advertising billboards set up a bicycle rental system in the city of Paris. The *Vélib'* system is claimed to be a success in terms of use. Close to 150,000 trips are made each day on weekends and more than half that amount on weekdays (Britton, 2007). Building on this success, the city is now planning to expand the project with about 4 000 self-service electric hire cars by the end of 2010 (Organisation for Economic Co-operation and Development, 2009, p. 69). Another example is the Chinese battery producer BYD diversifying into electric vehicles.
- When incumbent firms are threatened by new technologies, they will try to defend themselves by improving the existing technology. This is called the sailing ship effect. When steamships challenged sailing ships in the 1860s and 1870s, many improvements were made in sailing ships. To increase their speed, more masts and sail were added. To reduce labor costs, labor-saving machines were introduced, for example, to rig the sails. The improvement in incumbent technologies may delay the wide diffusion of new technologies. In the case of automobiles we see a similar dynamic. While showcasing electric concept cars on international fairs, car manufacturers are also improving the fuel economy of gasoline cars. In fact, the bulk of the investments is still in improved ICE vehicles (Chapters 6 and 10).
- Transitions may lead to Schumpeterian waves of creative destruction and the downfall of established firms. Because incumbent regime companies are locked in to technical competencies and particular consumer markets, they may be incapable of transforming themselves (Christensen, 1997). An example is that most established shipbuilders

did not make the shift from wood and sail to iron and steam, which required new skills (e.g., riveting and metal working) and new machine tools. These established wood and sail shipbuilders (gradually) disappeared from the market. Although General Motors recently has been rescued, it remains to be seen if it can make the transition to alternative propulsion cars requiring different competences. But one should not expect pioneering outsiders to always win and incumbent firms to go down. Because radical innovation takes a long time and many resources, pioneers often fail and 'burn out' (Olleros, 1986). They may even have to collaborate with incumbent firms if the latter posses 'complementary assets' such as specialized manufacturing capability, distribution channels and service networks (Teece, 1986; Tripsas, 1997). Because these 'complementary assets' are needed for the upscaling and broad commercialization of niche-innovations, green pioneers may need to collaborate with incumbent companies, something that seems to be currently happening in the car industry.

- Incumbent firms and old technologies may hold on in particular market niches for a long time, even after the new technology has become dominant, softening the pain for workers and for communities that depend on a particular employer. Although steamships broke through in the 1870s and 1880s, sailing ships continued to be used in bulk freight markets (e.g., iron, coal, rice, wool) well into the 20th century. Gasoline cars from Western Europe are having a second lifetime in Africa and Eastern Europe.
- Cultural framing is important because it affects the social legitimacy of old and new technologies, which in turn influences public acceptance, government protection and access to external capital (Lounsbury & Glynn, 2001). Proponents therefore try to link new technologies to wider cultural visions and values. Aircraft in the 1920s and 1930s, for instance, enjoyed great popular support because they were seen as means to a better world, the 'winged gospel' (Corn, 1983). Peak oil and climate concerns are currently legitimizing electric cars and delegitimizing fossil fuel cars. The formerly accepted negative features of gasoline/diesel cars such as pollution, CO2, noise and smell may become non-accepted.
- Opponents such as social movements may also try to delegitimize existing regimes through cultural criticism, negative discourses and negative symbols (Geels & Verhees, 2011). Climate change activists, for instance, label SUVs as "climate destroyer" and "Super Unpatriotic Vehicles." There have been, and still are, broader framing struggles about climate change. On the one hand, the environmental movement has developed catastrophe scenarios, which in some cases attracted much attention (e.g., Al Gore's movie in 2006). On the other hand, fossil fuel industries organized the Global Climate Coalition (GCC) in 1989 to challenge the climate science, pointing to the

lack of consensus among scientists and highlighting the uncertainties. It also promoted the views of climate skeptics in reports and press releases and ran advertisements about the costs and economic burden to the economy of CO_2 emission reductions (Levy & Egan, 2003). In the late 1990s, the GCC began to crumble when companies like Shell (in 1998) and Ford (in 1999) abandoned the Coalition because they felt that continued membership would damage their legitimacy and reputation. The discursive struggle then shifted from climate change denial to debates about responsibility, mitigation costs and the preferred kinds of policy instruments. The larger point is that sustainability transitions are accompanied by cultural discourse, which comes into play with markets, government support and control policies and technological development.

- Adoption and diffusion of new technologies also entail domestication and societal embedding. Consumption is more than buying, especially with regard to radically new technologies. New technologies have to be 'tamed' to fit in user contexts. This domestication involves: (a) symbolic work (see previous); (b) practical work, in which users integrate the artefact in their user practices; and (c) cognitive work, which includes learning about the artefact and developing new user practices (Lie & Sørensen, 1996). Mimi Sheller (Chapter 9 of this book) anticipates that the new ICT devices that enter cars will lead to new user practices as driving becomes interlaced with capacities for conversation, entertainment, information access, navigation, automation, tracking and surveillance. Likewise, BEVs may require people to change their driving behavior (e.g., more route planning to accommodate shorter range) and rearrange their garage to install recharging facilities. Geels and Schot (2010b) found that in the Netherlands the ANWB mobility organization played important roles in the articulation of user routines and the 'education' of the user. In the 1920s and 1930s the ANWB used its magazine to teach drivers to behave like gentlemen in traffic. These educational campaigns were accompanied by the creation of driving schools and license systems. Societal embedding is a complex phenomenon, which in the case of alternative fuel cars entails cultural acceptance and meanings, new infrastructures, new skills, special repair facilities and new policy programs and regulations (e.g., about end-of-life discharging of batteries to prevent pollution from heavy metals). Societal embedding is part of a process of mutual adaptation of society and technology. Policy support is often important for the diffusion of new technologies. For instance, national governments gave tremendous financial support to airline companies in the 1920s and 1930s and stimulated aviation by sponsoring research on aerodynamics, fuels and engines. This was possible because of societal support and cultural enthusiasm about aviation. Some examples of transport options that recently received substantial amounts of government support include automated vehicle guidance systems, high speed trains and Maglev trains and the high-speed "super bus." In the United States, the Department of Energy is investing more than $5 billion to electrify America's transportation sector.[6] It seems that high-technology options are able to garner more support than low-tech options such as bicycles (see also Chapter 14 on the preference of policy makers for large rail projects over bus programs).

- The political support for new technology can be part of wider power struggles, leading to accelerations or delay. In the early 20th century, American electric tram companies had a monopoly on urban mass transport and antagonistic relationships with city governments. When the automobile became a practical transport option in the 1920s and 1930s, city governments massively subsidized car transport through construction and improvement of roads. This was partly a response to demands from middle-class constituencies but also a move to undermine the strength of electric tram companies (Geels, 2005a).

3.6. A TRANSITION PERSPECTIVE ON POLICY

The role of policy in transitions is complex, in part because many issues are the focus of policy: innovation, competition, environmental protection, licenses, tax revenues and spending. Transport policy is made in multi-level systems of governance (see Chapter 5). The logic of the MLP suggests policy makers can follow two strategies if they want to influence transitions: (a) enhance the pressure on regimes through economic instruments and regulation (e.g., taxes, carbon emission trading, environmental legislation) and (b) stimulate the emergence and diffusion of niche-innovations. The second part of this strategy has been elaborated in the literature on Transition Management (TM) (Kemp et al., 2007; Loorbach, 2007; Rotmans et al., 2001) and strategic niche management (SNM; Hoogma et al., 2002; Kemp et al., 1998; Schot and Geels, 2008). Both policy approaches emphasize the importance of concrete demonstration projects to facilitate real-world learning processes, the build-up of social networks and the role of visions or expectations to guide niche developments and attract attention. But they also differ, with TM placing more emphasis on the creation of long-term visions in socalled 'transition arenas', and SNM giving more emphasis to experimental 'on-the-ground' learning through projects. Furthermore, the TM approach is probably more developed in terms of prescriptive operationalization. TM has also been tried out in practice in the Netherlands, where transition thinking was incorporated in the 4th Dutch National Environmental Policy Plan in 2001. TM has subsequently been applied in various programs and projects (Loorbach, 2007), especially by sections of the Ministry of Economic Affairs responsible for the so-called 'Energy Transition'.

In contrast to the attention given by the Ministry of Economic Affairs to the energy transition, there has been relatively little active support for TM within the Ministry of Transport, Public Works and Water Management (Nooteboom, 2006). Deeper analysis is needed on why this is the case, but it seems that TM does not fit with the transport engineering approach and devolution of responsibilities. Nevertheless, the idea of 'a transition to sustainable mobility' has attracted attention from Dutch business-representatives, NGOs and local policy makers. At the national policy level, 'transitions to sustainable mobility' mainly received some attention in the 'Energy Transition' program and in the context of the government-funded innovation program Transumo (Avelino et al., 2012; Kemp et al., 2011). So, the concept is being used in innovation programs but hardly in transport policy. This points to a general problem for transition policy, namely a preference for stimulating (niche) innovation rather than creating pressure on regimes. So, real-world policy makers (and TM and SNM scholars) give most attention to one dimension of the proposed two-pronged strategy.

The transition perspective does not offer a simple way out of transport problems through sustainability transitions but does help to diagnose the situation, for example, explaining how governments are part and parcel of transformations and experience lock-in. It is not the case that governments are doing nothing. For example, governments respond to technological promises, and when a car manufacturer wants to open a plant for a new type of vehicle they offer financial support for reasons of job creation and industrial policy. High-speed trains and light rail received support from government, which was necessary for those systems coming into existence. Urban authorities are responding to the demand for urban quality and sometimes create pedestrian-only areas in city centers and implement other car-restraining policies. Professions such as transport planning and transport engineering are also changing (see Chapter 7) and so are societal expectations and cultural aspirations. Car ownership appears to be in the process of becoming less important culturally for the young generation (see Chapter 4). Congestion problems are putting restraints on car use and increasing oil prices and climate change policies may put further strains on car-based mobility and alter the beliefs and mindsets that govern policy and travel behavior. So, despite the various types of embeddedness and stabilizing lock-in mechanisms, there are also developments that point in the direction of change; developments which partly work through policy. We will explicitly return to policy implications of the transition analysis in the concluding chapter.

3.7. SOME FINAL WORDS

In this chapter, we outlined and explained the transition perspective and suggested its relevance for studying transitions in auto-mobility. At the heart of the transition model is the MLP, which is grounded in (co-)evolutionary approaches of innovation, science and technology studies and social theory. The MLP has developed a specific way of looking at transition dynamics in terms of regimes, niches and landscape developments, leading to the discovery of certain patterns and mechanisms that help to explain what is currently happening, to speculate about what could happen, to anticipate outcomes of new developments and to identify useful interventions for working towards more sustainable systems of mobility.

The multi-level framework is applied differently in the different chapters. Because authors of different chapters work from different theoretical perspectives (e.g., critical theory, business studies, political science, cultural studies, innovation studies) and focus on different aspects of the transport regime, their interpretations and assessments of sustainability transitions differ. In the book project, we had discussions about whether the regime of automobility already is in transition. The answer depends on the significance given to real-world dynamics (whether these are irreversible) and depends on one's view of what counts as a transition, a niche and regime. The question of what constitutes a transition, how do you know they are happening and how do you determine start and end dates proved to be contentious and was not resolved. The answer centers around the boundaries of the system that is supposed to be in transition, the main elements of it and the problems for which we must find a solution. Are cars and technologies the main elements? Or are road infrastructures and traffic control the main elements? Or are the main elements fuel infrastructures and oil supply chains that extend to the Middle East and other regions? Are consumer behavior and mobility practices a (political) no-go area, or do these form the crucial dimensions of transport systems? Analysts hold different views about this, the same as real-world actors. Because the accepted definitions of problems and system boundaries change over time, analysts are studying a moving (or shape-shifting) target. So, we suggest that disagreement is inevitable because the struggle about interpretations and definitions of transitions (problems, directions, system boundaries) is part and parcel of transition processes.

Transitions defy an objective understanding. They are not a natural phenomenon. The same holds true for tipping points, a notion which is applied in some of the chapters and in the recent literature on mobility (for instance Dennis and Urry, 2009). Without wanting to go deeply into epistemic issues, we wish to say that developments can be cascaded by critical events, but an emphasis on successful trigger events ignores the importance of deeper, more gradual trends that bring the system towards the threshold. We prefer a middle-ground view in which particular events or actions can only have effects when preceding longer term processes have created the appropriate conditions. A further complication is that old ideas do not die out but live on. The predict-and-provide paradigm coincides with the predict-and-prevent paradigm. The gravity has shifted in the direction of the latter, but it can also shift back. In the Netherlands, the Balkende IV government engaged in a major road extension program. The support given to electric vehicles may also wane (with support shifting to other types of green vehicles). We agree with the authors of Chapter 10 that electric mobility has crossed a critical threshold but that this was not caused by a single event. Instead was the outcome of various developments building on each other (described in Chapter 10).

For this book, we hope the MLP can do three useful things: (a) enable a co-evolutionary analysis of transport systems that overcomes the dichotomy in transport studies between technology solutions and behavior change, (b) offer an analytical framework that helps scholars to investigate the conditions for transitions and provides ideas for transition policy and (c) offer a new vocabulary to discuss transformative change in the real world, for example, among experts, practitioners and policy makers, and to work towards it.

Three distinctive features of the transition perspective are, firstly, that it takes a longitudinal approach that allows a wider and historical perspective. Studies of transitions to sustainability should not start in the present but recognize deep path dependencies that may stretch back decades in time. Secondly, it draws attention not only to the factors that contribute to technological, social and political *change* but also to those factors that create powerful system *stability*. Transitions in automobility come about through the interdependencies and interactions between stability and change. Thirdly, the transition perspective provides an actor-based approach that helps in gaining a deeper understanding of shifting values, assumptions and mental models that underpin change and stability in behavior, policy, firm strategies, culture and infrastructure. This book aims to show how these distinctive features of a transitions approach provide added value compared to more established approaches in transport studies.

NOTES

1. The "transition to sustainable systems" literature is to be distinguished from the literature on economic transitions, dealing with transition problems and mechanisms involved in the transition from a planned economy to a market economy (Kornai, 1995; Roland, 2000).
2. The patterns are derived inductively from historical case studies. They are not an exhaustive list of possible patterns. More interactions could perhaps be differentiated on political, cultural, infrastructural and business dimensions. Developing a more differentiated view of niche-regime interaction seems generally a fruitful terrain for further research (see also Smith, 2007, on translations between niches and regimes).
3. http://abcnews.go.com/Technology/PCWorld/story?id=7067069
4. "The extent to which the old continues to dominate thinking about the new is nicely encapsulated in Thomas Edison's practice of regularly referring to his incandescent lamp as 'the burner'. Rather more seriously, in his work on an electric meter, a biographer reports, Edison for a long time attempted to

develop a measure of electricity consumption in units of cubic feet" (Rosenberg, 1986, p. 25).
5. http://www.guardian.co.uk/business/2010/mar/18/nissan-leaf-sunderland-factory-jobs
6. http://www.whitehouse.gov/files/documents/Battery-and-Electric-Vehicle-Report-FINAL.pdf

REFERENCES

Avelino, F., Bressers, N. and Kemp, R. (2012) 'Transition Management and Sustainable Mobility Policy: the Case of the Netherlands", in: Geerlings, H., Shiftan, Y. and Stead, D. (eds), *Transition towards Sustainable Mobility: The Role of Instruments, Individuals and Institutions*, Farnham (UK), Ashgate, forthcoming.
Britton, E. (2007) The new mobility agenda draft of 10-Nov-07, Available at http://www.ecoplan.org/library/3-velib-in-brief.pdf (accessed 11 August 2010).
Christensen, C. (1997) *The Innovator's Dilemma: When New Technologies Cause Great Firms to Fail*, Boston: Harvard Business School Press.
Corn, J.J. (1983) *The Winged Gospel: America's Romance With Aviation, 1900–1950*, New York: Oxford University Press.
Dennis, K. and Urry, J. (2009) *After the Car*, Cambridge, Polity Press.
Dyerson, R. & Pilkington, A. (2005) 'Gales of creative destruction and the opportunistic incumbent: The case of electric vehicles in California', *Technology Analysis & Strategic Management*, 17(4): 391–408.
Elzen, B., Geels, F.W. & Green, K. (eds) (2004) *System Innovation and the Transition to Sustainability: Theory, Evidence and Policy*, Cheltenham, England: Edward Elgar.
Fletcher, R.A. (1910) *Steam-ships: The Story of Their Development to the Present Day*, Philadelphia: Lippincott.
Geels, F.W. (2002) 'Technological transitions as evolutionary reconfiguration processes: A multi-level perspective and a case-study', *Research Policy*, 31(8/9): 1257–1274.
Geels, F.W. (2004) 'From sectoral systems of innovation to socio-technical systems: Insights about dynamics and change from sociology and institutional theory', *Research Policy*, 33(6–7): 897–920.
Geels, F.W. (2005a) 'The dynamics of transitions in socio-technical systems: A multi-level analysis of the transition pathway from horse-drawn carriages to automobiles (1860–1930)', *Technology Analysis & Strategic Management*, 17(4): 445–476.
Geels, F.W. (2005b) 'Processes and patterns in transitions and system innovations: Refining the co-evolutionary multi-level perspective', Technological Forecasting & Social Change, 72(6): 681–696.
Geels, F.W. (2005c) *Technological Transitions and System Innovations: A Co-evolutionary and Socio-Technical Analysis*, Cheltenham, England: Edward Elgar.
Geels, F.W. (2006) 'Co-evolutionary and multi-level dynamics in transitions: The transformation of aviation systems and the shift from propeller to turbojet (1930–1970)', *Technovation*, 26(9): 999–1016.
Geels, F.W. (2007) 'Transformations of large technical systems: A multi-level analysis of the Dutch highway system (1950–2000) ', *Science Technology & Human Values*, 32(2): 123–149.
Geels, F.W. & Raven, R.P.J.M. (2006) 'Non-linearity and expectations in niche-development trajectories: Ups and downs in Dutch biogas development (1973–2003)', *Technology Analysis & Strategic Management*, 18(3–4): 375–392.

Geels, F.W. & Schot, J.W. (2007) 'Typology of sociotechnical transition pathways', *Research Policy*, 36(3): 399–417.
Geels, F.W. & Schot, J.W. (2010a) The dynamics of transitions: A socio-technical perspective, in J. Grin, J. Rotmans, J. Schot, F.W. Geels & D. Loorbach (eds) *Transitions to Sustainable Development: New Directions in the Study of Long Term Transformative Change*, New York: Routledge.
Geels, F.W. & Schot, J.W. (2010b) 'Path creation and societal embedding in sociotechnical transitions: How automobiles entered Dutch society (1898–1970)'. Paper presented at international expert workshop on technological discontinuities and transitions, Eindhoven University, 14–16 May, 2010.
Geels, F.W. & Verhees, B. (2011) 'Cultural legitimacy and framing struggles in innovation journeys: A cultural-performative perspective and a case study of Dutch nuclear energy (1945–1986)', *Technological Forecasting & Social Change*, 78(6): 910–930.
Genus, A. & Coles, A.-M. (2008) 'Re-thinking the multi-level perspective of technological transitions', *Research Policy*, 37(9): 1436–1445.
Giddens, A. (1984) *The Constitution of Society: Outline of the Theory of Structuration*, Berkeley: University of California Press.
Grin, J., Rotmans, J., Schot, J, Geels, F.W. & Loorbach, D. (2010) *Transitions to Sustainable Development. New Directions in the Study of Long Term Transformative Change*, New York: Routledge.
Hendriks, C. (2008) 'On inclusion and network governance: The democratic disconnect of Dutch energy transitions', *Public Administration*, 86(4): 1009–1031.
Hendriks, C.M. & Grin, J. (2007) 'Contextualizing reflexive governance: The politics of Dutch transitions to sustainability', *Journal of Environmental Policy and Planning*, 9(3–4): 333–350.
Holz, G., Brugnach, M. & Pahl-Wostl, C. (2008) 'Specifying "regime": A framework for defining and describing regimes in transition research', *Technological Forecasting & Social Change* 75: 623–643.
Hoogma, R., Kemp, R., Schot, J. & Truffer, B. (2002), *Experimenting for Sustainable Transport Futures. The Approach of Strategic Niche Management*, London: Spon Press.
Huber, J. (2000) 'Towards industrial ecology: Sustainable development as a concept of ecological modernization', *Journal of Environmental Policy & Planning*, 2(4): 269–285.
Kaiser, F. G., Ranney, M., Hartig, T. & Bowler, P. A. (1999) 'Ecological behavior, environmental attitude, and feelings of responsibility for the environment', *European Psychologist*, 4(1): 59–74.
Kemp, R, Avelino F. & Bressers N. (2011) 'Transition management as a model for sustainable mobility', *European Transport/Trasporti Europei*, 47(1–2): 25–46.
Kemp, R., Rotmans, J. & Loorbach, D., 2007, 'Assessing the Dutch energy transition policy: How does it deal with dilemmas of managing transitions?' *Journal of Environmental Policy and Planning*, 9(3–4): 315–331.
Kemp, R., Schot, J. & Hoogma, R. (1998) 'Regime shifts to sustainability through processes of niche formation: the approach of strategic niche management', *Technology Analysis and Strategic Management*, 10(2): 175–196.
Kornai, J. (1995) Highway and Byways. Studies on Socialist Reform and Postsocialist Transition. Cambridge, MA: MIT Press.
Levy, D.L. & Egan, D. (2003), 'A neo-Gramscian approach to corporate political strategy: Conflict and accommodation in the climate change negotiations', *Journal of Management Studies*, 40(4): 803–30.
Lie, M. & Sørensen, K.H. (eds) (1996) *Making Technology Our Own: Domesticating Technology Into Everyday Life*, Oslo, Norway: Scandinavian University Press.

Loorbach, D. (2007) *Transition Management: New Mode of Governance for Sustainable Development*, Utrecht, the Netherlands: International Books.
Lounsbury, M. & Glynn, M.A. (2001) 'Cultural entrepreneurship: Stories, legitimacy, and the acquisition of resources', *Strategic Management Journal*, 22(6–7): 545–564.
Markard, J. & Truffer, B. (2008) 'Technological innovation systems and the multilevel perspective: Towards an integrated framework', *Research Policy*, 37(4): 596–615.
Meadowcroft, J. (2009) 'What about the politics? Sustainable development, transition management, and long term energy transitions', *Policy Sciences*, 42(4): 323–340.
Næss, A. (1973) 'The shallow and the deep, long-range ecology movement', *Inquiry* 16(1): 95–100.
Nooteboom, S.G. (2006) 'Adaptive networks. The governance for sustainable development', doctoral thesis, Erasmus University Rotterdam, the Netherlands.
Olleros, F. (1986) 'Emerging industries and the burnout of pioneers', *Journal of Product Innovation Management*, 1(1): 5–18.
Organisation for Economic Co-operation and Development (2009) *Eco-innovation in Industry. Enabling Green Growth*, Paris: Organisation for Economic Co-operation and Development.
Rip, A. & Kemp, R. (1998) Technological change, in S. Rayner & L. Malone (eds) *Human Choice and Climate Change*, Washington DC: Batelle Press.
Roland, G. (2000) *Transitions and Economics: Politics, Markets and Firms*, Cambridge, MA: MIT Press.
Rosenberg, N. (1986) The impact of technological innovation: A historical view, in R. Landau & N. Rosenberg (eds) *The Positive Sum Strategy: Harnessing Technology for Economic Growth*, Washington, DC: National Academy Press.
Rotmans, J., Kemp, R. & Van Asselt, M. (2001) 'More evolution than revolution. Transition management in public policy', *Foresight* 3(1): 15–31.
Schot, J.W. & Geels, F.W. (2008) 'Strategic niche management and sustainable innovation journeys: Theory, findings, research agenda and policy', *Technology Analysis & Strategic Management*, 20(5): 537–554.
Shove, E. & Walker, G. (2007) 'CAUTION! Transitions ahead: Politics, practice and sustainable transition management', *Environment and Planning A*, 39(4): 763–770.
Shove, E. & Walker, G., 2008, 'Transition management and the politics of shape shifting', *Environment and Planning A*, 40(4): 1012–1014.
Smith, A. (2007) 'Translating sustainabilities between green niches and sociotechnical regimes'. *Technology Analysis & Strategic Management*, 19(4): 427–450.
Smith, A. & Kern, F. (2009) 'The transitions storyline in Dutch environmental policy', *Environmental Politics*, 18(1): 78–98.
Smith, A., Stirling, A. & Berkhout, F. (2005) 'The governance of sustainable sociotechnical transitions', *Research Policy*, 34(10): 1491–1510.
Stern, N. (2006), *Review on the Economics of Climate Change*, London: H.M. Treasury.
Teece, D. (1986) 'Profiting from technological innovation: Implications for integration, collaboration, licensing and public policy, *Research Policy*, 15(6): 285–305.
Tripsas, M. (1997) 'Unravelling the process of creative destruction: Complementary assets and incumbent survival in the typesetter industry', *Strategic Management Journal*, 18(S1): 119–142.
Van Bree, B., Verbong, G.P.J. & Kramer, G.J., 2010, 'A multi-level perspective on the introduction of hydrogen and battery-electric vehicles', *Technological Forecasting & Social Change*, 77(4): 529–540.

Van Oosten, F.C. (1972) *Schepen onder Stoom: De Geboorte van het Stoomschip* [Ships Under Steam: The Birth of the Steamship], Bussum, the Netherlands: Unieboek, B.V.

Verbong, G.P.J. & Geels, F.W. (2007) 'The ongoing energy transition: Lessons from a socio-technical, multi-level analysis of the Dutch electricity system (1960–2004)', *Energy Policy*, 35(2): 1025–1037

Verbong, G.P.J., Geels, F.W. & Raven, R.P.J.M. (2008) 'Multi-niche analysis of dynamics and policies in Dutch renewable energy innovation journeys (1970–2006): Hype-cycles, closed networks and technology-focused learning', *Technology Analysis & Strategic Management*, 20(5): 555–573

Voß, J-P., Smith, A. & Grin, J. (2009) 'Designing long-term policy: rethinking transition management', *Policy Sciences*, 42(4): 275–302.

Unruh, G.C. (2000) 'Understanding carbon lock-in', *Energy Policy*, 28(12): 817–830.

Walker, W. (2000) 'Entrapment in large technology systems: Institutional commitments and power relations', *Research Policy*: 29(7-8), 833–846.

Part II
Stability and Regimes Pressures

4 The Dynamics of Regime Strength and Instability

Policy Challenges to the Dominance of the Private Car in the United Kingdom

Geoff Dudley and Kiron Chatterjee

4.1. THE CAR REGIME AND THE IMPACT OF SUSTAINABLE MOBILITY

Since the early 1990s, the concept of sustainable mobility has emerged from environmental niches and entered the mainstream of policy debates and content. It is a potentially potent idea that has offered significant challenges to the long time dominance of the car regime. For example, sustainable mobility can have major impacts on car design, the amount and type of fuel used, highway construction, the development and use of information technology and demand management policy instruments, together with affecting travel behavior itself. Yet at times it can also appear an extremely fragile concept, particularly when influential interests seize the opportunity to frustrate its implementation.

One key to understanding the varied fortunes of sustainable mobility is that it is subject to a wide degree of interpretation, and as such there are a number of different implications for the dominance of the car regime. For example, over the past two decades there has been a growing awareness of the impact of vehicle emissions on public health. More controversially, there is also the issue of the impact of 'greenhouse gases', such as carbon dioxide, on climate change. Albeit with some reluctance, the motor industry has been prepared to act on these issues, with such developments as the compulsory fitting on all new vehicles within the European Union of catalytic converters and in more recent years the wider availability of alternatively powered vehicles, such as petrol-electric hybrids. In these cases, although technological change may have significant economic implications for the industry, the car regime hitherto has been prepared to absorb the concept of sustainable mobility. Indeed, it can perceive environmental awareness as a significant asset in developing its public image, as a type of fit-stretch exercise in terms of adjusting to changing societal norms.

At the same time, sustainable mobility can offer a more direct threat to the dominance of the car regime, through such means as undermining the public expectation of expanding personal mobility, the promotion and development of alternative modes of transport and the use of demand management policy tools, such as congestion charging. In these instances, the car regime is likely to be considerably more resistant to change, with actors and institutions within the regime adopting an openly adversarial approach. The impact of the concept of sustainable mobility is therefore unpredictable, with complex interrelationships between niches, regimes and landscapes, to use the conceptual language discussed in Chapter 3.

In the United Kingdom, historically, the car regime developed relatively late, so that it experienced a rapid development from being a relatively weak niche in the early 1950s to a position of almost total dominance by the early 1960s. With the aid of the government itself becoming an integral member of the car regime, it retained this dominance for the next 30 years before the advent of the sustainable development concept, with its transport offshoot of sustainable mobility, led to an alliance of environmental interests that provided a significant challenge in the 1990s. The most significant tangible outcome of this period was the decision by the government to abandon its large scale road building strategy, changing from predict-and-provide to demand management (as discussed further in Chapter 6). Even more profoundly, some parts of the government defected from the car regime and launched a strategy based on sustainable mobility, including higher fuel taxes, promotion of alternative transport modes and plans for the introduction of congestion charging in large urban areas.

In the event, a protest against fuel prices in 2000 briefly threatened to bring the country to a standstill, leaving the government badly shaken by

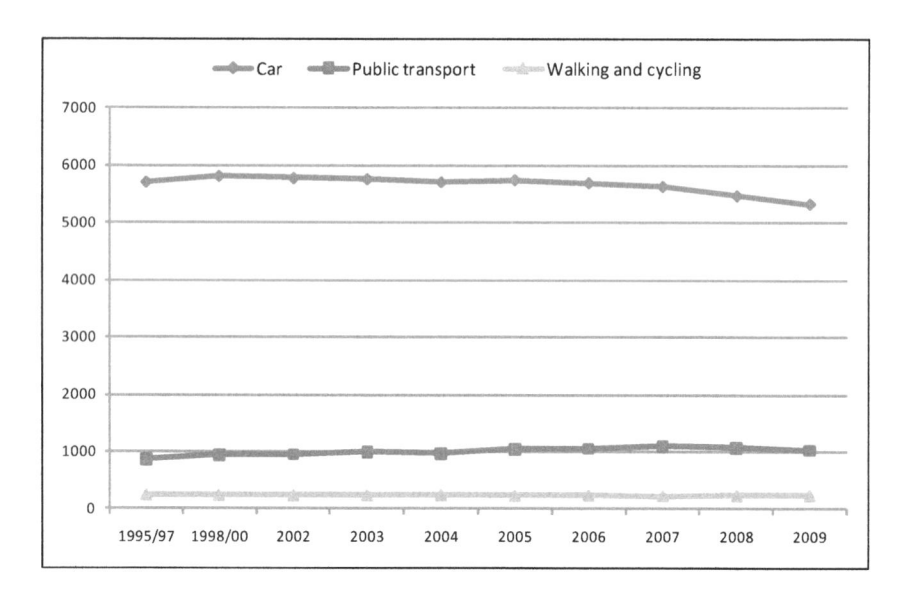

Figure 4.1 Average distance (in miles) traveled per person per year in Great Britain (Department for Transport, 2011).

the experience and reluctant to carry through measures that might provoke another revolt by the car regime. Niche developments such as the London Congestion Charge Scheme, introduced in 2003, have implemented the concept of sustainable mobility, but congestion charging has failed to win public support in other places in the United Kingdom, including Edinburgh and Manchester, where proposals were defeated in referendums. Nevertheless, there is evidence that personal car travel is no longer growing (see Figure 4.1).

A longitudinal study of the fortunes of the car regime in the United Kingdom, and of the impact on it of sustainable mobility, allows an opportunity to analyze how a powerful regime is constructed, the roots of its strength and also how it might be susceptible to change over time. The next section will therefore outline briefly the importance for regime stability and survival of the interrelationship between stability and change, together with the significance of the emergence of new ideas, such as sustainable mobility. The following seven sections then describe and analyze the dynamics of these processes, while the Conclusion examines the implications of the case study for assessing regime strengths and instabilities.

4.2. IDEAS AND TRANSITIONS

New ideas can act as powerful agents of change, in altering the ways in which technologies and regimes are framed. These shifts in framing can take a variety of forms, including the acquisition of knowledge, gaining fresh perspectives on new and old problems and changing the terms of debate on key issues. As Geels and Kemp illustrate in Chapter 3 (quoting Rosenberg, 1986) the fit-stretch pattern means that when radical new technologies appear, their impact can be severely handicapped by the tendency to think about them in terms of the old technology, such as in the case of the early cars being described as 'horseless carriages'. Alongside technological developments, therefore, car development involved a mental process of a leap in imagination that created new ideas about their potential. In doing so, their use was framed in a fresh way.

In more recent times, environmental ideas such as sustainable mobility can, like technologies, be developed within specialized niches and over time invade regimes and landscapes. In this context, ideas can be likened to viruses and have an ability to disrupt existing policy systems, power relationships and policies. A key issue for the entrenched interests (such as may be found in long-standing regimes) is the degree to which the new ideas and knowledge can be accommodated in existing and agreed policy frames, or whether completely new frames emerge (Richardson, 2000, pp. 1017–1018). Ideas can therefore have important but unpredictable effects on regimes and result in complex interrelationships with both niches and wider landscapes.

Geels and Schot (2007) distinguish four transition pathways with different relations between niches and regimes. These are transformation, de-alignment and re-alignment, technological substitution and reconfiguration. It could be said that the transformation path best describes how sustainable mobility has impacted upon the car regime. The transformation path occurs when there is moderate landscape pressure at a moment when niche-innovations have not yet been sufficiently developed, leading regime actors to respond by modifying the direction of development paths and innovation activities. In addition, Geels (2007) suggests that external pressure on regimes may also come from outsiders, such as social movements and scientists. Existing regimes may respond by adjusting some regime rules, which will then influence the direction of activities and trajectories.

As the UK case study of sustainable mobility and the car regime will illustrate, although ideas emerging from niches can be powerful agents of potentially major change, this type of change is not easily brought about and can take a considerable amount of time to pervade both regimes and landscapes. As a consequence of the discontinuities within the multi-level perspective, the forces of stability are likely to have the upper hand for much of the time and under the majority of conditions. It is only on the relatively rare occasions when "the idea whose time has come" (Kingdon, 1995, p. 1) is able to breach the walls of both regimes and landscapes that radical change takes place.

In terms of major change processes, therefore, the ability of an idea to pervade niches, regimes and landscapes is crucial. For example, the idea of sustainable mobility may have significant impacts at regime level, but if it has only a tenuous hold at landscape level, then this may limit its progress within the regime. In addition, opponents of the idea may themselves find new niches in which to mount a challenge. Thus sustainable mobility made significant progress from niches to regimes in the 1990s, only for its progress to be halted by a backlash from motoring interests that employed their own distinctive niche. In turn, this revolt appeared to win wide public sympathy.

After describing and analyzing these processes, the Conclusion will assess the implications for regime stability and change.

4.3. A LONGITUDINAL CASE STUDY OF THE BRITISH MOTOR CAR REGIME (1900–2010)

This section examines how, from relatively weak origins, by the late 1950s the car regime had established a position of enormous strength. It was only in the 1990s that this dominance was significantly challenged. Since that time, government has attempted to introduce sustainable mobility transport policies but with only limited success.

4.3.1. The Late Development of the Motor Car Regime (1900–1970)

It is significant to note that regimes themselves can often emerge from relatively long periods as niches in terms of their economic and political influence. Thus a structural niche may possess the potential to develop into something much more powerful that can supersede and eclipse established regimes and their associated interests but is hindered through a failure to project itself into a favorable situation. As such, the niche remains restricted in terms of the ability to 'stretch' into a strategic position that fulfills its apparent potential. For the first half of the 20th century, this was the situation in the case of the motor car in the United Kingdom. At the heart of the potential strength of the car regime was the economic strength of its constituent parts (including the motor industry, motoring organizations, the road construction industry and the oil industry), combined with the ability of the car to provide the individual with a high level of personal mobility that fulfills aspirations of economic and social well-being. The strength of the car regime can therefore be identified both in terms of its economic power and also its ability to reshape the way of life of society as a whole. In turn, this economic and social power meant that there were great political benefits to be gained for governments that supported and promoted the interests of the car regime. Given this formidable potential, it is surprising that for many years the car interests failed to impact significantly upon the prevailing regime.

The failure of the niche to break through was set in the early years of the 20th century, when a Road Board was founded by the government in order to aid local authority road improvement and maintenance and also initiate new motor roads, but in the event hardly any new roads were built (see Barker & Savage, 1974; Dudley, 1983; Dyos & Aldcroft, 1974; Hamer, 1987). When the Road Board was wound up in 1920, its powers were transferred to the newly created Ministry of Transport (MoT). However, during the 1920s and 1930s, the MoT was generally preoccupied with the state of the economically ailing railway companies, with the result that the motor industry remained largely a peripheral niche in terms of political attention. Nevertheless, one potentially significant shift in MoT responsibilities arrived in 1937, when the Ministry took direct control over some 4,500 miles of the most important national through roads, which were now legally defined as 'trunk roads'. Consequently, the MoT now assumed direct responsibility for constructing major roads and so provided the scope for a much closer relationship with the motor and road construction industries, and this now presented a context with the potential to move the industry from a niche to a more central and powerful regime. This potential, however, depended on the government committing itself to a major program of road building, but instead the Minister of Transport rejected the concept of a motorway network (events in the United Kingdom could be contrasted with what happened in countries such as Germany and Italy where road building became closely tied to the concept of economic growth and national regeneration (Charlesworth, 1984).

The motor interests remained in an economic and political backwater well into the 1950s. The austerity of the early post-war years led to a slow growth in the number of vehicles on the UK's roads, while industry became increasingly frustrated at its lack of progress. As Finer (1958) commented in the late 1950s: "Organizations with a special interest in roads form a vast complex of great social and industrial importance. Yet, for all this, the sums spent on roads since 1945 had been by any standard quite negligible" (p. 470).

This long-standing state of stagnation, however, was about to change radically and decisively. What had been lacking were crucial political leadership and an idea that could galvanize the government into action. From the late 1950s, therefore, road building became a key plank in the government's strategy of rebuilding post-war Britain, and this was shortly followed by the even more potent concept that increasing car ownership, and providing the roads for these vehicles to run on, could create a feel-good factor among the population that would reap political rewards. Thus the Conservative government was re-elected in 1959 using the slogan of 'you never had it so good'. As part of this strategy, the government made it a priority to complete the first major UK motorway, the M1 from London to Birmingham, by the time of the election. Hitherto, the growth in motor vehicles and the building of roads had been seen in terms of economic growth, but to this concept was now added the idea of cars and roads as a key element in a popular consumerist revolution (Dudley & Richardson, 2000, pp. 97–110). As Kingdon (1995) comments, "the greatest changes are likely to occur when the policy streams of problems, policies and politics converge and combine, particularly through the form of 'an idea whose time has come" (p. 1) and linking construction of the M1 to popular consumerism heralded the movement of the motor car from niche to enormously powerful regime.

From the late 1950s, the rapid rise of the car regime was also mirrored by the equally steep fall in the fortunes of the formerly powerful rail regime. This decline in rail was epitomized by the 1963 Report by British Railways Board Chairman Dr. Richard Beeching on the *Reshaping of British Railways* (British Railways Board, 1963). In particular, the Report recommended the closure of 2,000 stations that were considered to be uneconomic for the future running of the railway. The large majority of these closures were implemented over the following few years. Perhaps even more than the closures themselves, the Beeching Report indicated that the government now perceived rail as an industry in long term decline. Instead, it was the car regime that was seen as representing the future of transport, and every effort was to be made in promoting its rapid progress.

Continuing the trend toward popular consumerism, car ownership became closely identified with the phenomenon of the 'Swinging Sixties' in the United Kingdom. Thus while there had been a significant growth in the numbers of registered cars in Britain during the 1950s, from 2.0 million in 1950 to 4.9 million in 1960, this figure was dwarfed by the growth in the 1960s, from 4.9 million cars in 1960 to 10.0 million in 1970. These aggregate statistics do not tell the whole story, for while prior to the 1960s motoring was basically restricted to a (mainly male) social and economic elite, the 1960s saw car use become available across gender, age and social groups. Figure 4.2 illustrates how households with a car were a minority in 1955 (20%), but a majority by 1975 (56%). As motoring became a more classless activity, so cars became more diverse in size and design and also came to reflect, or even symbolize, the changing times. Perhaps nothing became more synonymous with the 'Swinging Sixties' than the Mini, the stylish car that was equally popular with Royalty, nouveau riche pop stars and young men and women owning their first vehicle (Society of Motor Manufacturers and Traders, 2007). Thus, although the Mini in itself represented a technological niche, it was influential in the 'fit and stretch' movement of the motor car to transcend social barriers and assume a place at the forefront of the consumer revolution. In this way, the car regime was in almost perfect harmony with the social and economic landscape. As Geels (2007, p. 126) comments, "transitions come about through alignments between processes at different levels".

It was also significant that the UK government rejected the idea of road tolls to pay for the motorways. Instead, they were free at the point of use by means of the Exchequer paying for their construction. This decision therefore encouraged the association of the motor car with a revolution in personal mobility, a concept that was embedded by the rapid growth in motorways during the 1960s, so that a 10-year target of 1,000 miles

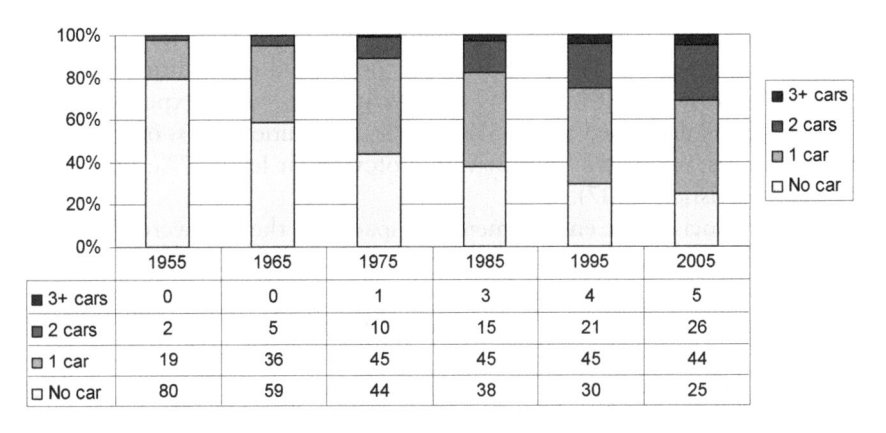

	1955	1965	1975	1985	1995	2005
3+ cars	0	0	1	3	4	5
2 cars	2	5	10	15	21	26
1 car	19	36	45	45	45	44
No car	80	59	44	38	30	25

Figure 4.2 Households with regular use of a car (Department for Transport, 2008).

of motorway by 1972 was achieved. The strength of the motor car regime was enhanced by the fact that the motorways were being built by the MoT itself, with the consequence that a powerful reciprocal relationship developed between the government and other members of the motor regime. By the early 1970s, this regime had achieved almost hegemonic power over UK transport policy (Dudley & Richardson, 2000, pp. 82–110), with apparently little threat to its dominance (see Chapter 6, however, for an account of the persistent challenge to the concept of urban road development, as opposed to the largely uncontested inter-urban development of motorways).

4.3.2 Early Environmentalism and Spasmodic Success (1970–1990)

The emergence and then domination of the car regime in the 1960s to the 1980s can be classified as a transition in terms of a major shift in a sociotechnical system, with co-evolutionary and multi-actor processes. In terms of the multi-level perspective, the car regime had clear functional goals; a variety of elements, including powerful economic, political and cultural factors; long-term stability; no single leader (although the central role of the government was clearly crucial); and a variety of dimensions within its component parts. The dominance of the car meant it was assumed that alternatives such as public transport were in inevitable secular long-term decline, so that a 'predict and provide' strategy was adopted with regard to road building, whereby road capacity would be expanded to accommodate the ever expanding number of vehicles. This meant that alternative regimes, such as that for public transport, were placed in a deeply subordinate position to that of the car regime, while few new technologies or ideas emerged from niches to challenge significantly the latter's almost unquestioned hegemonic supremacy.

In the wider landscape, the car embedded its position as a key element in the inexorable rise of popular consumerism. The percentage of households with a car increased from 56% in 1975 to 70% in 1995 (see Figure 4.2) with households having more than one car increasing from 11% to 25% over the same period. The proportion of household expenditure devoted to transport rose from 8% in 1957 to 16% in 2006, while expenditure on food and drink decreased from 33% to 15%. In addition, most of transport expenditure is associated with private motoring (at least 87%; Office for National Statistics, 2007).

Negative social and environmental impacts of the car were generally overlooked, with adverse effects framed in terms of solving traffic congestion problems through better accommodating the car. This policy solution was particularly evident in the highly influential 1963 Buchanan report on *Traffic in Towns,* that emphasized the need to adapt the urban environment to the growth of motor traffic through large-scale planning and design (although the report also advocated significant restraint on car use in large urban areas). It was only in the early 1970s that environmental impacts of the car were first given serious consideration, chiefly through the growth of environmental groups such as Friends of the Earth and Transport 2000. The latter campaigned for a multi-modal transport strategy, with particular emphasis on rail. These environmental groups for the first time challenged the assumptions framed in 'predict and provide', a process aided by the government being compelled in the mid 1970s to severely cut the road building program as part of the response to the economic crisis of the time.

The most effective challenge to the car regime during this period, however, came from a single individual, John Tyme, who traveled around the country organizing local residents to disrupt public inquiries held to consider a planned new road. In many cases this strategy proved highly effective, and it could be said that Tyme had discovered a procedural niche he could exploit in order to undermine the car regime (Tyme, 1977). In the late 1970s, the government abandoned the 'predict and provide' road building strategy in favor of a more flexible approach (Cmnd. 6836, 1977). However, the environmental groups did not constitute a regime that could challenge the car regime on a consistent long-term basis, and over the years that followed the car regime reasserted its power. This culminated in the 1989 White Paper *Roads for Prosperity* (CM 693, 1989), which effectively revived the concept of a strategic trunk roads plan and envisaged a doubling of expenditure.

4.3.3. Sustainable Development as an Effective Niche Idea (1990–1997)

In addition to its limited resources when compared with the car regime, it could be said that the environmental lobby had lacked a potent idea that could carry an effective public and political message and bring together the widest range of groups. This gap was filled through the medium of the 1987 report produced by the World Commission on Environment and Development, chaired by Gro Harlem Brundtland. Crucially, the report put forward the potent concept of sustainable development, defined as "development that meets the needs of the present without compromising the ability of future generations to meet their own needs" (Brundtland, 1987, 43).

For both environmental groups and governments, the adoption of policies that encompassed 'sustainable development' apparently combined a responsible attitude toward environmental issues with the virtues to be gained from economic growth. It therefore allowed the environmental coalition to maximize the spectrum of interests under its umbrella of 'sustainable development'. Thus the concept of a sustainable transport policy could bring together government and a wide range of environmental groups, while the discourse on 'sustainable development' gave environmental groups an effective means of undermining the car regime (Dudley & Richardson, 2000, pp. 141–162). Sustainable development therefore emerged from the niche of a specialized committee and by being adapted for transport purposes into the concept of sustainable mobility could become an active idea 'virus' in transport arenas.

The quest for sustainable mobility had particular resonance in the policy area of vehicle emissions, where the toxic substances involved were related to a variety of public health issues, including learning difficulties in children, the increasing incidence of respiratory diseases such as asthma and a potentially catastrophic warming of the earth's atmosphere through the production of carbon dioxide. Consequently, European Union directives set limits for lead in air and in petrol and in 1993 the requirement that all new cars be fitted with three-way catalytic converters. In the 1990s, therefore, new knowledge about the harmful effects of vehicle emissions provided a serious challenge to the values that produced the *Roads for Prosperity* strategy (Dudley, 1995). This was particularly evident in a 1994 report on transport by the influential Royal Commission on Environmental Pollution, which recommended a halving of expenditure on roads (Cm 2674, 1994).

For their part, environmental groups came together to form a multi-arena strategy designed to undermine the credibility of the roads strategy. This included a large number of direct action protests, whereby radical green activists and local residents united to occupy road construction sites. The widespread media attention provided to these protests further promoted the concept of sustainable mobility. Meanwhile, a report by the government sponsored Standing Advisory Committee on Trunk Road Assessment (1994) suggested that building new roads tended to have the effect of generating traffic, with the effect that this prevented them from solving congestion problems.

By permeating transport arenas, the concept of sustainable mobility had transcended its niche origins and became a powerful agent of policy change. In particular, by the late 1990s, the government had dismantled the *Roads for Prosperity* strategy, while for the first time in nearly half a century increasing car use was perceived as a serious policy 'problem,' rather than the chief 'solution'.

4.3.4. Optimistic Plans and High Ambitions for Sustainable Mobility (1997–2000)

The 1990s had seen a huge shift in the framing of transport policy, and by the end of the decade the concept of sustainable mobility had moved from its niche origins into the policy mainstream. Consequently, the policy makers perceived their chief task as translating this concept into practical policies. In this context, an early notable attempt was the 1991 Report, *Transport: The New Realism* (Goodwin, Hallett, Kenny & Stokes, 1991). The authors believed that, instead of the *Roads for Prosperity* (Cm 693, 1989) agenda, a new realism was called for involving a policy mix which would include

The Dynamics of Regime Strength and Instability   93

a substantial improvement in public transport, traffic calming, advanced traffic management systems and road pricing (see Chapter 6 for a detailed account of the construction and impact of *The New Realism*). In this new climate, road construction 'to meet demand' would no longer be the core of a transport strategy. For over 40 years, the enormously powerful motor car regime had become synonymous with the framing of UK transport policy, but now parts of government had separated themselves from the core automobile interests, through deserting the 'predict and provide' paradigm, and instead aligning policies with concepts of demand management. Consequently, the government was ready to implement the policy ideas expressed by *The New Realism*.

The Labour government elected in 1997 was particularly well placed to carry out this task, as the Transport and Environment Departments were merged to form the Department of the Environment, Transport and the Regions (DETR), with Deputy Prime Minister John Prescott appointed Secretary of State for the new Department. Unusually for transport, therefore, the Minister in charge was a powerful figure at the center of government, while Prescott himself had a personal commitment to place sustainable transport policies at the top of the policy agenda.

The first tangible product of this new era was the 1998 White Paper *A New Deal for Transport: Better for Everyone* (Cm 3950, 1998). In his Foreword to the White Paper, Prescott stressed that the main aim was to increase personal choice by improving the alternatives to car travel and so to secure mobility that would be sustainable in the long term. Consequently, the priority would be maintaining existing roads rather than building new ones and better management of the road network to improve reliability (Cm 3950, 1998, para. 3). Echoing the agenda set out in *The New Realism* (Goodwin et al., 1991), the White Paper saw the way forward as being through an integrated transport policy that included integration within and between different forms of transport, integration with the environment, integration with land use planning and integration with policies for education, health and wealth creation (Cm 3950, 1998, para. 1.22).

The White Paper set out its hopes and recommendations for improving the quality of public transport, as well as promoting walking and cycling, while particular emphasis in delivering these policies was placed on action at the local level, so that local authorities outside London would be required to produce Local Transport Plans that would set out their proposals for delivering integrated transport over a 5-year period (Cm 3950, 1998, para. 4.73). The implication therefore was that, although sustainable ideas had entered the policy mainstream, to a large degree their implementation would entail their returning to niches at the local level, whereby it would be hoped that over time they would spread through the country. This strategy had the political advantage for the government of passing responsibility for the implementation of sustainable mobility to the local authorities. At the same time, it also made the ambitious assumption that these same local authorities would have the resources and will to carry out these policies and that they would be successfully implemented to the degree that they would be widely taken up. However, the local authorities were not given many of the devolved powers to manage transport systems, particularly public transport, that would have matched the new onus of responsibility that was being placed on them.

These assumptions were particularly true in the case of controversial policies such as road user charging and workplace parking levies. A credible integrated transport policy required more than just the 'carrot' of promotion of alternative modes to the car. It also required the 'stick' element for discouraging car use, by means of such measures as fuel taxes and road pricing. In fact, the previous Conservative government had already made a significant move on fuel taxes, when in 1993 it introduced the fuel tax escalator, whereby fuel duty would increase by 5% per year above the rate of inflation, while in 1997 the new Labour government increased this figure to 6%. The 1998 White Paper potentially strengthened the scope of these regulatory policies by announcing that legislation would be introduced to allow road user charging pilot schemes to be implemented, either locally or on trunk roads and motorways (Cm 3950, 1998, para. 4.100). Similarly, legislation would enable local authorities to levy a new charge on workplace parking (Cm 3950, 1998, para. 4.107). Crucially, John Prescott won a concession from the Treasury that revenues from these schemes could be hypothecated to transport expenditure, so that they could help in providing the resources needed to implement an effective integrated transport policy. However, it was unclear if and when specific pilot schemes would be implemented.

The successor to the 1998 White Paper was a 10-year plan, *Transport 2010: The 10 Year Plan* (DETR, 2000). The plan set out an ambitious £180 billion investment strategy across the decade but also set a number of significant targets. For example, it was intended that congestion on interurban trunk roads would be reduced by 5% below current levels, compared with forecast growth of 28% by 2010 (DETR, 2000, p. 55). The target was also stated that there would be a 10% increase in bus passenger journeys by 2010 (DETR, 2000, p. 66).

With regard to larger urban areas, a target was set that congestion be reduced from a forecast growth of 15% by 2010 to an 8% reduction. In other urban areas, congestion growth would be reduced from 15% to 7% (DETR, 2000, p. 66). However, in addition to the problematic matter of the financial resources being available to meet the ambitious targets, the *10 Year Plan* (DETR, 2000) also made it clear that major assumptions were being made about the likelihood of the introduction of charging schemes. Thus the plan assumed that London and a number of local authorities would introduce local congestion charging or workplace parking levy schemes from 2004 to 2005 onwards. Net revenues from these schemes would total £1.5 billion in London and £1.2 billion to local authorities

The Dynamics of Regime Strength and Instability 95

in the rest of England. In turn, it was assumed that 8 of the largest towns and cities would introduce congestion charging schemes, and a further 12 would bring in workplace parking levies (DETR, 2000, pp. 104–105). It was expected also that up to 25 new light rail lines would be constructed by 2010 (although in the event only a handful of these lines have been implemented, with in many cases the escalating cost of financing them being given by government as the chief reason for withdrawal of support). The targets set out in the *10 Year Plan* (DETR, 2000) were therefore highly contingent on events running smoothly at niche level.

4.3.5. Problematic Implementation and a Severe Backlash from Motor Vehicle Interests (2000)

The difficulties in implementing sustainable mobility were great enough, even in the presence of a compliant motor regime. In the event, shortly after publication of the *10 Year Plan* (DETR, 2000), there was a backlash from the grassroots of the motor regime that for a period threatened to bring the country to a standstill and to severely undermine the credibility of the government (Lyons & Chatterjee, 2002). Consequently, just as there was a bottom-up process in the emergence of sustainable ideas, so the backlash also arose from vested interests. The trigger for the protests was the rise in fuel prices during the summer of 2000. Ironically, the government had scrapped the fuel tax escalator for that year in the budget of 2000, but the general rise in fuel prices highlighted the amounts taken in fuel duty by the government. To a significant extent, events in the United Kingdom took their lead from what was happening in France, where road hauliers blockaded oil refineries and fuel depots in an attempt to win concessions on fuel prices from the French government.

In September 2000, groups of farmers and hauliers embarked on similar action in the United Kingdom. Although it was estimated that at no time did the protesters number more than 2,500, the blockades proved highly effective in preventing fuel supplies being moved so that, when asked if Britain was facing a national crisis, Prime Minister Tony Blair conceded: "There's no point beating about the bush. Of course it is" (*Financial Times*, 14 September 2000). The breakdown in the fuel supplies threatened not only transport systems but also food supplies, health services and businesses generally. Nevertheless, the protesters won a high degree of public support and only called off their protests when they feared a breakdown in food supplies and health services would undermine that public goodwill.

From the perspective of implementing sustainable mobility policies, the fuel protests illustrated the continued strong hold on society of the motor car regime. As one commentator commented in the aftermath of the protests, the politicians simply underestimated the public's love affair with the motor car (*Financial Times*, 29 September 2000). The protests clearly rattled the government to its core, to the extent it became clear there were severe difficulties inherent in enforcing sustainable mobility policies within niches and also winning acceptance within the wider landscapes of society.

4.3.6 Central-Local Tensions and Achievements (2000–2008)

The separation of the government from the motor vehicle regime in the 1990s had the effect of placing it in a more exposed and vulnerable position. Ideas concerning sustainable mobility successfully permeated the policy making center, but the 'viruses' had not been so effective in inhabiting either the societal landscape or more local niches. The outcome in terms of implementing any kind of publicly controversial policies was something of a standoff, with government periodically expressing its good intentions on sustainable mobility, but lacking either the resources or the political will to carry them through. Instead, the onus was placed on local authorities to carry things forward, but here both public and political attitudes were highly problematic, with little chance of achieving any kind of consensus.

This standoff developed despite the fact that, at one point in the past decade, it briefly appeared that success in a significant niche would herald a major transition in terms of an idea cascading across niches. This was at the time of the successful implementation of the London Congestion Charge Scheme (LCCS) in February 2003. As we described previously, the *10 Year Plan* (DETR, 2000) envisaged that eight of the largest towns and cities would introduce congestion charging, with the revenues being used to make significant local improvements in public transport and provision for walking and cycling. The concept of congestion charging, in itself, represented an example of a fit-stretch process. Hitherto, urban road pricing schemes, for example those introduced in Norway, had been principally perceived as revenue raisers, rather than being used to reduce congestion. Thus the introduction of a Congestion Charge in a major city such as London was anticipated to be a landmark event that would cause road user charging to be perceived in a fresh light.

It was the election of Ken Livingstone as London Mayor in 2000 that transformed the LCCS into 'an idea whose time had come' (Kingdon, 1995). Livingstone chose to make the LCCS one of the principal ideas of his manifesto and pledged to introduce it during his term of office. Nevertheless, this political gamble required Livingstone to use the arena of the LCCS to shift the popular image of urban road pricing from an unwanted tax on the motoring public into a politically acceptable policy 'solution'. The political credibility of Livingstone therefore depended crucially on not only the technological and administrative feasibility of the LCCS and that it should achieve its principal policy objectives in terms of reducing congestion, but also in constructing a scenario of policy 'success' that would win public acceptability (Dudley, 2004). In the event, he successfully achieved these objectives. For example, the first official study of the LCCS found that 50,000 fewer cars a day were entering the zone, a reduction of 16%. Journeys within the zone were 14% quicker and traffic delays reduced by 30%, while accidents decreased by 20% (Transport for London, 2003). The LCCS had attracted widespread opposition from business and motoring interests prior to its implementation, but surveys of Londoners' attitudes toward congestion charging undertaken before and after the scheme was introduced showed an overall shift of opinion toward favoring the scheme, with four fifths of those who expressed an opinion considering that the scheme had been effective in achieving its objectives (Transport for London, 2004). In 2007, a western extension was added to the zone.

Despite its apparent commitment to congestion charging in the *10 Year Plan* (DETR, 2000), the government refused to endorse the LCCS prior to its implementation and preferred to let Livingstone take the full political risk. However, once it became clear that the LCCS had been deemed a policy success, then the government was prepared to explore the possibilities of building on this niche. Perhaps surprisingly, a 2004 feasibility study of road pricing found that it would ultimately be more effective to introduce a nationwide scheme than rely on a series of local schemes (Department for Transport [DfT], 2004). However, the government has continued to be highly reluctant to commit to a national scheme, as illustrated by the abandonment in 2005 of a planned national lorry road pricing scheme.

Instead, the focus has continued to be on local niches, but progress here has also stalled. Firstly, a proposal for dual cordon charging zones in Edinburgh was defeated in a referendum in 2005. Nevertheless, in 2004 the government brought forward a Transport Innovation Fund (TIF), where local authorities were able to bid for infrastructure investment funds from a budget of £1.4 billion for packages of schemes that overall would have a significant impact on road congestion or economic productivity. In most cases, TIF funding was dependent on the authorities introducing some type of congestion charging scheme. In the event, only the Greater Manchester local authorities proceeded with a specific charging proposal, but this was defeated heavily in a referendum of December 2008. With 78.8% of voters rejecting the proposal, it is considered that the public had a major concern that many motorists would have no viable alternative than paying road charges for essential journeys (such as journeys to work or the supermarket) that were well outside the most congested parts of Manchester. It also appeared that those in favor of the proposal were unable to convey the long-term advantages of the scheme. Thus a 'yes' vote could have led to a government grant of £1.5 billion from the TIF and £1.2 billion of local funding taken out as a 30-year loan and partly paid for by future revenues from the congestion charge. However, 'yes' campaigners conceded that people generally failed to grasp what the planned investment would mean for local transport and were more swayed by arguments that the congestion charge represented just one more tax on motorists (*The Guardian*, 12 December 2008).

In a similar manner to the impact of the fuel protests of 2000, the government's commitment to road pricing appears to have been weakened by a 2007 petition against charging placed on the Downing Street Web site and signed by 1.7 million citizens. At the time of the successful implementation of the LCCS, much weight was placed on the importance of political leadership, which makes it particularly surprising that so little attention has been paid to this factor by national government or in the case of the Edinburgh and Manchester proposals. Simply waiting for public opinion to move is unlikely to be effective; the public's understanding of the technology remains limited, and suspicions about the non-transport implications, such as the consequences for privacy and social justice, are easy to exploit in a context of uncertainty (Parkhurst & Dudley, 2008, pp. 68–69; see Chapter 6 for a further analysis of the feasibility of road pricing).

The *10 Year Plan* (DETR, 2000) also assumed that 12 towns and cities would introduce workplace parking levies, but here progress has been equally slow. Only the city of Nottingham has carried forward a significant proposal, where a workplace parking levy is due to be introduced in 2012, with the revenues used to pay for extensions to the light rail system. This proposal currently remains in place, despite fierce opposition from local business interests.

To a significant degree, the emphasis of government targets has switched to reducing carbon dioxide emissions through improving vehicle technology, rather than reducing congestion through reducing growth in car traffic. This is partly explained by reduced growth in car traffic in recent years (only 10.5% growth between 1997 and 2007 compared to 28.5% growth between 1987 and 1997). However, the government sponsored Eddington Report, published in 2006, forecasted that congestion would increase by 30% over 2003 levels by 2025 (Eddington, 2006), so congestion is expected to remain a central issue.

The 2008 Climate Change Act sets a demanding target for the United Kingdom that the net carbon account for all greenhouse gases in the year 2050 is at least 80% lower than the 1990 baseline. Greenhouse gas emissions from transport represent 21% of total domestic emissions with 58% of transport emissions attributable to cars. The government's analysis (DfT, 2009a) suggests that the main contribution to carbon reduction for road transport up to 2020 will be from working with the motor industry on technological improvements (especially engine efficiencies). The effect is therefore a tendency to work with the established car regime, rather than to fundamentally challenge its basic strengths.

4.3.7 Landscape and Local Trends (2000–2010)

Although major transport policy initiatives (such as congestion charging) have not been implemented in many towns and cities, it was noted earlier (see Figure 4.1) that average car travel per person in Great Britain has stabilized, and there has even been a slight decrease in recent years. Apart from

London, there have been decreases in traffic in a number of towns and cities, such as Birmingham, Manchester and Nottingham (where trends preceded the recent economic recession; DfT, 2009b). Manchester and Nottingham are notable as two cities in Great Britain where major investment has taken place in urban rapid transit. What is evident from examining these places is that other modes than the car are experiencing increases in use.

National investment is supporting local initiatives in some instances. The Cycle Demonstration Towns program between 2005 and 2009 involved six medium-sized towns in England and investment in cycling of about £10 per resident per year (10 times the average for England; Sloman, Cavill, Cope, Muller & Kennedy, 2009). Results indicate an average 27% growth in cycling trips across the 6 towns and 14% more adults cycling. The program has been extended now to a further 12 towns and cities. Cycling levels have also doubled between 2000 and 2006 in London. It is conceivable that cycling is emerging as more than a niche in some areas in the United Kingdom and perhaps has at least the potential to become an emerging regime.

Evidence is also emerging that the mix of mobility options being used by the public is increasing and that the car is reducing its dominance as a mode of transport in some areas (especially within large urban areas and for inter-urban travel where rail is increasing in popularity) so that some sub-groups of the population are increasingly managing without a car. For example, young adults under 30 years old have experienced a reduction in driving license holding. For 17- to 20-year-olds, the proportion of trips as car driver decreased from 26% in 1998 to 2000 to 25% in 2008, and for 21- to 29-year-olds it decreased from 45% to 40% (DfT, 2009c). Various explanations for this have been put forward, such as greater participation in higher education, more stringent driving tests and higher car ownership and use costs, but whatever the underlying causes, there is a changing landscape of mobility for this population group.

The previous evidence points to the prospect of the landscape developing, encouraged by policy actions, in such a way that the sustainable transport idea permeates the social landscape and that long-distance rail, urban rapid transit and cycling will apply pressure on the dominant motor car regime at a local level in some places and potentially transform the regime into something new.

4.4. CONCLUSION

Although the car regime in the United Kingdom has been significantly challenged by sustainable mobility in the past two decades, two factors in particular have been crucial in it retaining a significant amount of its strength. Firstly, there remains no true sustainable mobility regime that can match the car regime's assets. Thus the coalition of motor interests, including motor manufacturers, oil companies and motoring organizations, although less dominant than in previous decades, remains a highly influential political, economic and social force. In contrast, there is no equivalent identification of interests between environmental groups, bus and rail operators and cycling and walking groups. Instead, each tends to have its own agenda and priorities. Secondly, 'the popular consumerism' idea that underpinned the rise of the car regime in the late 1950s retains a significant degree of its strength in that the motor industry is still perceived as an important barometer of industrial and commercial prosperity in the United Kingdom (although there are no longer any domestically owned volume carmakers), while for many consumers individual mobility through car ownership remains an important element in personal identity and an indicator of social status and prosperity.

At the same time, there are several means by which the car regime has become less dominant over the past two decades, compared with the trends that underpinned its ascendancy from the 1950s to the 1980s. We can therefore identify a number of cracks in the strength and stability of the car regime.

First, an important political factor was the defection of government in the 1990s from its hitherto unquestioned adherence to the 'predict and provide' paradigm. However, the fuel protests of 2000, which formed a backlash from vested regime interests, illustrated that sustainable mobility had yet to truly permeate the societal landscape, with the result that government retreated from its commitment to sustainable mobility and an integrated transport policy. Nevertheless, it cannot be said that government has rejoined the car regime. Consequently, the subsequent outcome has been something of a stalemate, with niche successes for sustainable mobility such as the LCCS countered by the fear of the government to precipitate another major rebellion similar in character to that of the 2000 fuel protests. At the same time, the long-term defection of government from its ranks does mean that the car regime can no longer assume that policy framing will be in its favor.

Second, perhaps even more significantly, we have seen that, over the past decade, there is evidence within the societal landscape that a limit has been reached in personal mobility through motor vehicle use. Figure 4.1 showed that the average distance traveled by car per person has not increased since 1995/97. There are a variety of possible explanations for these changes, many of them not connected directly to transport causes, including changes in working patterns, technological developments, shifts in locations of home and work, demographic factors and changing attitudes and behavior within sections of society. Over time, however, when these factors are added to more direct transport factors, such as congestion and a rise in the real cost of fuel, then the strength and stability of the car regime can appear less secure.

Third, other modes of transport, including bus, rail, light rail and the bicycle, have become relatively stronger and more influential since the early 1990s. Rail in particular was perceived as being in steep and terminal decline in the 1960s and 1970s, but in recent years rail patronage has risen sharply (an increase from 42 billion passenger kilometers in 1994 to 59 billion passenger kilometers in 2007 (DfT, 2008), while bus use and cycling have also experienced significant growth, at least in some areas. As we noted previously, there is still no unified passenger transport and cycling regime that could significantly challenge the car regime. Nevertheless, these alternative modes to the car are now generally accepted as significant elements that are likely to grow in significance, in representing at least part of the 'solution' to problems posed by road congestion, air quality and climate change. This is exemplified by the recent emergence of a wide political consensus that a high speed rail link is required between London and Scotland. However, it should also be noted that politically it is much easier to offer the 'carrot' of infrastructure investment than wield the 'stick' of demand management policies, such as road pricing. It is perhaps only with the widespread implementation of the latter that the car regime will be challenged more fundamentally.

Fourth, over the past two decades there has been a major re-evaluation of the role of the car in urban areas. Consequently, the concept that town and city centers must be remodeled to accommodate the car has been widely replaced by the introduction of large pedestrianized, and generally car-free, areas. This cascading trend also has the effect of framing urban planning and transport issues in a fresh way, so that access to these urban areas is perceived by society much more in multi-modal terms. These largely local and incremental changes may not attract sensational headlines, but over time have the potential to shift significantly perceptions in society about the status of ubiquitous car use.

In examining the car regime, therefore, perhaps more can be learned from assessing its conditions for continued stability, rather than those of change itself, for it is in the dynamics of stability that change may find opportunities to break through. Consequently, in order to maintain its strength and stability, a regime must not only seek to construct a unity of purpose among its components parts but must also adjust to, and absorb, changes and challenges emerging from not only niches but also the wider social and economic landscape. It could be said that, for a regime, the dynamics of stability can have many gradations, from at one extreme being the equivalent of an individual strolling down a wide and firm road to, at the other extreme, being that of a tightrope walker progressing along a wire, while being buffeted by varying air currents from different directions.

For the car regime in the United Kingdom, it was the 'wide road' experience of stability that pervaded the 1960s to the 1990s. Since that time, the advent of sustainable mobility has meant that the road has become somewhat narrower and the ground a little more unstable, but still a basic stability has been retained. In the late 1990s, the government set ambitious targets in an attempt to introduce sustainable mobility policies, but the achievements have been limited. The fuel protests of 2000 acted as a deterrent in carrying these policies forward, with implementation of controversial demand management policies delegated to the local level, but here the necessary economic and political resources were generally lacking. In this context, political leadership was required at both national and local levels but with one or two exceptions has been conspicuous by its absence.

At the same time, the 'tightrope' type of stability for the car regime would arrive if societal landscape changes meant that the link to 'popular consumerism' was to be broken so that a significant number of individuals no longer identified personal mobility (associated with status and identity) with car ownership. At the moment, there are some small, but significant, signs of a weakening of that link. The great challenge for the car regime is to ensure that it progresses no further. The challenge for niches is to see if they can transform or substitute the car regime.

REFERENCES

Barker, T.C. & Savage, C.I. (1974) *An Economic History of Transport in Britain*, London: Hutchinson.
British Railways Board (1963) *The Reshaping of British Railways*, London: HMSO.
Brundtland, G.H. (1987) *Our Common Future. The Report of the World Commission on Environment and Development*, Oxford, England: Oxford University Press.
Buchanan, C.D. (1963) *Traffic in Towns: A Study of the Long-Term Problems of Traffic in Urban Areas*, Reports of the Steering and Working Groups appointed by the Minister of Transport, London: HMSO.
Charlesworth, G. (1984) *A History of British Motorways*, London: Thomas Telford.
Cm 693 (1989) *Roads for Prosperity*, London: HMSO.
Cm 2674 (1994) *Transport and the Environment*, Royal Commission on Environmental Pollution, Eighteenth Report, London: HMSO.
Cm 3950 (1998) *A New Deal for Transport: Better for Everyone. The Government's White Paper on the Future of Transport*, London: HMSO.
Cmnd 6836 (1977) *Transport Policy*, London: HMSO.
Department of the Environment, Transport and the Regions (2000) *Transport 2010. The 10 Year Plan*, London: Department of the Environment, Transport and the Regions.
Department for Transport (2004) *Feasibility Study of Road Pricing in the UK*, London: Department for Transport.
Department for Transport (2008) *Transport Statistics Great Britain 2008*, London: Department for Transport.
Department for Transport (2009a) *Low Carbon Transport: A Greener Future—A Carbon Reduction Strategy for Transport*, London: Department for Transport.
Department for Transport (2009b) *Road Traffic Statistics for Local Authorities: 1993–2008*, London: Department for Transport. Available at http://www.dft.gov.uk/pgr/statistics/datatablespublications/roadstraffic/traffic/rtstatisticsla/ (accessed on 26 June 2010).
Department for Transport (2009c) *Transport Statistics Bulletin—National Travel Survey: 2008*, London: Department for Transport.
Department for Transport (2011) *National Travel Survey*, London: Department for Transport. Available at http://www.dft.gov.uk/pgr/statistics/datatablespublications/nts/ (accessed on 19 March 2011).
Dudley, G. (1983) The road lobby: A declining force? in D. Marsh (ed) *Pressure Politics*, London: Junction.

Dudley, G. (1995, February/March) 'What will Europe dare to do about the car?' *European Brief*: 50–56.
Dudley, G. (2004) 'Constructing policy 'success' and a shift in political 'image': Urban road pricing and implementation of the London congestion charge,' *Local Governance*, 30(3): 109–124.
Dudley, G. & Richardson, J. (2000) *Why Does Policy Change? Lessons From British Transport Policy 1945–99*, London: Routledge.
Dyos, H.J. & Aldcroft, D.H. (1974) *British Transport. An Economic Survey from the Seventeenth Century to the Twentieth*, London: Pelican.
Eddington, R. (2006) *The Eddington Transport Study: The Case for Action: Sir Rod Eddington's Advice to the Government*, London: HM Treasury.
Financial Times, 'Blair Admits Crisis as Army Tankers Move In,' 14 September 2000.
Financial Times, 'Blair Tries to Stay on Track,' 29 September 2000.
Finer, S.E. (1958) 'Transport interests and the road lobby', *Political Quarterly*, 1: 47–58.
Geels, F.W. (2007) 'Transformations of large technical systems: A multilevel analysis of the Dutch highway system,' *Science, Technology & Human Values*, 32: 123–149.
Geels, F.W. & Schot, J.W. (2007) 'Typology of sociotechnical transition pathways,' *Research Policy*, 36(3): 399–417.
Goodwin, P., Hallett, S., Kenny, F. & Stokes, G. (1991) *Transport: The New Realism*, London: Rees Jeffreys Road Fund.
The Guardian, 'Manchester Says no to Congestion Charge,' 12 December 2008.
Hamer, M. (1987) *Wheels Within Wheels. A Study of the Road Lobby*, London: Routledge and Kegan Paul.
Kingdon, J.W. (1995) *Agendas, Alternatives and Public Policies* (2nd ed.), New York: HarperCollins.
Lyons, G. & Chatterjee, K. (eds.) (2002) *Transport Lessons From the Fuel Tax Protests of 2000*, Aldershot, England: Ashgate.
Office for National Statistics (2007) *Family Spending 2007 Edition*, Basingstoke, England: Palgrave Macmillan.
Parkhurst, G. & Dudley, G. (2008) Roads and traffic: From 'predict and provide' to 'making best use', in I. Docherty & J.Shaw (eds) *Traffic Jam. Ten Years of 'Sustainable' Transport in the UK*, Bristol, England: Policy Press.
Richardson, J. (2000) 'Government, interest groups and policy change,' *Political Studies*, 48(5): 1006–1025.
Rosenberg, N. (1986) The impact of technological innovation: A historical view, in R. Landau & N. Rosenberg (eds) *The Positive Sum Strategy: Harnessing Technology for Economic Growth*, Washington DC: National Academy Press.
Sloman, L. Cavill, N., Cope, A., Muller, L. & Kennedy, A. (2009) *Analysis and Synthesis of Evidence on the Effects of Investment in Six Cycling Demonstration Towns*, Report for the Department for Transport and Cycling, England.
Society of Motor Manufacturers and Traders (2007) *Motor Industry Facts 2007*, London: Society of Motor Manufacturers and Traders.
The Standing Advisory Committee on Trunk Road Assessment (1994) *Trunk Roads and the Generation of Traffic*, London: HMSO.
Transport for London (2003) *Congestion Charging Six Months On*, London: Transport for London.
Transport for London (2004) *London Congestion Charge—Impacts Monitoring Second Annual Report, Section 5: Social and Behavioral Impacts*, London: Transport for London.
Tyme, J. (1978) *Motorways Versus Democracy*, London: Macmillan.

5 The Governance of Transport Policy

Iain Docherty and Jon Shaw

5.1. INTRODUCTION

Governments all over the world are involved in directly providing, regulating and funding the provision of transport infrastructure and services, but the precise role of the state and its policy structures and instruments vary over time according to the wider economic and social context in which government finds itself. Several longstanding and deepening processes of governmental restructuring continue to influence the development of transport policies across all modes and sectors, subtly altering how the state chooses to intervene in the market for transport. Policies, and the governing frameworks responsible for them, are not just developed in response to *transport* problems—congestion, pollution, lack of accessibility and so on—but also according to the broader and much more complex political landscape in which the governing policy regime operates.

This chapter explores the impacts of changing forms of governance on the formulation and implementation of transport policies, focusing on two jurisdictions within the United Kingdom, namely Scotland and London. We begin by exploring why the state is involved in the transport sector at all and then go on to analyze how the roles of different institutions of governance in developing and managing modern transport systems have evolved. To do this, we explore two domains of governmental restructuring that have been of such profound importance to the United Kingdom that they can be considered as transitions in themselves:

- The *marketization transition*, that is the functional restructuring of the powers and competencies of the institutions and networks of governance through processes including privatization and deregulation and the reactions to these neoliberal perspectives generated by the response to the global financial crisis
- The *devolution transition,* which is the transfer of power from the UK central government to London and the Celtic 'nations' of Scotland, Wales and Northern Ireland, which has been driven by a coalition of constitutional and administrative reform agendas

In discussing these processes, we draw upon examples of changing passenger transport policy formulation and delivery. First is the marketization of the transport sector that began under the Conservative governments of the 1980s and culminated in the difficult and controversial privatization of British Rail in the mid 1990s. Since then, something of a gradual retreat from this free-market posture has become apparent, with more regulation and the active management of the market and its competitive processes returning as a policy objective. Second is the unfolding of the devolution 'project', in which responsibility for transport along with other areas of domestic public policy in London, Scotland, Wales and Northern Ireland has been (largely) transferred away from the central UK state to new elected assemblies and administrations, which have often chosen to make the shift from inherited policies toward new priorities.

The implications for transport policy of these two transitions are many and various: Key issues to date have included changing the balance between economic, environmental and social priorities for transport policy, how it might be possible to ensure equity in provision between different territories and socio-demographic groups and the role of innovative transport policies in defining and legitimizing the new institutions of governance. But these two transition processes have also substantially altered the context against which the move to greener mobility will take place. Our analysis demonstrates that although marketization and devolution have generated plenty of innovation in transport policy, this has not always made the strategic transition to a more sustainable mobility system in the future more likely.

5.2. TRANSPORT, GOVERNANCE AND THE STATE

The management of mobility has always been at the heart of state intervention in transport, and any understanding of the contemporary governance of transport policy must emerge from this perspective. At the fundamental level, the state has long been involved in the regulation and delivery of transport services because the conditions rarely exist for a genuinely free market in transportation to evolve. The peculiarities of transport—such as natural monopolies of infrastructure provision, the complexity of the distribution of transport benefits and a socio-political belief in the 'right to mobility'—together create an environment of less than perfect competition (Voight, 1960).

In other words, the landscape against which government transport policy is developed and played out has for a century or more been one of addressing market failure, since only state intervention has seemed capable of achieving the levels of security, social welfare and (relative) equity of access to mobility demanded by liberal societies. As early as 1909, Joseph Schumpeter emphasized the potential of state intervention as a means to maximize this kind of 'social value', that is the various 'welfare' benefits that were unlikely to be expressed through private profit in the market (Schumpeter, 1909).

In transport terms, this means that public provision of key infrastructure and services is potentially more efficient and equitable than the market alternative. Indeed, public provision is often essential since the risks involved mean the market is frequently reluctant to provide such infrastructure at all. In circumstances where the market is active, state intervention can avoid wasteful competition and duplication of assets (such as separately owned, parallel road or railway routes) and maximize the value obtained from the limited capital resources available to the economy by managing the planning and implementation of transport investment carefully. Finally, since transport infrastructure investment is particularly 'lumpy' in terms of the significant capital investments required to construct large schemes, the costs are usually amortized over long time periods that are relatively unattractive to market financiers.

The power of Schumpeter's argument was reflected in the fact that it became orthodoxy for policy regimes engaged in transport for over 50 years. The near elimination of private competition for transport infrastructure delivery in Europe had obvious, immediate appeal to the Left, which was quick to point to the wave of bankruptcies across European private railway companies in the 1920s as support for a state-led investment policy (see Chapter 7). But this approach was also attractive to many on the Right, albeit for entirely different reasons. Most important was that the potential step-change in the level of mobility of firms and private individuals made available by the private car and van would only be achieved if the necessary road infrastructure and support for the vehicle industry was made available (Foster, 1981; Glaeser, 2004; Yago, 1984).

In addition to economic considerations, (welfarist) transport policy has always sought to achieve a range of societal goals, ranging from maximizing transport safety to ensuring access to mobility for disadvantaged groups. The first challenge to this policy landscape—which triggered the profound restructuring of governing systems, structures and objectives that is still unfolding today—was the so-called 'crisis of mobility' that came about almost immediately after the oil price shocks of the 1970s. It seemed that overnight, policy regimes had to shift their attention from constructing new transport infrastructure in order to cater for rapidly increased travel demand to the new imperative of reducing public spending. Against this environment, a number of radical challenges to policy orthodoxy emerged, including the rise of the Green movement in continental Europe and, at least at a rhetorical level, the promotion of energy conservation and fuel efficiency and their importance as part of a wider transition to a (more) sustainable future (Heldmann, 2002; Willeke & Verbeek, 1977).

But the most important outcome of this period was that it created the conditions for the emergence of a new approach to governance, which represented a radical break with the previous orthodoxy and which had immediate and long-lasting effects. The combination of stagnating economies, 'big' government and economic uncertainty provided the conditions in which a new generation of political leaders, most importantly Margaret Thatcher and Ronald Reagan, could put into practice the neoliberal policies advocated by their favored economists and philosophers such as Milton Friedman, Friedrich von Hayek, William Niskanen and Charles Tiebout.

5.3. THE MARKETIZATION TRANSITION

As a political and economic philosophy, or the "kind of operating framework or 'ideological software' (needed) for competitive globalization" (Peck & Tickell, 2002, p. 380), neoliberalism countered the prevailing assumptions underpinning the welfarist economies of much of the developed world by stressing the need to reduce state intervention in the economy and promote market forces in order to encourage competition, enterprise and individual self-reliance (Harvey, 2005). Transport was profoundly affected by this shift in political philosophy since, from the neoliberal perspective, transport is the "same as any other good, subject to market forces and the rigours of competition" (Sutton, 1988, p. 132). The private car regime as we know it today was largely created in this period: The car was venerated as a symbol of freedom and progress—occupants can within reason go anywhere at any time—and governments deemed steady increases in personal mobility facilitated by the car necessary to achieve the kinds of structural adjustments in the labor and housing markets essential for increased economic growth (Meyer & Gomez-Ibanez, 1981; Pucher & Lefevre, 1996).

As a result, the position of car manufacturers as central to national industrial policy was entrenched (see Chapter 6), and large road building programs reappeared on the policy agenda in many countries as soon as the financial climate allowed, designed to match increasing traffic volumes in an approach that became known as 'predict and provide' (Charlesworth & Cochrane, 1997; Goodwin, 1999; see also Chapter 7). But as the state found it difficult to keep up the desired pace of infrastructure development, the procurement and financing of infrastructure also became more dependent on the private sector. Keen to continue delivering increased capacity but also equally alive to their wider political commitments to reduce taxes, public spending and debt liabilities, governments innovative mechanisms such as the Private Finance Initiative through which the procurement of infrastructure could be achieved without adding to public debt, were used extensively (Acerete, Shaoul, & Stafford, 2009; Debande, 2002; Shaoul, Stafford & Stapleton, 2006).

On top of the change of approach to the provision of transport infrastructure, radical governments also had the large public corporations that had evolved in many states to regulate and operate transport services firmly in their sights. Heavily unionized and often with very large workforces, these organizations became a clear target for downsizing by neoliberal governments across the world, albeit at different rates and in different styles (Butler & Pirie, 1981). In the United Kingdom, public transport—primarily bus and train services—was managed by large municipal or nationalized concerns and so were deemed ripe for reform. Intellectually, the neoliberal perspective perceived these modes as part of a single transport market that included the car. Therefore, the level of public transport service provision should be flexible and determined by evolving user choice rather than policy-led assumptions about efficiency or wider planning objectives as had until then been the case (Baumol, 1967; Foster, 1981; Yago, 1984).

Successive UK governments therefore championed a range of reforms including the removal of competition restrictions—the process of deregulation—in an attempt to open up markets previously serviced only by the public sector. The downsizing, administrative fragmentation and in many cases outright privatization of state owned transport undertakings was rolled out first across the bus sector and then latterly on the railways (Knowles & Abrantes, 2008; Preston, 2008). The logic was therefore clear: Sustained increases in personal mobility facilitated by the car were essential for economic growth, and, following on from this, since "private car use would increase", it was necessary to increase road capacity. "And public transport use would decline, therefore it would be logical to reduce service levels" (Goodwin, 1999, p. 659). As might be expected, each stage of this marketization transition from a public to private transport-based approach to economic development increased the dominance of the private car regime—which Margaret Thatcher famously referred to as 'The Great Car Economy'—considerably.

There is little doubt that many of these market reforms in Britain succeeded in realizing their intended cost savings, improvements in process efficiency and reductions in bureaucracy. Equally, though, many of them "did not bring about most of the positive effects forecast" in terms of improving the scope and quality of transport services (Pucher & Lefevre, 1996, p. 133). Adverse effects such as service contraction, reductions in public transport ridership and worsening traffic congestion became increasingly apparent (Roberts, Clearly, Hamilton & Hanna, 1992; see also Chapter 3). There were, of course, exceptions, the phenomenal growth in passenger aviation being a particularly notable example (Graham, 2008), although this sector benefitted greatly from deregulation and increased competition coinciding with the completion of the European Single Market and the 1990s economic boom that substantially increased demand for international leisure travel. As Peter Mackie (2008) has noted, the deregulation of the aviation market presented "the opportunity to move from The Great Car Economy to the Great Car and Plane Economy," a development which recent UK governments welcomed with open arms (Graham, 2008).

Critiques of the neoliberal era of transport governance and its unstinting support for the car regime inevitably focus on traffic congestion, since this was the most immediate and visible negative consequence of deregulation and a laissez faire approach to the rapid expansion of private car use. Despite their desire to do so, governments (or private sector holders of road building concessions) were simply unable to build themselves out of traffic congestion, although many to this day continue to try. Partly this was because the financial, practical and political limitations on road building were real no matter how 'free' the neoliberal enthusiasts hoped and believed the transport market could be. But it was also not lost on other politicians from the Right that accommodating continued increases in mobility through road building was paradoxical since it implied (very) high levels of government spending no matter how successful accounting tricks such as the Private Finance Initiative became. As the Organization for Economic Development and Cooperation (an organization not noted for advocating radical state intervention in the market) had warned with remarkable foresight, a strategy focused on road building to match increasing demand for mobility was likely to fail sooner or later: "Since further extension of the road infrastructure to meet growing demand for car use is not everywhere possible for urban planning and financial reasons, nor desirable from environmental, energy and often social policy standpoints, the only remaining transport policy option is to swing modal split in favour of public transport by investment and/or pricing policy measures" (Organization for Economic Development and Cooperation, 1979, p. 149).

The realization that there is more to transport policy than trying to build new roads to meet the demand for travel, especially in urban areas where both the problems and the physical challenges for road builders are most severe, has been an important driver in the re-appraisal of the role of regulation in the transport market. Many economists argue that some form of demand management needs to be put in place once the level of traffic congestion reaches a certain threshold, otherwise the costs of congestion will jeopardize the overall welfare of the economy, and the benefits of infrastructure investment will not be maximized (Glaister, 2004; Thomson, 1974). Ironically, this again positions the state in the role of arbitrator between private and public interests as envisaged by Schumpeter, since in order to ensure the overall welfare of the economy, it becomes necessary to restrict individual mobility, either through pricing or by other measures. The modern day problems of congestion thus provide a new context for a welfarist type of transport economics to re-emerge as the management of the supply of mobility becomes increasingly important given that building new roads is ever more difficult.

A further significant trend informing the re-engagement of the state in transport policy was also predicted by the Organization for Economic Development and Cooperation's (1979) report. Despite its critical importance to the performance of the economy, mobility management is now only one of (at least) three principal public policy concerns that governments seek to address through their transport strategies and programs. Although it took until the late 1980s to really impact on general consciousness, it was the realization of the potential scale of any impending environmental crisis, and transport's contribution to it, that marked the environmental 'watershed' in transport policy to which the state had to respond (Goodwin, 1999).

Economic theories have always attempted to incorporate the environment in their definitions of wealth, although whether its true value has ever been accurately reflected in the models and assumptions used is open to considerable doubt (Hanley & Spash, 1993; Ison & Wall, 2007). Economists have traditionally conceived of the environment as a store of 'natural resources', and their interest has largely been confined to the impacts of the scarcity of materials on their market values and thus short term economic performance, rather than on any more fundamental conceptual or philosophical perspective.

The emergence of, and development of the debates surrounding, the concept of sustainability have completely altered this position. As a critical introductory passage from the Brundtland Report noted, there is a need for concerted government action to tackle the impending environmental crisis, since there exists "a growing realization . . . that it is impossible to separate economic development issues from environment issues . . . and environmental degradation can undermine economic development" (as quoted in World Commission on Environment and Development, 1987, p. 3). Not only has an extensive literature developed in the field of environmental economics that attempts to formalize and quantify some of these relationships (see, e.g., Baumol & Oates, 1988; Ison, Peake & Wall, 2002; Pearce, Markandya & Barbier, 1989), but also the case for government intervention in the transport market has become much stronger, as the scale and international scope of transport-related pollution and emissions have become apparent (Commission for Integrated Transport, 2007; European Conference of Ministers of Transport, 1989; Greene & Wegener, 1997). Indeed, the environmental crisis has now established itself as the most significant challenge to the supremacy of the car regime, with the development of the narrative concerning the *global* impacts of the car's carbon emissions underpinning the central concern of this book, specifically whether we will attempt to meet future mobility demands by 'greening' the car or attempt a more comprehensive reorganization of the transport system across modes.

Added to the environmental case for increased state intervention was the challenge to the neoliberal stance from the social policy perspective (Schaeffer & Sclar, 1975). At the same time as attention was being focused on the environmental costs of a car-dominated transport regime, the potentially significant social costs of these policies also became increasingly apparent (see Chapter 3). With the continued contraction of public transport services resulting (at least in part) from marketization having become sufficiently advanced that vulnerable and disadvantaged groups such as the elderly, the young, the unemployed and the infirm were 'mobility deprived', calls for a

The Governance of Transport Policy

more interventionist approach to safeguard at least some degree of equity in transport provision grew louder (Buchan, 1992; Torrance, 1992). In other words, the traditional political balancing act of economic efficiency versus social equity was reasserting itself in the transport policy arena: Policy makers in all societies have values or objectives that inform the making of social judgments and hence guide the making of social decisions. For most societies, "these are likely to include attaining an efficient use of scarce resources and the promotion of an equitable or just distribution of those resources" (Le Grand, 1991, p. 1).

Many states have therefore, at least at the rhetorical level, begun to adapt their transport strategies and policy tactics in response to the "changing connections and inter–relations between social, political and cultural factors" that characterize the modern internationalizing and dynamic political economy (Painter, 1995, p. 276). Some, including the United Kingdom, have sought to follow a 'third way' combining elements of market discipline in service procurement and delivery with stronger state regulation for social and/or environmental reasons (Giddens, 2000). But the extent to which any of this suggests that governments are genuinely prepared to tackle the externalities of the car regime is highly uncertain. The recent global financial crisis is a case in point. Governments across the world rushed to bail out collapsing national car manufacturers citing their importance to the industrial base, high technology research, skills and employment and so on (Moulton & Wise, 2010; see also Chapter 4). Almost overnight, many carefully constructed narratives about the need to reduce dependence on the car were sacrificed at the altar of short-term economic stability and *increasing* car production and sales. The UK Government was no exception: Manufacturers had their customer finance divisions underpinned by government guarantees, and a scrappage scheme to incentivize new car purchase was introduced, although this was not, as in France for example, limited to new cars below a certain carbon emissions threshold.

5.4. THE DEVOLUTION TRANSITION

As the marketization transition became increasingly established, a second strategic trend has become apparent—that of changing institutions and processes of governance. In many countries, complex, multi-layered systems of governance have evolved as a response to challenges from 'above'— the internationalization of markets, the global environmental imperative and so on—and from 'below', that is the growing demands from regionalist or nationalist challenges to the authority of the nation state in an increasingly globalized world (Rodriguez-Pose & Gill, 2003). At the same time, hierarchical systems of administration are being replaced with structures that rely more on networks and alliances of stakeholders, designed to promote greater innovation in the development and implementation of policies across a range of sectors (Morgan, 1997; Stoker, 1999). The governance of transport policy has therefore changed profoundly as the spaces of governance themselves have evolved. In the United Kingdom, the most important transformation of governance has been that of administrative devolution.

Devolution in the United Kingdom is characterized by its highly complex and 'asymmetric' nature. This means that each of the four devolved territories—the countries of Scotland and Wales, Northern Ireland and London—have been afforded different powers. Scotland has a full Parliament capable of determining law over most areas of domestic policy—including the majority of transport issues—whereas the Assemblies in Wales and London do not have primary legislative powers. Northern Ireland, with its difficult recent history and highly contested constitutional status, has potentially the greatest devolved autonomy, but this has been subject to substantial periods of suspension in recent years due to various political crises.

5.4.1. Devolution and Transport Policy

British devolution has been supported by different political groups for different reasons but undoubtedly offers the potential both for real policy innovation and transformation. 'Localist' advocates of devolution—usually those from political perspectives that remain committed to the future of the United Kingdom as a nation state—emphasize the potential of devolution to create 'local solutions to local problems', and thus introduce a (controlled) level of policy divergence between different parts of the country in order to improve the outcomes of government intervention (Greer, 2003; Hazell, 2001; Jeffrey, 2002). 'Nationalist' voices, on the other hand, point to the potential of more radical policy transformations that become possible with high degrees of autonomy or outright independence from the UK state (see Murkens, Jones & Keating, 2002).

At first glance, transport appears to be rather removed from many of the larger theoretical and constitutional debates that drive transitions in governance such as devolution. Yet mobility is a critical facilitator of economic and social development, and although the rather prosaic tasks of policy delivery suggest that transport is an unlikely starting point for political arguments in favor of greater regional autonomy, the reality of transport policy-making in the devolved United Kingdom has neatly highlighted how powerful the devolved institutions can be in practice and how they might increase this power in future.

Reflecting its 'domestic' function, the main transport responsibility of each devolved territory is the development and management of roads infrastructure, and so the devolved institutions have since day one had an important part to play in the future development of the car regime in their territories. In the Celtic countries, the longstanding view that road infrastructure has suffered from decades of under-investment quickly translated itself into substantial programs of inter-urban road building and upgrading,

The Governance of Transport Policy 113

especially in Scotland and Northern Ireland. While this may seem at odds with the environmentalist rhetoric that has been increasingly adopted by all governing parties (see next chapter), the (perceived) quality of road infrastructure is a powerful political issue, especially in more remote rural regions with a high degree of dependence on the car (Gray, Farrington, Shaw, Martin & Roberts, 2001).

Governance of the railways is much more complex since the network hosts a mix of local, regional and inter-regional services and is subject to a strong regulatory regime controlled across Great Britain. In Scotland, further rail powers including responsibility for the terms of the ScotRail franchise and the funding of rail infrastructure were transferred, respectively, in 2001 and 2005, the latter event being widely regarded as "the most significant transfer of powers since devolution" (Scottish Executive, 2005).

Devolution in London reinforced the longstanding tradition of autonomy in transport matters, which dates from the creation of the original London Passenger Transport Board in 1933. The Mayor of London and Greater London Assembly (GLA) now have control over the Underground, London Buses, taxis, river boat services, light rail and key roads, all of which are managed by the dedicated agency Transport for London (TfL). Added to these are strong planning powers, which enabled the introduction of the globally significant congestion charge despite very considerable hostility to the policy from the UK central government. Unlike the Scottish Parliament and the other devolved administrations, the GLA can also issue bonds to the market to finance capital investment in transport. All of this, coupled with the 5-year funding agreement with central government that TfL was able to secure, means that funding for transport investment in London is both at a greater scale and much more secure than that elsewhere in the United Kingdom (Passenger Transport Executive Group, 2010).

5.4.2. Contrasting Leadership, Contrasting Outcomes

Comparison of the development of key transport policies in Scotland and London immediately highlights important differences both in the kinds of interventions both administrations were able to enact, and in the extent to which each territory positioned its system of governance to address strategic questions on the future of the car-based mobility regime. The story of transport in London since devolution is dominated by one issue—the Congestion Charge—and the importance of the commitment to the policy of the first elected Mayor, Ken Livingstone. Livingstone's political and leadership skills were widely acknowledged by supporters and opponents alike (Borraz & John, 2004). A tactically and strategically adept politician having been both the leader of the Greater London Council—an earlier incarnation of cross-London government—and a Member of Parliament for 13 years, Livingstone's path to the Mayoralty included not just winning the election itself but first having to position himself to do so by resigning from the Labour Party, since it was clear that the then Prime Minister Tony Blair would do everything in his power to block his selection as the Labour candidate.

In order to win as an independent, Livingstone deployed all of his political nous, differentiating himself from the official Labour candidate by stressing his commitment to, and deep knowledge of, London society and the problems facing it, especially transport. He also realized, unlike most other commentators who were keen to flag up the potential pitfalls of the Congestion Charge, that relatively few Londoners drive frequently in the central zone and that, as a result, the political reaction from *his* electorate could be made positive by using the Charge revenues to transform the bus network across London to the benefit of many, in turn bolstering support for his transport policies and the office of Mayor more generally.

London's was unquestionably the most radical and high profile transport policy agenda in the United Kingdom in the 2000s. Less than 2 years into his first term, the Mayor's *Transport Strategy* (GLA, 2001) strengthened the transformational rhetoric by declaring "the biggest single problem for London is the gridlock of our transport system" (GLA, 2001, p. i). Since then, the story of transport in London has been one of unprecedented investment: in the Underground addressing decades of underinvestment in many lines; other local rail services—several of which have been transferred from the National Rail network to TfL to become the 'Overground'; in the frequency and coverage of the bus network; and in the Oyster smart ticketing scheme.

Taken together with the Congestion Charge, these initiatives place London in a unique position in the United Kingdom, one in which the role of the car is highly managed as one component of a genuinely multi-modal strategy. There are limits to this, however—the extension of the Congestion Charge zone to the west of the central area met with significant opposition, and Boris Johnston, the Conservative who defeated Livingstone at the 2008 Mayoral elections, abolished the extension in January 2011. Nonetheless, the legacy of the devolution transition in London is important—TfL has cemented its position as a world-class professional delivery-focused organization (Marsden & May, 2006), and it continues to oversee a large capital investment program aimed at providing sustainable mobility, especially on the heavy rail network. Alongside further expansion of the Overground concept by upgrading and enhancing services on existing lines, construction work on the £16 billion Crossrail regional metro scheme has begun. Crossrail is a major new east–west tunnel under central London that will combine existing radial services into a new high frequency and high capacity line greatly enhancing accessibility to key nodes in the central area and relieving overcrowding on the Underground.

In Scotland (as in Wales and Northern Ireland) transport did not enjoy anything like the prominence afforded to it in London. To some extent this is because transport policy is a lesser concern for governance organizations in the Celtic nations with their wide-ranging responsibilities as compared to London where issues such as congestion and its constraints on economic performance are much larger and more immediate. But it is also because, outside London, the early period of devolution was notable for the lack of a high-profile political champion for transport. In Scotland, this was made worse by the succession of different transport ministers appointed from 1999 to 2004. This meant that little leadership was evident to help make the case for a transition to more sustainable mobility, and that as a result, transport policy delivery in Scotland in the first two terms of devolution (1999–2007) became particularly susceptible to charges of favoring projects for reasons of local and short-term political expediency (Docherty & Hall, 1999; Docherty, Shaw & Gray, 2007).

The first example of this came less than 5 months into the life of the Scottish Parliament, when the results of the new devolved administration's *Strategic Roads Review* were presented to Parliament (Scottish Executive, 1999). The roads program that resulted was a curious mix, reflecting neither Labour's own policy position that new roads should be built "as a last resort rather than a first" (Department of the Environment, Transport and the Regions, 1997, p. 1), nor the economic appraisals continued in the *Review* itself, which demonstrated that some of the rejected schemes had a very strong benefit: cost ratios. Instead, the package of schemes chosen for implementation was an assemblage of smaller projects unlikely to arouse strong objection from environmentalists and a few larger schemes that would prove popular in local areas politically favorable to the new administration.

The fact that the choice of those transport projects to be advanced under the new administrative arrangements was so politicized from the outset affected transport policy in Scotland throughout the first two terms of devolution. A policy based on a process of Ministerial 'pick and choose' not only reflected the lack of experience of many of the politicians in the first Scottish Parliament—hardly any had held significant political office outside local government—but also raised the importance of lobbying within the decision-making architecture. Transport policy and strategy were deeply discredited by the ensuing rush by many public- and private-sector stakeholders to have their preferred schemes put back in the program, spurred on by the evidence of the government's own economic appraisals that many of them were better value for money than the projects actually going ahead.

The most important example of this process was the M74 completion project in Glasgow, an urban motorway interconnector scheme that had been first proposed in the 1960s. The then Scottish Executive did not actually cancel the road, but instead said that the local authorities benefitting from it should shoulder the circa £500 million cost, which was clearly impossible for them. Given that the M74 project had a very strong benefit: cost ratio performance—anything from 5 to a remarkable 17 depending on the assumptions made—a concerted, evidence-based lobbying effort was put in place by local interests determined to have the motorway reinstated in the transport capital program. This was successful but not without problems; by the time the new Scottish National Party government signed the final contract for the road in February 2008, its total cost had increased by a factor of three, even though its capacity had been reduced by around a third from the scheme outlined in the *Strategic Roads Review* (Scottish Executive, 1999).

The second key early decision shaping the context for transport policy in the devolved Scotland came with the passing of the *Transport (Scotland) Act 2001*, which was designed in an attempt to effect the kind of transition toward sustainable mobility alternatives that is at the heart of the book's concerns. Sadly, the same combination of political inexperience, combined with vociferous lobbying from political and other interest groups, again exposed the lack of leadership in transport and precipitated another missed opportunity. At the same time as the London Mayor was pressing ahead with congestion charging, Scottish Ministers decided at the 11th hour not to include powers for charging on their road network in the legislation, which were seen as too politically dangerous and 'anti-motorist'. Instead, in a repeat of the buck-passing seen over the original M74 decision, local councils were given the power to implement their own schemes if they could demonstrate local support.

In an attempt to replicate London's success, the City of Edinburgh Council tried to introduce congestion charging in the Scottish capital. But the Edinburgh scheme failed ignominiously after a public referendum on the proposals—which was undertaken to meet the requirement of demonstrating public support contained in the 2001 Act—rejected them by a majority of 3:1 (Gaunt, Rye & Allen, 2007). This was unsurprising on several counts: First, the scale of the concerted lobbying against the proposal illustrated the strength and influence of certain interest groups in the city and its conservative political culture; second, public transport enhancements such as the proposed tram scheme were not in place *before* charging was to be introduced as was the case in London; third, there was no leadership figure actively outlining the benefits of the scheme to those who had a vote on its introduction, such leadership being especially important when radical policies are being implemented in the face of populist opposition (see Hambleton & Sweeting, 2004). Although central government leadership was lacking in both the London and Edinburgh cases, in London, the Mayor's political skill more than compensated. In Edinburgh, there was no alternative champion once government ministers had abandoned the idea.

None of this is to say that there has been an absence of public transport policy successes in Scotland since devolution. For example, both the first two Labour/Liberal Democrat coalition administrations and the minority Scottish National Party Government from 2007 gained plaudits from many observers for actively pursuing the extension and enhancement of the rail network. The motivations behind this pro-rail stance are complex:

Alongside transport objectives, including improving the accessibility of regional centers such as Inverness and Aberdeen by reducing journey times to the central belt and supporting the labor market depth of Edinburgh and Glasgow by maximizing the range of both cities' commuting zones, there are also structural governance considerations. Most important of these is that rail development can be funded by the infrastructure company, Network Rail, rather than the Scottish Government itself. Although the devolved Scottish Government has no borrowing powers of its own, it can enter into arrangements to pay back third-party borrowers such as Network Rail through an annual payments regime, and so substantial elements of rail development in Scotland have been funded in this manner.

Overall, then, transport investment in Scotland has been more pragmatic than genuinely visionary: Government has seized the opportunity of the growth in public expenditure in the 2000s to address the historic backlog of investment in infrastructure, especially roads, and used the available alternative funding mechanisms that exist to supplement this by developing the rail network. Rail expansion allows the devolved administration to make two political points: first that it is investing in the key alternative mode to the car and thus can claim that it is fulfilling its promises on making transport more sustainable (despite the very low actual contribution of rail to Scotland's overall mobility system), and second that this state of affairs is a triumph of devolution since outside of London there has been virtually no rail expansion in England, where devolution has not been introduced.

While infrastructure investment in the quality of both rail and road networks is welcome, the strategic outcome of the approach to transport investment in Scotland—"more of everything"—is that the overall characteristics of the mobility regime have not fundamentally altered. Unlike in London, there has been no real challenge to the car regime, and the more the policy narrative of expansion of mobility across all modes becomes entrenched, the harder it becomes to introduce more powerful restrictive measures on car use in the future. In part this is due to the country's geography and the political importance of the rural north, but it is also due to the politicization of the decision-making process and the broader failure of political leaders to grasp the opportunities afforded by devolution to put in place policy "sticks" such as road pricing as well as the "carrots" of improved infrastructure and services.

5.5. CONCLUSIONS

The marketization and devolution transitions have been argued to be highly significant changes to the system of governance in the United Kingdom. They have clearly had a number of profound political and policy effects, re-energizing debates on the kind of mobility regime that will predominate in the future. At the heart of this book are the questions of whether the car-dominated mobility regime will react to the sustainability imperative by focusing on the greening of vehicles, or whether it might evolve through a more fundamental transformation into a new regime built upon a more balanced and diverse definition of sustainable mobility in which alternatives to the car have a much more prominent role. Devolution and marketization are now both sufficiently established to send important signals about how and why each of these potential mobility transformations may or may not come about.

The fact that the devolved London institutions were able to adopt radical policies such as congestion charging does indeed suggest that—at least to *some* extent—it is possible for politicians with sufficient mandate and leadership skills, operating inside sufficiently strong and robust institutional environments, to initiate the transition to a more sustainable mobility system in which the car plays a smaller and a more highly managed role than has been the norm since the marketization transition fully took hold around 30 years ago. But it should not be automatically assumed that this can be easily replicable elsewhere: For many decades, London has had a highly developed public transport system, and the systems and structures of governance to manage and invest in it, which together mean that the city has for some time enjoyed a much more balanced pattern of mobility than anywhere else in the United Kingdom. So while London has been held up as something of a trailblazer for more sustainable transport choices (White, 2008) there are particular local issues at play, not least the UK central government's more generous approach to transport investment in London, especially compared to that which it has been prepared to fund in the larger provincial cities of England (Passenger Transport Executive Group, 2010).

The experience since devolution in Scotland—where the Scottish Parliament also has very considerable transport powers—demonstrates that the adoption of rhetoric in favor of more sustainable mobility is not always easy to implement given the competing demands of everyday politics (Mackinnon, Shaw & Docherty, 2008). For much of the first three devolved Parliaments, relatively generous resources for transport (as for other public investment) have been available. The ensuing "more of everything" approach to investment is perhaps an 'easy' political choice as it satisfies most geographical and modal constituencies, but it is also defensible given decades of underinvestment in infrastructure compared to benchmark countries in continental Europe and strong desires to engage in 'nation building' by distributing investment around the country.

The outcome of this situation has been a slow but steadily increasing disconnect between policy rhetoric on the one hand—most notably the Scottish Government's much trumpeted Climate Change Act with its "world leading" emissions reduction targets (Scottish Government, 2009)—and an overall policy architecture that, although containing very significant rail investment of the kind not seen for decades, nonetheless ignores the stronger available tools that might aid the wider transition to a less car-dominated mobility regime.

The different experiences in Scotland and London also underline the resilience of the marketization transition in transport. Although at one level London has been demonstrably more radical in its approach to challenging the car regime than has Scotland (and the other devolved territories), there have been clear limits to this. It has not been easy to repeat the early success of congestion charging, since the extension of the chargeable zone has become politically controversial and continued financial support for the bus network at recent levels will be extremely difficult in the new economic climate. Perhaps the most important legacy of the marketization transition is that the complexity of the deregulated transport market makes it even more difficult to implement transport policies aimed at delivering environmental (or social) objectives such as greener mobility. Devolution has provided the opportunity for different jurisdictions to try and direct their policies toward particular local objectives including social and environmental priorities. But the underlying strength of the marketization transition and its weakening of the state's influence over transport in general mean this is far from easy to achieve. Although London leads the way in demonstrating what a genuinely multi-modal mobility system might look like, the wider urban region remains as dominated as any other by the car regime. Even here it seems that the future transition path for mobility is unclear.

REFERENCES

Acerete, B., Shaoul, J. & Stafford, A. (2009) 'Taking its toll: The private financing of roads in Spain', *Public Money & Management*, 29(1): 19–26.
Baumol, W. (1967) 'Macroeconomics of unbalanced growth: The anatomy of urban crisis', *The American Economic Review*, 57(3): 415–426.
Baumol, W. & Oates, W. (1988) *The Theory of Environmental Policy*, Cambridge, England: Cambridge University Press.
Borraz, O. & John, P. (2004) 'The transformation of urban political leadership in Western Europe', *International Journal of Urban and Regional Research*, 28(1): 107–120.
Buchan, K. (1992) Enhancing the quality of life, in J. Roberts, J. Clearly, K. Hamilton & J. Hanna (eds) *Travel Sickness*, London: Lawrence and Wishart.
Butler, E. & Pirie, M. (eds) (1981). *Economy and Local Government*, London: Adam Smith Institute.
Charlesworth, J. & Cochrane, A. (1997) Anglicising the American dream: Tragedy, farce and the 'postmodern' city, in S. Westwood & J. Williams (eds) *Imagining Cities: Scripts, Signs and Memories*, London: Routledge.
Commission for Integrated Transport (2007) *Transport and Climate Change*, London: Commission for Integrated Transport.
Debande, O. (2002) 'Private financing of transport infrastructure: an assessment of the UK experience', *Journal of Transport Economics and Policy*, 36(3): 355–387.
Department of the Environment, Transport and the Regions (1997) 'New roads as a last resort—Strang', Press Release 216, London: Department of the Environment, Transport and the Regions.
Docherty, I. & Hall, D. (1999) 'Which travel choices for Scotland? A response to the government's white paper on integrated transport in Scotland', *Scottish Geographical Journal*, 115(3): 193–209.

Docherty, I., Shaw, J. & Gray, D. (2007) 'Transport strategy in Scotland since devolution', *Public Money and Management*, 27(2): 141–148.
European Conference of Ministers of Transport (1989) *Transport Policy and the Environment*, Paris: European Conference of Ministers of Transport/Organization for Economic Development and Cooperation.
Foster, M. (1981) *From Streetcar to Superhighway: American City Planners and Urban Transportation*, Philadelphia: Temple University Press.
Gaunt, M., Rye, T. & Allen, S. (2007) 'Public acceptability of road user charging: the case of Edinburgh and the 2005 referendum', *Transport Reviews*, 27(1): 85–102.
Giddens, A. (2000) *The Third Way and Its Critics*, London: Polity Press.
Glaeser, E. (2004) *Four Challenges for Scotland's Cities*, Allander Series, Glasgow, Scotland: University of Strathclyde.
Glaister, S. (2004) *London—On the Move or in a Jam?* London: Development Securities PLC.
Gray, D., Farrington, F., Shaw, J., Martin, S. & Roberts, D. (2001) 'Car dependence in rural Scotland: Transport policy, devolution and the impact of the Fuel Duty Escalator', *Journal of Rural Studies*, 17(1): 113–125.
Greater London Authority (2001) *The Mayor's Transport Strategy*, London: Greater London Authority.
Greer, S. (2003) 'Policy divergence: Will it change something in Greenock?' in R. Hazell (ed.) *The State of the Nations 2003: The Third Year of Devolution in the United Kingdom*, Exeter, England: Imprint Academic.
Goodwin, P. (1999) 'Transformation of transport policy in Great Britain', *Transportation Research Part A*, 33(7/8): 655–669.
Graham, B. (2008) UK air travel: Taking off for growth? In I. Docherty & J. Shaw (eds) *Traffic Jam: Ten Years of 'Sustainable' Transport in the UK*, Bristol, England: Policy Press.
Greene, D. & Wegener, M. (1997) 'Sustainable transport', *Journal of Transport Geography*, 5(3): 177–190.
Hambleton, R. & Sweeting, D. (2004) 'US-style leadership for English local government?' *Public Administration Review*, 64(4): 474–488.
Hanley, N. & Spash, C. (1993) *Cost-Benefit Analysis and the Environment*, Cheltenham, England: Edward Elgar.
Harvey, D. (2005) *A Brief History of Neoliberalism*. Oxford, England: University of Oxford Press.
Hazell, R. (2001) Conclusion: The state of the nations after two years of devolution, in A. Trench (ed) *The State of the Nations 2001: The Second Year of Devolution in the United Kingdom*, Exeter, England: Imprint Academic.
Heldmann, H. (2002) *50 Jahre Verkehrspolitik in Bonn: ein Mann und zehn Minister*, Bonn, Germany: Kirschbaum-Verlag.
Ison, S., Peake, S. & Wall, S. (2002) *Environmental Issues and Policies*, Harlow, England: FT Prentice Hall.
Ison, S. & Wall, S. (2007) *Economics*, Harlow, England: FT Prentice Hall.
Jeffery, C. (2002) Uniformity and diversity in policy provision: insights from the US, Germany and Canada, in J. Adams & P. Robinson (eds) *Devolution in Practice: Public Policy Differences Within the UK*, London: Institute for Public Policy Research and the Economic and Social Research Council.
Knowles, R. & Abrantes, P. (2008) Buses and light rail: Stalled en route? in I. Docherty & J. Shaw (eds) *Traffic Jam: Ten Years of 'Sustainable' Transport in the UK*, Bristol, England: The Policy Press.
Le Grand, J. (1991) *Equity and Choice*, London: HarperCollins.
Macki, P. (2008) Personal Communication, November 25.
Mackinnon, D., Shaw, J. & Docherty, I. (2008) *Diverging Mobilities? Devolution, Transport and Policy Innovation*, Oxford, England: Elsevier.

Marsden, G. & May, A. (2006) 'Do institutional arrangements make a difference to transport policy and implementation?' *Environment and Planning C: Government and Policy*, 24(5): 771–789.
Meyer, J. & Gomez-Ibanez, J. (1981) *Autos, Transit and Cities*, Cambridge, MA: Harvard University Press.
Morgan, K. (1997) 'The learning region: Institutions, innovation and regional renewal', *Regional Studies*, 31(5): 491–504.
Moulton, S. & Wise, C. (2010) 'Shifting boundaries between the public and private sectors: Implications from the economic crisis', *Public Administration Review*, 70(3): 349–360.
Murkens, J., Jones, P. & Keating, M. (2002) *Scottish Independence: Legal and Constitutional Issues: A Practical Guide*, Edinburgh, Scotland: Edinburgh University Press.
Organization for Economic Development and Cooperation (1979) *Report of the Seminar on Urban Transport and the Environment*. Paris: Organization for Economic Development and Cooperation.
Painter, J. (1995) Regulation theory, post-fordism and urban politics, in D. Judge, G. Stoker & H. Wolman (eds) *Theories of Urban Politics*, London: Sage.
Passenger Transport Executive Group (2010) *Transport Works: A Case for Investing in the City Regions*, Leeds, England: Passenger Transport Executive Group.
Pearce, D., Markandya, A. & Barbier, E. (1989) *Blueprint for a Green Economy*, London: Earthscan.
Peck, J. & Tickell, A. (2002) 'Neoliberalising space', *Antipode*, 34(3): 380–404.
Preston, J. (2008) Is labour delivering a sustainable railway? in I. Docherty & J. Shaw (eds) *Traffic Jam: Ten Years of 'Sustainable' Transport in the UK*, Bristol, England: The Policy Press.
Pucher, J. & Lefevre, C. (1996) *The Urban Transport Crisis in Europe and North America*, London: Macmillan.
Roberts, J., Clearly, J., Hamilton, K. & Hanna, J. (eds) (1992) *Travel Sickness*. London: Lawrence and Wishart.
Rodriguez-Pose, A. & Gill, N. (2003) 'The global trend towards devolution and its implications', *Environment and Planning C: Government and Policy*, 21(3): 333–351.
Ryley, T. (2005) 'Lessons from the Edinburgh congestion charging referendum', *Traffic Engineering+Control*, 46(4): 130–131.
Ryley, T. (2005) "The propensity for motorists to walk for short trips: Evidence from West Edinburgh", *Transportation Research Part A: Policy and Practice*, 42(4): 620–628.
Schaeffer, K. & Sclar, E. (1975) *Access for All: Transportation and Urban Growth*, London: Penguin.
Schumpeter, J. (1909) 'On the concept of social value', *Quarterly Journal of Economics*, 23(2): 213–232.
Scottish Executive (1999) *Travel Choices for Scotland: Strategic Roads Review*. Edinburgh, Scotland: Scottish Executive.
Scottish Executive (2005) 'Transfer of rail power to executive', Press Release, 17 October, Edinburgh, Scotland: Scottish Executive.
Scottish Government (2009) 'Climate change bill passed', Press Release, Edinburgh, Scotland: Scottish Government
Shaoul, J., Stafford, A. & Stapleton, P. (2006) 'Highway robbery? A financial analysis of Design, Build, Finance and Operate (DBFO) in UK roads', *Transport Reviews* 26(3): 257–274.
Stoker, G. (ed) (1999) *The New Politics of British Local Governance*, Basingstoke, England: Macmillan.

Sutton, J. (1988) *Transport Coordination and Social Policy*, Aldershot, England: Avebury.
Thomson, J. (1974) *Modern Transport Economics*, Harmondsworth, England: Penguin.
Torrance, H. (1992) Transport for all: Equal opportunities in transport policy, in J. Roberts, J. Clearly, K. Hamilton & J. Hanna (eds) *Travel Sickness*, London: Lawrence and Wishart.
Voight, F. (1960) *Die volkswirtschaftliche Bedeuteung des Verkehrssystems*, Berlin, Germany: Dunker & Humblot.
White, P. (2008) Transport for London: Success despite Westminster?' in I. Docherty & J. Shaw (eds) *Traffic Jam: Ten Years of 'Sustainable' Transport in the UK*, Bristol, England: Policy Press.
Willeke, R. & Verbeek, F. (1977) Verkehr und Umwelt (Traffic and environment), in B. Reihe (ed.) *Deutsche Verkehrswissenschaftliche Gessellschaft*, Schriftenreihe 35, Köln Germany.
World Commission on Environment and Development (1987) *Our Common Inheritance*, Oxford, England: Oxford University Press.
Yago, G. (1984) *The Decline of Transit: Urban Transportation in German and US Cities, 1900–70*, Cambridge, England: Cambridge University Press.

6 The Nature and Causes of Inertia in the Automotive Industry
Regime Stability and Non-Change

Peter Wells, Paul Nieuwenhuis
and Renato J. Orsato

6.1. INTRODUCTION

Those interested in understanding socio-technical system change often, unsurprisingly, focus on precisely that: change. There is a tendency to underplay, therefore, the degree to which change does not happen and, importantly, why socio-technical systems can exhibit inertial resistance to change. Indeed, the wider business and management literature also reflects this inherent bias toward the relentless search for newness and difference, and of course success, at the expense of sufficiently accounting for stability, sameness and continuity. The business history literature too has seen a debate about this issue in recent years, whereby Chandler (1977, 1990) is charged with emphasizing stability too much by, notably, Louçã and Mendonça (2002) who instead put their emphasis on continuous change.

Yet, regimes, once established, can be enduring. This chapter therefore examines the historical basis of an obdurate regime, the automotive industry and its attendant cultural manifestation, automobility, in order to arrive at an understanding of the scope for transition within and without the existing regime. This chapter has taken as its starting point the matter of transition (Kemp, Rip & Schot, 2001; Struben & Sterman, 2008) but given emphasis to why such transition may not occur. Transition is said to involve multiple parties making more or less synchronous steps such that there is a co-evolution of infrastructures, vehicles and supporting frameworks such as legislation and incentives (Melaina, 2003; Struben & Sterman, 2008). These matters have been extensively discussed elsewhere (Pilkington & Dyerson, 2006), but it is evident from the discussion here that the scale, complexity and embeddedness of the automotive industry and its constituent members is mirrored by the degree to which the prevailing culture of automobility is embedded in modern societies. Hence, the lack of transition even in the face of a compelling need to change may be anticipated arising out of the concrete and historically specific characteristics of the automotive industry.

The case is presented that the fundamental architecture of the existing regime was created during the period 1914 to 1925 in the United States, emergent as a phenomenon out of the technical, business and

124 *Peter Wells, Paul Nieuwenhuis and Renato J. Orsato*

social innovations that embedded mass production of the contemporary car within an over-arching landscape that became remarkably enduring. Figure 6.1 illustrates the landscape, regime and niche levels for the automotive industry. As the characteristic components of the landscape became more deeply grounded, and spatially extensive, the automotive industry regime became increasingly embedded and resistant to fundamental change. Thus with the development of road infrastructures and the spatial organization of cities predicated upon private car ownership and use, along with other reinforcements such as the cultural status of the car as the symbolic and real manifestation of personal freedom, so the landscape level became shaped by and in turn shaped the form of the automotive industry (at the regime level) evident from the 1920s onward. In turn, the regime level has multiple positive feedback loops that all act to preserve the regime, or at least contain the rate, character and direction of change.

Thereafter the chapter analyzes the pressures at a landscape level on the existing automotive industry regime, and the extent to which internal contradictions at the regime level have become manifest, leading to some weakening of the positive feedback loops. In view of these pressures, the following section provides an account of the extent and degree of change or non-change in the automotive industry. As Diamond (2005) has so poignantly argued, the danger is that societies may be unable to adjust sufficiently far and sufficiently fast in the face of impending disasters. Hence,

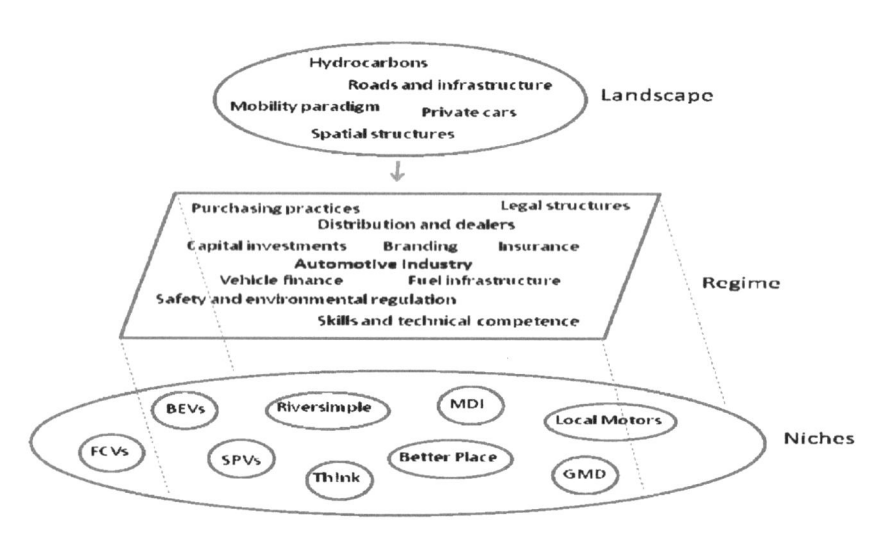

Figure 6.1 The automotive industry landscape, regime and niches (the niche-level developments are for illustrative purposes only). FCVs = fuel cell vehicles; SPVs = special purpose vehicles; BEVs = battery electric vehicles; MDI = Motor Development International; GMD = Gordon Murray Design.

The Nature and Causes of Inertia in the Automotive Industry

a key consideration in the conclusion is the pace of change that is achievable within the automotive industry and automobility regime. This theme is explored in the remaining sections of the chapter.

6.2. THE FOUNDATIONS OF AUTOMOTIVE INDUSTRY REGIME STABILITY

Mass production of the car was a US invention. Ford took the original concept of the car and adapted it to mass production: at least in terms of the manufacture and assembly of the mechanical, engine and chassis elements (Duncan, 2008). Edward Budd developed technologies enabling the mass manufacture and assembly of stamped and welded all-steel bodies with multiple secondary advantages (Courtenay, 1987; Nieuwenhuis & Wells, 1997, 2003, 2007; Palmer, 1913). Finally, a new business model was created by Alfred P. Sloan that embedded mass production and enabled mass motorization (Flink, 1988).

We then have three key elements that drive the current car mass production and consumption paradigm: mass manufacture and assembly of internal combustion engines and other vehicle components; mass manufacture, assembly and painting of vehicle bodies; and the business operational systems and concepts for bringing these mass produced elements to market, contributed respectively by Ford, Budd and GM. Thereby a trajectory was initiated that replaced labor with capital; created large, centralized assembly facilities requiring inbound and outbound logistics networks of great complexity; separated manufacturing from retail and distribution; and supplied customers with more or less standardized products optimized for mass manufacturing at low per-unit cost.

The purpose here is not to argue that the automotive industry is simply a monolithic, enduring entity in which individual vehicle manufacturers are undifferentiated. On the contrary, as the 'varieties of capitalism' school of thought has cogently argued, there are distinct cultural and operational differences that matter (Mikler, 2008, 2009). Moreover, it is evident that the fundamental business model for the industry has evolved since the early days of Fordism (Batchelor, 1994), with for example a substantial shift in the degree of vertical integration at the level of the vehicle manufacturers. Despite these observations, the framework described in the foundations of the existing automotive industry regime remain broadly in place, and it is pertinent therefore to ask why that may be so.

As long as access to automobility is defined in terms of the as-new purchase cost of a single vehicle, it is likely that the existing regime will prevail because it is powerfully adept at least cost-per-unit manufacturing, even if the lifetime cost of 'ownership' or 'usership' could be lower with a different business model and mobility paradigm. This is the fundamental challenge faced by those seeking a transition to a new paradigm. It is virtually impossible to compete with the existing regime on its own terms, and therefore the terms of competition need to be changed. However, because the existing automotive industry is so deeply embedded into legal, social, cultural and economic practices it is extremely difficult to find the space within which to initiate and subsequently develop an alternative way of working.

Furthermore, as the automobile penetrated deeper into society, so entire lifestyles and attendant spatial structures came to be predicated upon automobility, setting up a positive feedback loop of mutual dependency between consumers and producers within the prevailing socio-technical paradigm. Physical infrastructures and social norms such as those applied by the car insurance industry are exceedingly enduring and simultaneously expensive or difficult to replace with alternatives. Similarly, social practices that have accreted around the existing socio-technical paradigm are often resistant to change. Hence, in parallel with the automotive industry attempts at tactical or operational measures to absorb pressures for change, there are social changes that enable the current socio-technical regime to remain viable. We may include in this category much of what passes for regulation of the industry, intended as it is to ameliorate the worst of the external costs imposed on society by this paradigm.

In other words, a broader form of path dependency (David, 2007) has been created whereby the type of mobility (Urry, 2007) consumed in contemporary society is derived from and reinforces the distinct framework of automotive industry. One only has to consider the wider socio-technical embedding of the existing regime via rules, norms and behaviors and practices such as safety standards, cultural expectations of the car and physical infrastructures to begin to appreciate that mutual dependency is a powerful force limiting change in the automotive industry. To take a simple example, vehicle impact standards are defined in terms of what can be achieved by an all-steel vehicle body, and these vehicles have come to predominate on the roads of the world. Some very lightweight alternatives that could have superior environmental performance are unable to achieve a sufficient level of safety performance in comparison and are excluded from the market (Riley, 1994). Similarly, emissions regulation is predicated on the use of internal combustion engines, and as a result, current trends toward increasing use of electric power train to some extent meet a regulatory void, particularly in the context of carbon dioxide emissions. This is a particular category of barrier to entry, of which there are many in the automotive industry. Hence the process of creating path dependency is somewhat one of creating barriers to entry to those that might disrupt the paradigm. Such barriers can be technical, as in the case of vehicle impact or vehicle emissions standards. They can also be social. Hence in the same vein as impact standards, there is a degree of social acceptance that in a country like the United States upwards of 45,000 people per annum are killed on the roads.

The barriers to entry can also be cultural. Over the years consumers have been 'taught' that certain attributes in the vehicle are important (such as range, the rate of acceleration or the extent of top speed) in a manner that clearly speaks to the capabilities of product design and performance within the existing paradigm. Even in comparatively trivial matters such as the appearance of the paint finish on new vehicles the 'rules' of performance and qualitative expectations are informed by that which is achievable with steel body structures (Andrews, Nieuwenhuis & Ewing, 2006a, 2006b). Further barriers to entry include the sheer capital cost of establishing a high volume vehicle manufacturing operation and, as importantly, the cost of creating a brand and dealer network that consumers will have sufficient confidence in.

6.3. Pressures for Change on the Existing Regime

To understand the pressures for change on the existing regime it is necessary to appreciate that in both economic and environmental terms the prevailing regime is steadily exhausting the efficacy of traditional solutions, while simultaneously facing pressures for an increased pace of change. The contradictions that accumulate within the regime may indeed be resolved or deferred or may contribute to the ultimate demise of the regime. The automotive industry is heavily implicated in carbon emissions for example, but it is not clear whether this will be sufficient to compel radical shift. In 2004, US cars and light trucks contributed an estimated 314 million metric tons carbon-equivalent such that the United States with 5% of the world population, and some 30% of the world's automobiles, contributed 45% of the world's automotive CO_2 emissions (DeCicco, Fung & An, 2006). With the progressive spread of automobility, the share of the United States will fall, but this may just have the effect of diffusing the political blame. Greater understanding of anthropogenic climate change (Lynas, 2007; World Energy Council, 2007) provides potent arguments for developing stronger regulation of the automotive industry, while concerns over resource scarcity and the uneven geographical distribution of oil provides an international context for developing alternatives to hydrocarbon fuels (International Energy Agency, 2009). In those nations that are particularly dependent on imported oil such as the United States, the problem is viewed as acute and of national strategic importance (Aleklett, 2007), a perspective that may well extend to other countries and for some is the basis for further global conflict (Heinberg, 2005).

Despite these comments, it appears unlikely that, given current circumstances, there is sufficient pressure on the regime at a landscape level to enact major regime change. One issue is the sheer scale of the industry. According to Organization Internationale des Constructeurs d'Automobiles (OICA, 2009), in 2005 if automotive manufacturing was counted like a country, it would be the sixth largest in the world with an equivalent turnover of US$2 trillion, it supported an estimated 9 million direct jobs and 50 million supplier jobs, and it contributed US$400 billion to governments in tax. Quite simply, there is too much to lose for radical change to be contemplated.

What is likely is 'greening of cars' without substantial and enduring change to the regime itself. The evidence for this statement comes from recent history, in which governments around the world sought to offer rescue and market stimulation packages to vehicle manufacturers, in some instances virtually amounting to nationalization, rather than see them fail. Many of the global vehicle manufacturers suffered from a shortage of finance in the post-crisis environment from mid-2008 onwards. Governments had to step in to provide 'soft loans' and loan guarantees (Lampinen, 2009) because so many new car purchases are made via various finance schemes rather than outright purchase by customers. In the first half of 2009 the European Union European Investment Bank provided €7 billion (US$9 billion) for vehicle manufacturers (Proctor, 2009). In the case of GM Acceptance Corporation (GMAC) in North America, for example, new vehicle consumer financing contracts during the first quarter of 2009 fell significantly to US$3.4 billion from US$13.1 billion in the first quarter of 2008. At the end of 2008 new contracts had collapsed, and it was only with an investment from the US Treasury's Troubled Asset Relief Program in late 2008 that GMAC expanded its new North American retail car financing activities (AutomotiveWorld, 2009). These rescues were conspicuous for their lack of connection with any sustainability agenda, or even the amelioration of carbon emissions, apart from some very weak measures. Put simply, in the choice between stimulating (environmentally malign) economic recovery and the longer term option of designing a new industry to meet the pressures for sustainability, it is evident that short-termism triumphed.

As events during the economic crisis that unfolded across North America and Europe after 2007 have shown, the automotive industry was able to garner (and indeed required) state intervention and support. Hence, even 'failures' in the existing business model have not been exposed as fully or as ruthlessly as a 'pure' market-forces approach would entail. Interestingly, this means that the automotive industry also shows the rather neglected significance of barriers to exit. It is expensive and difficult to withdraw from the automotive industry given that the capital intensive assets and much of the industry-related skills are of limited transferability. Simultaneously, the post-2007 era illustrates that the intervention of the state has been crucial to sustaining the existing regime. What is intriguing about the events of the period since mid-2007 is that the historically specific moment is quite distinct in combining both an economic crisis with an enduring environmental crisis and that both are intimately connected. Previously the automotive industry had experienced growing regulatory pressures regarding environmental performance but had done so from the comfort of relative prosperity as markets experienced a prolonged boom from the latter 1990s onward. By 2009 however the industry had become seriously weakened just as the environmental imperatives for change intensified. This twin pressure on the prevailing paradigm is in this respect unprecedented, and hence the historical performance of the industry in terms of resisting change may not be such a good guide to future expectations (Wells, 2010a). Fundamental changes in external operating conditions can provide the basis for regime failure and, in extreme circumstances, regime collapse. It is a judgement as to whether those conditions will prevail in the automotive industry from 2010 onward'

Institutional isomorphism and risk aversion has tended to result in individual vehicle manufacturers seeking to emulate the 'best practices' uncovered elsewhere (Hoed, 2004), an approach most evident with respect to the ways in which the Toyota Production System was codified and promulgated by the proponents of Lean Production during the early 1990s (Ichijo & Kohlbacher, 2007; Womack, Jones & Roos, 1990). From the perspective of transitions, the issue then is whether and to what extent optimization practices are able to sustain the existing regime in the face of external threats and internal contradictions. The same comments apply to the practices by which crisis and potential failure are managed. In the automotive industry, as with many others, there are several adoptable (and co-determining) tactics to deal with impending crisis (Wells, 2010a) including the use of purchasing power to pass cost pressures downwards onto the supply base (Humphrey, 2003), the use of shared engineering platforms across different models and brands, and mergers and acquisitions that enable duplicated back office activities such as logistics to be rationalized. The result is the emergence of groups such as the VW Group under which multiple brands and productive facilities are more or less integrated, deriving ever-greater economies of scale and hence lower per-unit costs. On the other hand, in the period since 2000 there has been an equally visible collection of high-profile failures of such consolidation measures including the catastrophic Daimler-Chrysler merger and the de-consolidation of the Ford Group and the GM Group. The abysmal failure of the merger between Daimler and Chrysler is a classic case (Han & Kleiner, 2003; Weber & Camera, 2003). In 1998 Daimler paid US$36 billion for Chrysler; while in 2007 it sold the majority of the business (80.1%) to Cerberus Capital for just US$7.4 billion, having endured major losses in the previous 2 years. Dieter Zetsche, chief executive of DaimlerChrysler reportedly said in 2007 that "We obviously overestimated the potential synergies." The failed merger was part of a wider strategy adopted by Daimler under the then Chairman Jürgen Schrempp in the mid-1990s termed the 'Welt AG' plan under which the company would seek a dominating role in the key market regions of the world through a series of mergers and alliances including Chrysler, Mitsubishi and Hyundai. Paradoxically, the strong leadership of Schrempp could have been the cause of the merger failure (Stadler & Hinterhuber, 2005) as he took the company into a series of relationships that the rest of the management found difficult to make operational. These failures point to the limits of tactics to sustain the existing paradigm and crucially to the impending significance of diseconomies of scale as the automotive industry becomes truly global in scope. What the failures appear to have discovered is that sometimes the costs of integration can be higher than the expected benefits and that in the longer term such multi-brand constellations can become unwieldy while harboring the atrophy of brand distinctions. Unfortunately, what also tends to be discovered is that de-consolidation is also expensive, time-consuming and with uncertain outcomes: A classic case in 2009 was the inability of GM to sell Saab to any other interested party except Spyker, a company which made 43 cars in 2009 (Clark, 2010), and Ford's prolonged attempts to sell Volvo, which was ultimately sold to Chinese manufacturer Geely. Crucially therefore, the existing regime continues to sustain itself, despite repeated crises, by the ability to uncover new levels of efficiency and productivity and hence restore profitability to a level sufficient for retaining credibility in the markets by reducing costs. Simultaneously, geographic market expansion and production internationalization has enabled volume growth and hence revenue growth, albeit unevenly (Dunford, 2009) and leaving behind dysfunctional locations (Bailey, Koyabashi & MacNeill, 2008). Tables 6.1 and 6.2 illustrate how industrial concentration in the automotive industry changed between 1998 and 2008.

Whether considering all vehicles (Table 6.1) or just cars (Table 6.2), it is evident that the share of the leading 5 and leading 10 vehicle manufacturers fell over the 10 years to 2008. The reduction of concentration was most marked when all vehicles are concerned, but even in the case of cars the Top 5 lost two percentage points, and the Top 10 lost eight percentage points.

It should be recalled that corporate failure and the tectonic shifts in the geography of production and consumption also play a part in sustaining the prevailing orthodoxy. The removal of a company such as MG Rover

Table 6.1 Production Volume and Percentage Share of Total for All Motor Vehicles, 1998 and 2008

	1998		2008	
	million	%	million	%
Top 5	28,669	54.1	33,275	47.9
Top10	40,843	77.0	47,919	68.9
Total OEMs	52,987	100.0	69,561	100.0

Note: Derived from OICA (2009). OEMs = original equipment manufacturers.

Table 6.2 Production Volume and Percentage Share of Total for Cars, 1998 and 2008

	1998		2008	
	million	%	million	%
Top 5	19,321	50.9	27,117	48.6
Top10	29,678	78.3	39,335	70.4
Total OEMs	37,925	100.0	55,846	100.0

Note: Derived from OICA (2009). OEMs = original equipment manufacturers.

simultaneously removed a theoretical 500,000 units of competing capacity, thereby releasing market space for the remaining companies. The gradual process whereby existing plants are closed and new ones opened also allows for a degree of change to happen, notably in terms of removing more expensive production locations or older plants with inefficient layouts (as has been the story in North America for example as GM, Chrysler and Ford closed capacity from the latter 1980s onward). The changes that thereby ensue do not amount to a regime shift, but rather to an increase in efficiency (or reduced cost) within the prevailing regime and therefore can be seen as yet another mechanism whereby more fundamental change is diverted; the same is true of the introduction of 'Toyotist' practices in the wake of the lean production movement. Over a period of 20 to 30 years changes of this type can be quite substantial and yet remain 'manageable' for the companies concerned and for the wider socio-political process. Neither do these changes challenge the prevailing mode of automobility. Moreover, geographic integration of markets may help to erode protected 'islands' of differentiation that might have nurtured the seeds of transition. Iceland was until recently a potential example of a literal and figurative island of experimentation in fuel cell vehicles, but the impact of the financial crisis in that country has been profound enough to undermine the attempt to introduce the world's first hydrogen economy. Instead the country is now adopting battery electric vehicles on an experimental basis, working with Mitsubishi. On the other hand, as the automotive industry becomes more streamlined and 'efficient' so the numbers of direct and indirect jobs attributed to the industry declines (Andera, 2007). In turn, this means that the bargaining power of the industry in terms of its economic importance is somewhat reduced, and so the end result may be that policy makers and society at large will envisage the social costs of radical change to be less than was previously the case.

In order for the existing regime to remain viable it is necessary for profitability to be retained, either by reducing costs or increasing revenues. It is also necessary to quash emergent and potentially threatening niches or to subsume them. The argument here is that the automotive industry has enacted a series of tactical, operational measures to achieve these objectives.

A further interesting aspect of the ability of the incumbent regime to resist socio-technical shift is that of absorbing or subsuming potential threats. An early example of both the potential and limits of this strategy by the established vehicle manufacturers was the case of Ford when the company purchased the struggling niche electric vehicle manufacturer Th!nk. The story is a long and complex one (see Orsato, Wassenove & Wells, 2008) and is still ongoing, but essentially the traditional vehicle manufacturer had failed to create such an innovative electric vehicle, needed to comply with California's ZEV Mandate and compensated for this by buying the company. A similar approach was adopted by DaimlerBenz with the purchase of a significant shareholding in the emergent US electric vehicle manufacturer

Tesla (Motavalli, 2009) and more recently with BYD of China. This speaks to the extent to which it is possible to achieve transition within or without the existing regime, a theme that is elaborated upon further in the following section.

6.4. THE SCOPE FOR TRANSITION WITHIN AND WITHOUT THE EXISTING REGIME

When contemplating the scope for change in a socio-technical regime, given the pressures for change noted previously, there are two broad alternatives. The first is that the change process will come from within the regime and hence that the prevailing members of the regime along with the socio-cultural practices that help define the regime will, over a period of time, adopt new practices, technologies and rationalities. The alternative is that the regime is effectively 'stranded' by change beyond the control and influence of the members as new practices, technologies and rationalities come to the fore. Over time, therefore, if the regime is stranded in this manner it may be expected to wither away and become marginal or parochial. The impetus for this latter change may be expected to come from the niches below the regime, including of course the electric vehicles and fuel cell vehicles covered elsewhere in this book.

In an industry as complex and pervasive as the automotive industry it is unlikely that socio-technical transition will proceed uniformly in all places and at all times. In addition, the theoretical distinction between internal and external regime shift may be less precise in reality. Rather, we may expect to see a variety of more or less successful socio-technical experimentation with alternatives to the mainstream paradigm, and indeed degrees of change within the mainstream paradigm, proceeding at differential rates and in historically specific settings. One interesting theoretical issue is whether the point is reached whereby the external costs of the regime will overwhelm the ability of the regime to resist or absorb change and if so where the impetus for change will be manifest. That is, it might be the case that the financial collapse of the existing automotive industry will be the basis for widespread and rapid change as the socio-technical paradigm exhausts the possibility of creating and extracting surplus value. Alternatively, it might be that regulation from government is driven by an imperative greater than that which calls for the continued nurture of the existing automotive industry and results in interventions that force transition. It might be that the declining utility of the current pattern of automobility, combined with ever-escalating costs, comes to reverse spatial decentralization of social structures and compel the re-concentration of the social activities that provide the current demand for automobility.

The existing socio-technical paradigm manages the imperative for change, using mechanisms such as technology roadmaps to both define the

The Nature and Causes of Inertia in the Automotive Industry   133

technological trajectory of the industry along a future timeline and also to create consensus among participants in the regime as to the direction and pace of change (Petrick & Echols, 2003). A typical example, for the UK case, is that presented by Parker and McGinity (2006). Moriarty and Honnery (2008) argue that the search for technological solutions to the negative externalities of transport is insupportably optimistic in terms of the scale and pace of improvement that can be achieved. Hence mobility itself and the future of the passenger car as a concept are to be questioned. In this analysis, the central task at a social level is in fact to reduce the need for mobility in all its forms and hence to design low-mobility into future transport systems. Thereafter the remaining demand for mobility is to be met in very different ways. The prevailing socio-technical regime can therefore embody degrees of resistance to change but also embodies a capacity to restrain the qualitative impact of change into manageable forms that fall short of catastrophic change. Put simply, the prolonged technology roadmap, which essentially preserves the primacy of automobility, is to be preferred to the more radical low-mobility future, at least as far as the participants are concerned.

What is evident here is a potential for quite distinct outcomes despite similarities in procedure for both neo-conservative technology mapping policies and the attempts to create more sustainable societies via socio-technical transitions. Transition management is not concerned with policy instruments as such but involves different ways of interacting, changes to the mode of governance and consensus goal seeking. Innovation and learning are important aims for transition management (Rennings, Kemp, Bartolomeo, Hemmelskamp & Hitchens, 2003), enabling progress from one stage through to the next. Through transition management, the endeavor toward more sustainable systems is supposedly institutionalized. Transition management therefore aims to help increase the chance of achieving a transition toward more sustainable systems in energy, transport, agriculture and food and equally aims to help achieve greater socio-technical diversity (Vellema & Loorbach, 2006). The end-state is not fixed but open ended (Rennings et al., 2003). However, the rather vague content of transition management renders it a process liable to co-option by existing powerful interests as appears to be the case with the automotive industry.

In addition, the notion of partial transition is one that the existing incumbents appear more prepared to contemplate. It is illustrative, for example, that in Germany the government in mid-2009 promised €500 million for electric vehicle infrastructures and technology development with the intention of making the country the biggest 'electro-mobility' market in the world. Interestingly this would result in an electric vehicle fleet of just 5 million units by 2030, compared with about 50 million petrol and diesel cars on the road in 2009. There are a great many such national and regional initiatives in the automotive industry, all of which appear designed to maximize the potential capture of economic wealth generation from a resurgent industry constructed around green cars, rather than a radical reappraisal of mobility as a whole. In all these instances, the pace of change seems to be measured in decades rather than months and years, such that as far as change from within the regime is concerned, there is only the envisaged slow process of greening of cars with the gradual replacement of high carbon emission vehicles with lower carbon emission vehicles.

Nevertheless, it is clear that at one level, the industry does take the 'greening' agenda seriously. The evidence is there. BMW has in recent years shown what can be achieved in terms of CO_2 reduction by combining relatively simple and low-cost technologies in an existing product range under the EfficientDynamics label. The combination of these technologies allows the MINI Cooper D to return CO_2 emissions comparable to a Toyota Prius. The resulting improvements have been implemented throughout the BMW and MINI range. This technology push does not come for free. During the principal development and production planning phase, BMW's research and development costs rose by nearly 40% to €835 million, and in turn the expenditures dominated the second quarter results of 2007, making BMW's net profit fall 4.3% to €753 million (Ciferri & Franey, 2007). In effect, the promotion of low-carbon but non-radical technologies by BMW represents a strategic decision as to the position of the brand and of the perceived customer base, though of course the company is also pursuing electric vehicles for the future.

Similarly, at the 2007 Frankfurt motor show, Mercedes showed a concept car which gives an indication of what could be achieved with technologies currently under development (Kable, 2007). The F700 is a large luxury saloon, which is nevertheless powered by a small 1.8 litre engine. The engine uses a combination of diesel and Otto cycles to produce 258bhp, but with CO_2 emissions of only 127g/km for a car of 5.17 meters in length and a weight of around 1700kg (Mercedes-Benz UK press release, 11 September 2007). This performance is achieved by combining the internal combustion engine with a hybrid power train, while the engine itself has two-stage turbo-charging and optimized internal combustion technology. By going some way to meet the low-carbon agenda, it could be argued that the industry in Europe has somewhat been able to defuse the pressure for more radical or more rapid change. As a result of such incremental measures, the European Union new car fleet of 2009 emitted significantly lower levels of CO_2 than 10 years ago (see Table 6.3). Much of this reduction in overall fleet CO_2 output has been achieved by a greater reliance on diesel engines. The message sent by the established industry through these developments is quite clear: Environmental problems can be solved by the existing industry, using existing technologies, and while remaining firmly rooted within the prevailing regime—there is no need for radical change.

While these technological developments took, the vehicle manufacturers were also engaged in a vigorous lobbying effort in a partially successful attempt to reduce the impact and effectiveness of carbon reduction legislation.

The Nature and Causes of Inertia in the Automotive Industry   135

Table 6.3   Low Carbon Vehicles Available in 2009 (< 100 g/km CO_2 on NEDC)

Vehicle manufacturer	Model	Variants	CO_2 g/km	Powertrain
Ford	Fiesta	1.6 TDCi Econetic	98	Diesel
SEAT	Ibiza	1.4 TDI Ecomotive	98	Diesel
Smart	ForTwo	0.8 cdi	88	Diesel
	ForTwo	0.8 cdi Passion	88	Diesel
Tesla	Open Roadster	R'str/Signature Edition	0	Battery Electric
Toyota	IQ	1.0	99	Petrol
	Prius	1.8 VVT-i T3	89	Hybrid
		1.8 VVT-i T4/T Spirit	92	Hybrid
Volkswagen	Polo	1.4 TDI 80 Bluemotion	99	Diesel
	Golf	1.6 TDI 105 Bluemotion	99	Diesel
Volvo	C30	1.6 D DRIVe	99	Diesel

Note: Derived from Autocar (2009); NEDC = New European Driving Cycle.

The lobbying process ultimately led to Regulation 443/2009, which established a CO_2 emission target of 130 g/km for the fleet average of new cars sold by 2015. While such developments may indeed bring us cars that emit significantly less carbon, this seems to be the limit of ambition for change from within the regime level as constituted by the contemporary automotive industry. It is however important to recognize that the automotive industry is less monolithic than often assumed; although industry representative bodies such as Association des Constructeurs Européens d'Automobiles (ACEA)attempt to represent a common position, in reality interests and preferences vary even within the existing dominant regime. Those manufacturers with lower CO_2 emitting fleets, for example, initially lobbied in favor of tighter regulation. This included Fiat and PSA Peugeot-Citroën, who gained the support of their respective governments. In time, however, it was the luxury car manufacturers of Germany and the United Kingdom who with their governments ultimately carried more weight in the debate. The leading role of Toyota and Honda in seeking to commercialize vehicles with more radical alternative power train—notably hybrid and fuel cell—should also be noted. At the same time it is important to emphasize that even they are firmly entrenched within the dominant regime.

6.5. CONCLUSIONS

While it is the case that transitions theory envisages change as occurring over periods of perhaps 50 years, there is scant evidence thus far that—despite low profitability now being endemic in the industry—the fundamental structure and operating practices of the automotive industry are changing. It is also the case that much of the technological effort of the automotive industry is going into areas other than power train or body architecture but in fact into aspects of occupant comfort, safety, entertainment and communications. Ironically, this may indeed be the start of a transformation of the car, from a device essentially used to deliver personal mobility and freedom of travel, to a device to protect and connect people within an increasingly urbanized and congested spatial environment. The emphasis of much of this new wave of technology is on vehicle—infrastructure interaction, presented to users as enhanced efficiency through route planning, the identification of parking spaces, proximity to recharging points and so forth, all of which cocoon the individual further within the shell of the vehicle (Mitchell, Borroni-Bird & Burns, 2010). This illustrates that there is also an inexorable path dependency within the automobile toward electric power train, making the car over time less of a mechanical and more of an electric and electronic device (Nieuwenhuis, 2009).

New technology in product design and manufacturing process that can be traced to the imperative to improve environmental or sustainability performance may create the space for new entrants and/or new business models (Borroni-Bird, 2006), and these new entrants may presage a genuine transition in automobility from the outside. However, this need not necessarily be so. It is true that smaller, more niche emergent suppliers of electric vehicles have emerged including companies such as Tesla, Miles Electric Vehicles, ZAP (Zero Atmospheric Pollution), ZENN (Zero Emission, No Noise) and Fisker, all of which have developed or are developing their own versions of electric vehicles (Connell, 2009). On the other hand, these companies are either doomed to obscurity or in danger of being absorbed by the incumbents. Even more on the fringe are micro-niche socio-technical experiments with innovative companies and social partners offering alternatives to traditional conceptualizations of the automotive industry alongside innovative technology. Instances such as Riversimple, Local Motors and MDI Air Car can be found but remain entirely irrelevant within the vastness of the industry.

Equally, small-scale or localized developments in terms of infrastructure to support alternative mobility practices are under way as cities and regions attempt to capture both the economic benefit of the emergent automobility paradigm as well as the more obvious environmental and transport benefits. Hence there has been a multiplication of 'hydrogen highways' and electric vehicle re-charging networks across North America, Europe, Japan and increasingly in emergent markets such as China. Such instances of diverse development trajectories suggest that the monolithic character of the prevailing socio-technical paradigm is also under threat as localities seek specific and place-appropriate solutions to sustainable mobility (Wells, 2010b). Whether such locations are also the living embodiment of strategic niche management as defined by some of the theorists of socio-technical transition (Hoogma, Kemp, Schot & Truffer, 2002) is something that can only be determined empirically.

REFERENCES

Aleklett, K. (2007, December) 'Peak oil and the evolving strategies of oil importing and exporting Countries; Facing the hard truth about import decline for OECD countries', Discussion Paper No. 2007–17, OECD-International Transport Forum, Paris.

Andera, J. (2007) Driving Under the Influence; Strategic Trade Policy and market Integration in the European Car Industry, in *Lund Studies In Economic History*, 42, Stockholm: Almqvist & Wiksell.

Andrews, D., Nieuwenhuis, P. & Ewing, P. (2006a) 'Black and beyond—Colour and the mass-produced motor car', *Optics and Laser Technology*, 38(4–6): 377–389.

Andrews, D., Nieuwenhuis, P. & Ewing, P. (2006b) 'Living systems, total design and the evolution of the automobile: the significance and application of holistic design methods in automotive design, manufacture and operation', *Design and Nature*, Vol. 2: 381–446, ISSN: 1478-0585.

Autocar (2009) New Cars A-Z, 6th June, pp. 90–107.

AutomotiveWorld (2009) US: GMAC financial services reports US$676m Q1 loss. Available at http://www.automotiveworld.com/news/oems-and-markets/76341 (accessed 19 June 2009).

Bailey, D., Koyabashi, S. & MacNeill, S. (2008) Rover and out? Globalisation, the West Midlands auto cluster, and the end of MG Rover. *Policy Studies*, 29(3): 267–279.

Batchelor, R. (1994) *Henry Ford: Mass Production, Modernism and Design*, Manchester, England: Manchester University Press.

Borroni-Bird, C. (2006) The reinvention of the automobile, in P. Nieuwenhuis, P. Vergragt & P. Wells (eds) *The Business of Sustainable Automobility: From Vision to Reality*, Sheffield, England: Greenleaf.

Chandler, A.D. (1977) *The Visible Hand: The Managerial Revolution in American Business*, Cambridge MA: Harvard University Press.

Chandler, A. D. (1990) *Scale and Scope: The Dynamics of Industrial Capitalism*, Cambridge MA: Harvard University Press.

Ciferri, L. & Franey, J. (2007, 20 August) 'BMW will spend big to keep green lead', *Automotive News Europe*, 12(17): 6.

Clark, A. (2010, 27 January) 'Saab saved from closure in $400m buyout by Dutch sports car maker', *The Guardian*: 28.

Connell, D. (2009) New generation auto suppliers: Enova Systems, Inc. Available at http://www.automotiveworld.com/news/components/77892 (accessed 16th August 2009).

Courtenay, V. (1987) *Ideas That Move America . . . The Budd Company at 75*, Troy, MI: The Budd Co.

David, P. (2007) 'Path dependence: A foundational concept for historical social science', *Cliometrica*, Vol. 1(2): 91–114.

DeCicco, J., Fung, F. & An, F. (2006) *Global Warming on the Road: The Climate Impact of America's Automobiles*, New York: Environmental Defense.

Diamond, J. (2005) *Collapse*, New York: Viking Press.

Duncan, J. (2008) *Any Colour—So Long as it's Black; Designing the Model T Ford 1906–1908*, Auckland, New Zealand: Exisle.

Dunford, M. (2009) 'Globalization failures in a neo-Liberal world: The case of FIAT Auto in the 1990s', *Geoforum*, 40(2): 145–157.
Flink, J. (1988) *The Automobile Age*, Cambridge, MA: MIT Press.
Han, N. & Kleiner, B.H. (2003) ,The effective management of mergers', *Leadership and Organization Development Journal*, 24(8): 447–454.
Heinberg, R. (2005) *The Party's Over: Oil, War and the Fate of Industrialised Societies*, Cabriola Island, British Columbia, Canada: New Society Publishers.
Hoed, R. van den (2004) *Driving Fuel Cell Vehicles: How Established Industries React to Radical Technologies, Design for Sustainability Program*, Publication No. 10, Delft, the Netherlands: Delft University of Technology.
Hoogma, R., Kemp, J., Schot, J. & Truffer, B. (2002) *Experimenting for Sustainable Transport. The Approach of Strategic Niche Management*, London: Spon Press.
Humphrey, J. (2003) 'Globalization and supply chain networks: the auto industry in Brazil and India', *Global Networks*, 3(2): 121–141.
Ichijo, K. & Kohlbacher, F. (2007) 'The Toyota way of global knowledge creation: The 'learn local, act global' strategy', *International Journal of Automotive Technology and Management*, 7(2/3): 116–133.
International Energy Agency (2009) World Energy Outlook 2009 factsheet: Global energy trends, Paris: International Energy Agency. Available at www.iea.org (accessed 15 September 2009).
Kable, G. (2007, 12 September) 'F700 shines light on next S-class', *Autocar*: 12–13.
Kemp, R., Rip, A. & Schot, J. (2001) Constructing transition paths, in R. Garud & P. Karnoe (eds) *Path Dependencies and Creation*, London: Lawrence Erlbaum Associates.
Lampinen. M. (2009) Japan: Toyota Financial Services seeks state-backed bank loan. Available at http://www.automotiveworld.com/news/oems-and-markets/75138 (accessed 19 June 2009).
Louçã, F. & Mendonça, S. (2002) 'Steady change: The 200 largest US manufacturing firms throughout the 20th century', *Industrial and Corporate Change*, 11(4): 817–845.
Lynas, M. (2007) *Six Degrees: Our Future on a Hotter Planet*, London: Fourth Estate.
MBUK (2007), F700 a luxury car for the future, Mercedes-Benz UK press release, 11 September 2007.
Melaina, M.W. (2003) 'Initiating hydrogen infrastructures: Preliminary analysis of a sufficient number of initial hydrogen stations in the US', *International Journal of Hydrogen Energy*, 28(7): 743–755.
Mikler, J. (2008) 'Framing responsibility: national variations in corporations' motivations', *Policy and Society*, 26(4): 67–104.
Mikler, J. (2009) *Greening the Car Industry: Varieties of Capitalism and Climate Change*, Cheltenham, England: Edward Elgar.
Mitchell, W., Borroni-Bird, C. & Burns, L. (2010) *Reinventing the Automobile; Personal Urban Mobility for the 21st Century*, Cambridge, MA: MIT Press.
Moriarty, P. & Honnery, D. (2008) 'Low-mobility: The future of transport', *Futures*, Vol. 40(10): 865–872.
Motavalli, J. (2009) Wheels: Daimler takes a stake in Tesla Motors. Available at http://wheels.blogs.nytimes.com/2009/05/19/daimler-takes-a-stake-in-tesla-motors (accessed 21 May 2009).
Nieuwenhuis, P. (2009, August) 'Triple path dependency in automobiles; its effect on the introduction of fuel cell vehicles', in *Innovation, Markets and Sustainable Energy; The Challenge of Hydrogen and Fuel Cells*, Symposium conducted at the US Academy of Management Conference, Chicago.
Nieuwenhuis, P. & Wells, P. (1997) *The Death of Motoring? Car Making and Automobility in the 21st Century*, Chichester, England: Wiley.

Nieuwenhuis, P. & Wells, P. (2003) *The Automotive Industry and the Environment—A Technical, Business and Social Future*, Cambridge, England: CRC Press.
Nieuwenhuis, P. & Wells, P. (2007) 'The all-steel body as a cornerstone to the foundations of the mass production car industry', *Industrial and Corporate Change*, 16(2): 183–211.
Organization Internationale des Constructeurs d'Automobiles (2009) Economic contributions, Paris: Organization Internationale des Constructeurs d'Automobiles. Available at http://oica.net/category/economic-contributions/ (accessed 16 June 2009).
Orsato, R., Wassenove, L. & Wells, P. (2008) *Eco-entrepreneurship: The Bumpy Ride of TH!NK. INSEAD Teaching Case04/2008-5485*, France: INSEAD Business School.
Palmer, C. (1913, 22 January) 'Body features revealed at the New York shows', *The Horseless Age*: 189.
Parker, S. & McGinity, B. (2006) *Vision for the UK Automotive Industry in 2020 Focusing on Supply Chain and Skills and Technology*, Leamington Spa, England: Ricardo Ltd.
Petrick, I.J. & Echols, A.E. (2003) 'Technology roadmapping in review: A tool for making sustainable new product development decisions', *Technological Forecasting and Social Change*, 71(1–2): 81–100.
Pilkington, A. & Dyerson, R. (2006) 'Innovation in disruptive regulatory environments', *European Journal of Innovation Management*, 9(1): 97–91.
Proctor, T. (2009) Europe: European Investment Bank to lend €7bn to OEMs in first half 2009. Available at http://www.automotiveworld.com/news/oems-and-markets/74915 (accessed 19 June 2009).
Rennings, K., Kemp, R., Bartolomeo, M., Hemmelskamp, J. & Hitchens, D. (2003) Blueprints for an integration of science, technology and environmental policy (BLUEPRINT), Final report of 5th Framework Strata project. Available at http://www.insme.info/documenti/blueprint.pdf (accessed 13 March 2008).
Riley, R. (1994) *Alternative Cars in the 21st Century: A New Personal Transportation Paradigm*, Warrendale PA: Society of Automotive Engineers.
Stadler, C. & Hinterhuber, H.H. (2005) 'Shell, Siemens and DaimlerChrysler: Leading change in companies with strong values', *Long Range Planning*, 38 (5): 467–484.
Struben, J. & Sterman, J. (2008) 'Transition challenges for alternative fuel vehicle and transportation systems', *Environment and Planning B*, 35(6): 1070–1097.
Urry, J. (2007) *Mobilities*, London: Polity Press.
Vellema, S. & Loorbach, D. (2006) 'Strategic transparency between food chain and society', *Production Planning and Control*, 17(6): 624–632.
Weber, R.A. & Camera, C.F. (2003) 'Cultural conflict and merger failure: An experimental approach', *Management Science*, 49(4): 400–415.
Wells, P. (2010a) *The Automotive Industry in an Era of Eco-austerity*, Cheltenham, England: Edward Elgar.
Wells, P. (2010b) 'Sustainability and diversity in the global automotive industry', *International Journal of Automotive Technology and Management*, 10(2/3): 305–320.
Womack, J. Jones, D. & Roos, D. (1990) *The Machine That Changed The World*, New York: Rawson.
World Energy Council (2007) Energy and climate change study, London: World Energy Council. Available at http://www.worldenergy.org/documents/wec_study_energy_climate_change_online.pdf (accessed 12 February 2009).

7 Providing Road Capacity for Automobility
The Continuing Transition
Phil Goodwin

7.1. OBJECTIVE OF THE CHAPTER

The problem of infrastructure in a transport regime dominated by cars is that cars inherently require large amounts of road space, compared with either the more compact requirements of public transport or the modest requirements of walking and cycling. Broadly speaking, we can say that if traffic increases more than road capacity, congestion will increase (in intensity, duration or geographical spread). It was to avoid this that for much of the 20th century, a planning philosophy known as 'predict-and-provide' attempted to deliver sufficient road space for automobility by a formal process of predicting the future volumes of traffic in time and space and designing roads of sufficient capacity to allow more or less unfettered movement.

However, it would not remotely have been possible to increase the length of the road system at the same rate as the growth in traffic, and the gap was bridged, at least for a period, by expanding the capacity of the road network by two further policy instruments, namely the construction of bigger roads, especially motorways, rather than more roads and the use of traffic management methods (traffic signals, one-way systems, etc.) intended to make maximum use of the road space available.

Figures 7.1 and 7.2 show the progress of road building in Great Britain, which over a century produced a nearly 40% increase in the length of roads overall and the construction of a completely new network of motorways now carrying some 20% of the total traffic volume.

Provision of enough road capacity to cope with forecast increases in traffic underpinned the creation of large national road networks providing for interurban movement, but for urban areas it never held unchallenged dominance, being contested throughout the automobile age. But in both the contested and the uncontested areas, the conflict between the demands of the car for road space and the inability to satisfy those demands was a recurrent flaw in the success of automobility to deliver its promise.

In the United Kingdom, at the end of the 1980s and beginning of the 1990s, there was a major transition in policy which Owens (1995, 42) described as "From 'predict-and-provide' to 'predict-and-prevent'," a move away from

Providing Road Capacity for Automobility 141

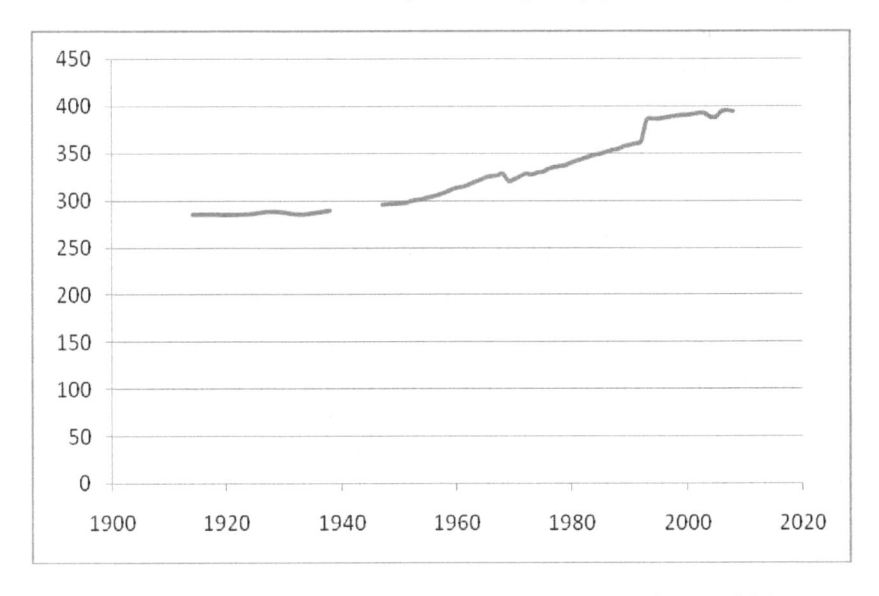

Figure 7.1 Length of road in Great Britain, 1914–2009 (in thousand kilometers). Some of the apparent discontinuities and jumps in the diagram are due to the shifts in definitions and measurements in long-term time series. They are not material to the argument (con-structed from Department for Transport, 2009, Table 7.6).

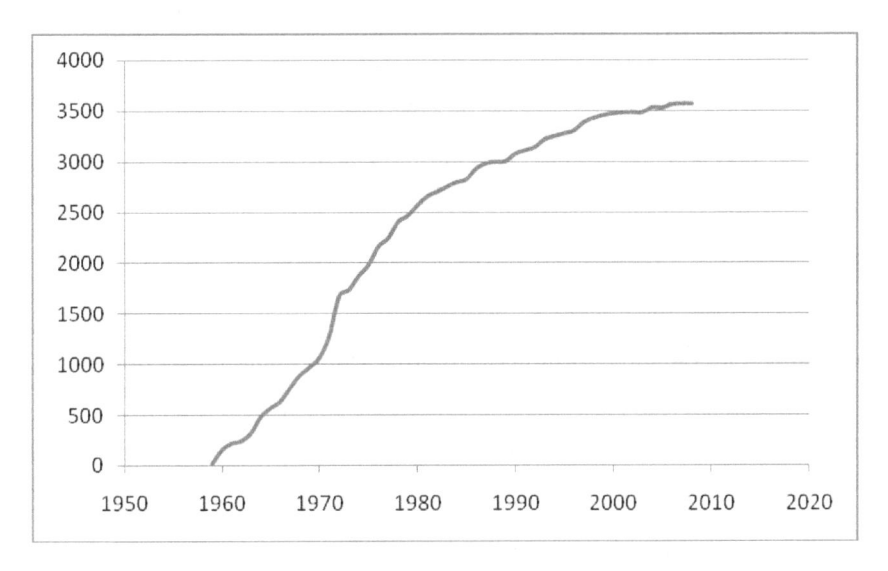

Figure 7.2 Length of motorway in Great Britain, 1950–2009 (in kilometers). Some of the apparent discontinuities and jumps in the diagram are due to the shifts in definitions and measurements in long-term time series. They are not material to the argument (constructed from Department for Transport, 2009, Table 7.6).

providing enough road capacity to match forecast increases in traffic, toward deliberately reducing the amount of traffic to a level which would be compatible with road, economic and environmental constraints. This raised the question of whether we are now no longer in the phase of an unstoppable transition to automobilism but a difficult transition away from it.

7.2. A CENTURY OF UNRESOLVED DISPUTE

Although both the increases in car use and the provision of roads for that increase undoubtedly happened, it would not be true to say that there was an unchallenged acceptance of either, even during the periods of greatest change. A characteristic feature of the argument about automobility and how (or whether) to cope with it is the very long period over which ideas, immediately recognizable as 'modern', have been stated and contested. The point of the following historical account is to demonstrate a surprisingly long continuity of thought, spanning virtually the entire age of the car, of intellectual division and challenge.

The core continuing thread was the argument about whether to provide for, or to restrict, car use. In the following account only the most influential voices are reported, being either official publications or the most widely cited and substantial representatives of professional thinking.

The early case for building roads to accommodate the traffic was asserted by two of the leading thinkers of early 20th century socialism, Sidney and Beatrice Webb (1913, 254):

> We cannot doubt that—whatever precautions may be imposed for the protection of foot passengers and whatever constitutional and financial readjustments may be necessary as between tramways, omnibuses and the public revenues—the roads have once more got to be made to accommodate the traffic, not the traffic constrained to suit the roads.

Both strands were seen throughout the 1920s and 1930s. A conflict between car traffic and local quality of life was explicitly recognized: Alker Tripp (a very influential though now largely forgotten senior police officer in London, who defined the principles of traffic engineering from the point of view of those responsible for enforcement of regulation and later described as the 'father of traffic calming') had the view that motor traffic will never and can never mix safely with pedestrians and pedal cycles. His broad idea will be to give traffic a free run on the sub-arterials and a very slow and awkward passage if it attempts to take a short cut through the precincts.

He was uncompromising. "There shall be three generic types of road and three only," (41) he wrote in 1942. On arterial roads, exclusively for fast long distance traffic, "there must be no frontages, no loading and unloading, no standing vehicles—and no pedestrians." (42)

At the other end of the spectrum, he was equally uncompromising about local roads.

> They will be so designed as to discourage through traffic of any kind from entering them at all. The only traffic to be coped with in these roads will be traffic having business in the particular locality. (p43)

It was the building of an entirely new network of main highways which became the predominant approach in discussions during and immediately after the War, focusing on inter-urban travel and given added momentum by an influential report by Glanville and Smeed (1958). This developed considerable support and indeed was built largely as planned.

But the argument about provision and use of infrastructure in urban areas continued unresolved. Colin Buchanan (MOT 1963), lead author of the very influential report 'Traffic in Towns', returned to Tripp's idea of separation of car from residential areas, though that report subsequently became used to justify their arguments both by those who wanted to build enough roads to cater for all the forecast traffic and by those who wanted to restrict traffic growth substantially. Buchanan—although typically and carefully ambiguous rather than uncompromising—was much influenced by Tripp, stating: "Basically, however, there are only two types of roads—distributors for movement and access roads to serve the buildings." (44)This was possible because the report itself is guardedly ambiguous on the topic, the words tending to stress the negative impacts of cars more and the pictures tending to illustrate large scale road building generally in a rather positive style. The Steering Group of that study wrote an introduction saying "Distasteful though we find the whole idea, we think that some deliberate limitation of the volume of road traffic in our cities is quite unavoidable." (para 6)

The opposing thread was seen in an early example of the new forms of mathematical analysis being brought over from the United States (where the strength of opposition to large scale urban road building was much more muted than in Europe and sometimes non-existent). A major study by consultants Freeman Fox (1966) suggested that a very large scale urban road building program should be carried out in London, providing a road network of sufficient capacity to serve the predicted increases in car use and a residual public transport system for a declining number of people unable to use cars. This was highly controversial, with much campaigning both for and against especially in the early 1970s, and the most ambitious road building plans were rejected.

The point of this account is that throughout most of the 20th century, in European urban areas it would be very difficult to describe one approach or the other as a 'dominant' set of ideas and practices in the sense used in transition studies, namely representing a nearly unchallenged consensus.

Rather, it was a continual battleground, with victories and defeats on both sides: Car ownership and use continued to grow throughout, but the

144 Phil Goodwin

'predict and provide' school of infrastructure response to it was contested. Many road expansion schemes were opposed and abandoned, especially in cities. The same was less true for road building between towns where the motorway programs, already conceived during the war and launched in the 1950s and carried out throughout the 1960s and early 1970s, were at first more widely (though still not unanimously) welcomed. They were initially seen as great successes, but became a focal point of great public dissension: While car use continued, the same public resisted willing the means to provide capacity for it.

The result of this was that the road building programs substantially slowed. This slowing occurred for a number of reasons, of which three seemed decisive at the time. First, what started out as a rather popular program generated increasing public resistance as the scale of environmental and local effects became apparent and as the promised 'solution' to increasing congestion failed to deliver open uncongested roads. Secondly, the cost of the program was substantial and increasing. McCulloch (2001), in a paper tellingly titled 'Is the motorway network complete', suggested that there had been a progressive increase in the cost of providing new capacity over time. There had been a tendency to adopt a building sequence of choosing easier or cheaper sections of the network first, other things being equal, and leaving the difficult ones to later. But then the increasing public resistance to road projects, together with associated environmental constraints and requirements, resulted in pressures for higher design standards which cost more. Examples include more extensive provision to control water runoff, design elements that minimize the impact of the road on the landscape, soundproofing and other environmental mitigation measures. Third, at the border between professional and public concerns was what traffic forecasters and road designers called 'generated' (later 'induced') traffic, namely the additional traffic which was attracted by the improved conditions which temporarily at least followed the expansion of road capacity. The existence and size of induced traffic had been intermittently recognized and debated since the 1930s: At some periods this was reflected in an official distancing from the idea that building roads would solve congestion (e.g., during the 1930s and the late 1960s), and at other periods this caveat tended to disappear (e.g., during the late 1950s and early 1990s). The official but independent technical advisory body, the Standing Advisory Committee on Trunk Road Appraisal (SACTRA), reconsidered the evidence in the late 1980s, and its somewhat monumental report (SACTRA, 1994) demonstrated that new roads, built in the context of already existing or expected future congestion, had the effect of increasing the volume of traffic to a greater level than would have existed without the construction. The conclusion, summarised in a special issue of the journal *Transportation* largely written by the report's authors (Goodwin, 1996, p. 51), was:

An average road improvement, for which traffic growth due to all other factors is forecast correctly, will see an additional [i.e., induced] 10% of base traffic in the short term and 20% in the long term.

The SACTRA report was described (though not by its authors) as 'the last nail in the coffin' of the road building program. The issue of induced traffic was one of the key technical issues contributing to a decline in the unchallenged authority of building roads to meet demand, since road building would itself add more traffic, at least eroding (and in some circumstances possibly completely offsetting) the relief from congestion the roads were intended to achieve and always adding to the environmental impact[1].

The other instrument for increasing infrastructure capacity, traffic management, also ran out of steam for a number of reasons, one of which was a loss of confidence that it was in fact the right thing to do: Methods intended to operate a road network at levels close to its absolute maximum capacity resulted in a vulnerability to random variations, a loss of essential redundancy and higher overall traffic volumes with associated disruptive effects.

Thus even taking account of the increase in total length of roads, the construction of a new network of motorways and the use of traffic management techniques to maximum throughput, the total capacity of the road system still increased more slowly than the growth in traffic over the period as a whole. This became the key fault line in the discussion of a policy transition in the 1990s.

7.3. THE TRANSITION IDENTIFIED? BRITAIN IN THE 1990s

Pivotal change in UK transport planning came in 1989, with key but contradictory developments in UK discussion on transport infrastructure and management. It was these contradictions which constituted the start of a transition in the relative weight of the two strands of argument described previously. Four specific developments were critical:

(a) Publication of revised official National Road Traffic Forecasts by the Department of Transport. These noted that traffic growth had already exceeded the upper bounds of the previous forecasts and suggested that, in the first quarter of the 20th century, traffic (mostly, but of course not only, private cars) would overall be of the order of double the 1988 figures. These were described as a 'watershed'.
(b) Publication of a very large government program of planned road building, called 'Roads to Prosperity', and famously described as 'the largest road program since the Romans'. It was intended to provide enough road capacity to cope with the large forecast traffic increases.

146 *Phil Goodwin*

(c) A report by the Institution of Civil Engineers—the professional body whose members are primarily responsible for the design and construction of roads—saying that road infrastructure could not keep pace with traffic growth, and it was necessary to 'take the strain off the road system' by measures to reduce traffic.

(d) A conference of European Ministers of Transport in Paris, chaired as it happened by the UK Minister for Roads (though only because it was his turn) which adopted the first collective statement at multi-governmental level in favor of reducing car use for economic and environmental reasons.

In the following 2 years, local authorities in particular confronted a technical problem: If the national traffic forecasts were correct, what did that imply for their own area? The problem was that even if the proposed national road program was successfully and completely realized (which was not certain) there would then be even heavier congestion on the local roads, including urban roads, which were their responsibility, and there was no possibility at all of expanding the local road system to match the national one. There was a considerable ferment of discussion, study groups, re-statements of policy and argument, summarized in an account of the time under the heading 'Transport: the New Realism' (Goodwin, Hallett, Kenny & Stokes, 1991).

This brought together policy statements from a wide range of institutions, professionals, pressure groups, public opinion and government. It suggested not just that a different policy course was desirable, but that there was already, as it were spontaneously, the basis for a consensus around that course. The agencies discussed included the Institution of Civil Engineers whose 1989 report had 'set the tone'; the Chartered Institute of Transport (mainly representing transport operators); the Royal Institute of Chartered Surveyors (important in the road design process); the Royal Institute of British Architects; a new think-tank called the Institute for Public Policy Research (later seen as an important influence on Labour thinking); several organizations representing local government bodies in London, other cities, districts and counties; the two main motoring organizations, the Automobile Association and Royal Automobile Club; bodies representing employers, senior managers and big private companies; the Freight Transport Association; academics' discussion papers; several different public opinion surveys; and a new government White Paper in 1990 were included. The conclusion was summarized as follows:

> Since the publication of the revised traffic forecasts, there developed a radically new situation. The new feature was that, for the first time, there grew a universal recognition that there is no possibility of increasing road supply at a level which approaches the forecast increases in traffic. That is of central importance, because it logically follows that:

a) Whatever road construction policy is followed, the amount of traffic per unit or road will increase, not reduce: i.e., all available road construction policies only differ in the speed at which congestion gets worse, be it in intensity or in spread.
b) Therefore demand management will force itself to centre stage as the essential feature of future transport strategy, independently of ideology or political stance. (Goodwin et al., 1991, p. 111)

The policy conclusions which followed will now seem rather familiar, and indeed many had been suggested for many years previously. They included:

- A substantial improvement in the quality and scale of public transport
- Traffic calming both as a set of detailed engineering techniques and as a general strategy to tilt the balance of advantage in favor of pedestrians and sometimes cyclists
- Advanced traffic management systems but with a revised statement of objectives, that is, *not* to achieve maximum throughput of vehicles (which had been the previous assumption, the phrase 'best' and 'maximum' being used interchangeably, as a way of attempting to match capacity to demand by management methods when construction could not do so) but to provide a deliberate margin between traffic levels and capacity and priority for favored classes of vehicles notably buses, delivery lorries, emergency services and disabled travelers

This list of policy instruments concluded as its *last* element:

In this logic, assessment of the need for new road construction is seen to follow from a consideration of how much traffic it is desirable to provide for, which will be influenced by the combined effect of the policies described. There will still be occasions when new road construction is clearly justified—for example in connecting a new industrial or residential development to the network—but construction 'to meet demand' is no longer the core of a transport strategy. (Goodwin et al., 1991, p. 166)

Listing road construction last was deliberate, and indeed the phrase 'road construction is the last measure we should consider' was later frequently used, with a telling ambiguity. The point however was an analytical one: Since you could not decide the design standards and locations of road building without calculating or assuming the level of traffic which those roads were intended to provide for, it followed that the policy measures having the effect of changing traffic levels (whether intended or not) needed to be detailed before the roads could be assessed.

148 Phil Goodwin

The proposition was not just that this was a better transport policy than the preceding one, but that there was already sufficient consensus that it was better to make it happen. It was not worded in terms of 'a transition is necessary' but rather that 'a transition has happened'. The strength of that analysis was that, in context, it appeared to offer a way out of endless repeated arguments. The weakness was that, in focusing on the ideas of transport planning rather than the concrete reality of what was happening on the ground, it presupposed that the practice would follow the ideas and then be effective.

The transition in government thinking was identified during the last few years of the then Conservative government, but it was articulated explicitly in the Transport White Paper 'A New Deal for Transport', (DETR 1998), one of the early initiatives of the newly elected 'New Labour' government. The White Paper itself used the language of change and transition, though Docherty and Shaw (2010) argue that the transition in ideas (which in their view would be very desirable) has not in fact been converted into real outcomes: Rather it was an illusion, a 'decade of displacement activity' (224) undertaken by a government using the rhetoric of transition to obscure a lack of real commitment.

7.4. ISSUES IN IMPLEMENTATION OF A NEW PLANNING REGIME

Once there has been a formal abandonment of the legitimacy or feasibility of providing enough road space to keep pace with unrestricted traffic growth, it follows that the main policy instruments will be those of demand management, provision of alternative modes of transport, land-use planning and behavior change. A full review of the transport policy initiatives that would be necessary to replace 'predict-and-provide' is outside the objectives of this chapter, but from a transition point of view there are three areas which it is useful to discuss. These are road pricing, 'softer' methods of achieving behavior change, and some specific aspects of information technology applied to transport.

7.4.1. Road Pricing—Forever a Niche?

There has been a long tradition (at least since 1964 in terms of UK government interest) in the idea that pricing the use of roads is or could be the primary method of managing the efficient use of the infrastructure, based on an economic argument that essentially demonstrates that the benefits of reduced congestion is greater than the disbenefit of some lost trips. If it were implemented, it has the dual effects of raising a lot of money for some agreed purpose (the debate on the use of the money being long-standing) and also saving money from the reductions in road building that would then be seen as necessary. In a little noticed technical report by HM Treasury (2006) but reporting work carried out by DfT officials, it is calculated that the size of the inter-urban road building program that would appear to give good value for money as they measured it, if there were an efficient system of road pricing, would be cut by over 60% compared with the road building program without road pricing.

The UK 1998 White Paper, and two pieces of primary legislation that followed it, provided for local authorities to take the lead in establishing local pricing schemes, either road pricing or workplace parking charges. It was assumed that within 10 years around 20 of the biggest towns and cities would have chosen to take up this policy, making use of the funds generated to support public transport and other improvements. In the event, central London was implemented with great publicity (and reasonable popularity and success), but 2 other cities voted against it in referenda, and many others buried the idea in studies and controversies. The government had taken a view—first stated in the Smeed (1964) Report and repeated intermittently since then—that the appropriate technology for road pricing would be available in about 10 years time, which still seems to be the assessment:

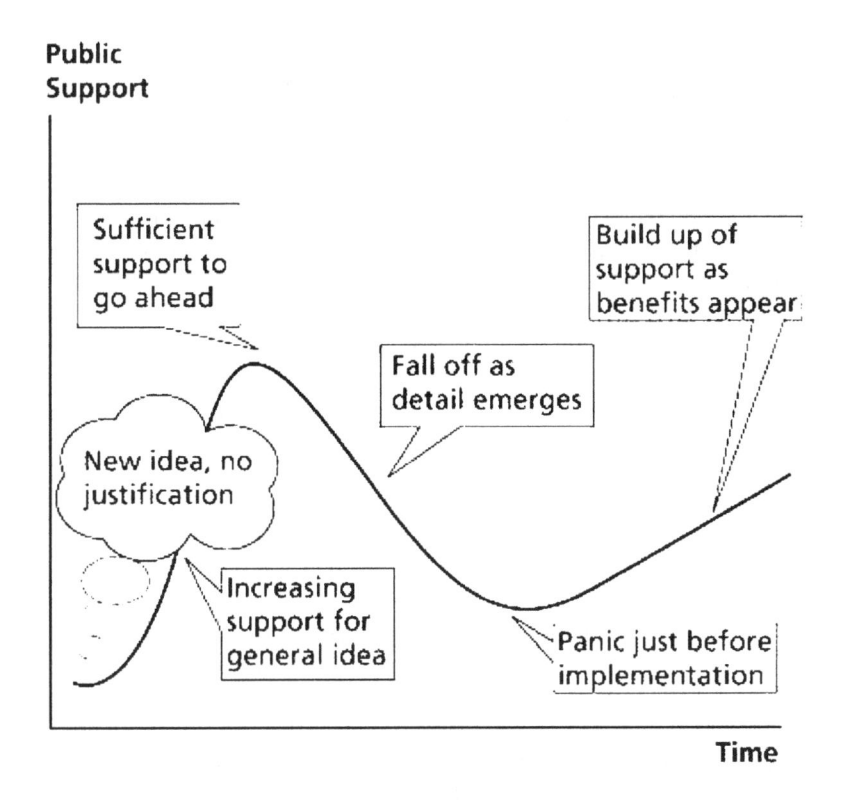

Figure 7.3 The evolution of public support for a new policy idea (see footnote 2).

10 years is an important period of time in political discussions because it means outside the scope of the current or the next government.

There has been much discussion on this, and it is probably fair to say that there is a very wide degree of consensus (subject to one caveat discussed later) that the development of road pricing as a way of managing the use of transport infrastructure is neither led nor constrained by technical innovation. The technologies available in the 1960s would have provided a crude system, the technologies now a quite refined one, and in each successive 10-year period we can be confident the technologies would be more flexible, exact, small and efficient. The constraints for road pricing to break out of its niche are not technical but political.

In this context, it is interesting to note the use of a diagram with some superficial similarities to the Gartner 'hype cycle' cited in Chapter 3, though its origin[2] and interpretation are different.

The proposition is that road pricing will not happen, ever, as a result of the spontaneous appearance of entirely satisfactory technology. It will happen if there is a conjunction of political will and public acceptability. That inevitably means temporary coping technology in the first phase and the evolution of better technology as it develops. At the beginning, there will be little or no public support for road pricing. This is not just an empirical observation but an important inherent characteristic: There is no spontaneous constituency or pent-up latent market demand thirsting for the idea. It is a creation of professionals, policymakers and academics, derived from pure thought, not personal desires.

A second phase is characterized by a buildup of support as constituencies are created. From past research, this support has been connected with recognition of the existence of a problem of congestion, environmental damage, quality of local life or other effects of excessive traffic, combined with a rejection of the traditional solution—build more roads—as ineffective or undesirable. (If that is not accepted, there is little basis for the mental jump to demand management.) At some point along the trail, a decision is taken that there is sufficient public support to merit detailed plans and schemes. The experience unfortunately so far is that at this point there will be a fall-off of public support, especially among some of the organized stakeholders whose support was always contingent on it being implemented in a particular way that meets their own objectives. Everybody can be kept happy while the discussion is still about principles but not when the devil of the detail emerges.

This is the stage at which, very often, there is a stepping back by the lead agency, in national or local government, and the policy is abandoned or delayed. However, if there is sufficient steadiness of purpose to keep moving, with a cleverly designed implementation program that keeps clear of election periods, the process is brought to completion with a vesting date defined.

During this phase things get hot and complex. Paradoxically, the fears engendered by a media-inspired panic are one of the reasons why the outcome is not as bad as feared, and there is a swift release of tension, followed by a building up of support, perhaps over many years as the promises of improvement are actually, more or less, delivered.

7.4.2. Managing the Use of Infrastructure by Targeted Behavior Change

In a context where road pricing is seen as efficient but unpopular, there has been a very rapid growth in the use of policy instruments formerly called 'soft measures', and in the United Kingdom now retitled 'smarter choices', following a report by Cairns et al. (2004) and an update by Sloman et al (2010). Smarter choices are a broad collection of different initiatives aimed at enabling people to choose improved standards of accessibility with less car use. They include workplace and school travel plans, personalized travel planning, public transport information and marketing, travel awareness campaigns, car clubs and car sharing; teleworking, teleconferencing and home shopping; residential and leisure travel plans; and initiatives on cycling and walking.

Because the instruments have no natural opposition and are suitable for local implementation, there are by now some hundreds of specific studies and initiatives, with a case-study literature whose main significance is perhaps to give measures of whether car-based behavior patterns are too deeply embedded to change or are subject to flexibility and alteration. It is interesting to note that while the literature on road pricing is largely theoretical with only limited empirical case studies, exactly the opposite is true of smarter choices. There is wide variation in experience from place to place.

- Workplace Travel Plans. A large case study literature had already accumulated by 2002. Subsequent new evidence on best practice in 20 organizations indicated an average reduction in car driver trips of 18%, equivalent to 14 fewer cars arriving per 100 staff. Similar figures had been found in the United States and the Netherlands. In further research in seven areas, we found that about 16% of the 2.2 million employees in the areas worked for companies already engaged in travel plans. Results available from 28 workplaces also showed an average reduction of 18% in car driver trips.
- School Travel Plans. Six reviews of evidence published between 2000 and 2003 found some reductions in car use of up to 50% but mostly in the range 10% to 20%. We carried out further study in 8 areas, detailed work in 3 areas and had access to results from 23 areas generated by a parallel Department for Transport project. The results suggested that local authorities might expect to achieve an average cut in school car use of 8% to 15% across all engaged schools, with substantial reductions at some schools (up to 50%) but with relatively little effect at others, particularly those where generating a travel plan document was not matched with subsequent travel initiatives.

- Personalized Travel Planning. A small but growing number of specialist agencies conduct programs of interviews and guidance aimed at the specific travel patterns of individuals and households. Reductions in car use of 2% to 15% had been reported in 17 applications, mostly in Australia and Germany, by the early 2000s. We added detailed information on three case studies and reviewed several other newly emerging ones. Overall, there were typical reductions in car driver trips of 7% to 15% in urban areas and less robust indications of 2% to 6% in rural areas. There has since been a very rapid growth of interest in this technique, and three 'Sustainable Travel Towns' were designated to try out the combined effects of several different measures. Sloman et al. (2010) report that the results were somewhat less than the 2004 estimates but only in proportion to the somewhat smaller initiatives which were put together.

Allowing for other effects and initiatives not summarized here, the overall 2004 estimate was of a potential reduction in car use of 10% to 20% in different conditions, and experience since then has reinforced this assessment.

7.4.3. Information Technology: The Idea of Functional Transition

One insight that follows from the shift in thinking about infrastructure is that new technologies do not have a uniquely defined function: Technologies devised for one purpose can find themselves used for a quite different, even opposite, purpose (Goodwin, 2006a). For example, in the 1980s an advanced form of cruise control was seen as a way of allowing platoons of very fast, very closely packed cars, avoiding collision by computer control with reaction speeds many times faster than humans can manage. In the event, however, there has been more interest in the opposite function, that is, to slow traffic down by externally triggered speed limiters set to prevent illegal or temporarily unsafe speeds—a sort of electronic traffic calming, offering all the benefits of speed bumps without their intrusiveness and inflexibility. A second example is the technology of integrated traffic signals, initially thought of as giving a cheap way of expanding the capacity of a road network to allow more vehicles to use it but now increasingly being used to *reduce* the flows at specific locations in order for the network to operate more reliably overall, for example, in ramp control.

But the most interesting example of such a transformation was first (as far as I know) devised by a European Union official in charge of managing the DRIVE research program in the late 1980s and early 1990s. He proposed what he named the 'soft implementation strategy' for road pricing. This observes that there is a genuine potential market in driver information, typically the sort of information useful during the course of a journey and focused mainly on route choice. This requires some sort of in-vehicle equipment for which drivers would pay, together with some form of communication system with an information-gathering center set up to provide information to drivers about traffic conditions, route advice and so on. Because the most efficient source of this information would include the population of drivers themselves, this would involve a two-way communication. Such a system would be popular among drivers, who would pay the progressively reducing cost, initially as a voluntary subscription, subsequently as standard equipment included at the factory.

The resulting system would have all the technical attributes necessary to be used as the basis for a road pricing system; therefore one would simply wait until a sufficiently large proportion of the vehicle fleet were equipped with the necessary equipment, all installed at the owners' own expense, and then switch on its latent pricing mode at much lower investment cost than if a complete pricing system had to be funded from scratch.

There is an added dynamic that would assist this process, due to an unfortunate feature of information systems of this sort. Various simulation studies come to the common view that the biggest individual advantage of receiving traffic information arises when that information is privileged and only shared by a minority of drivers (say around 15%). At this level the owner is significantly better informed than the market as a whole and can find better routes than drivers without that information. However, at this point two dynamics set in. First, the disadvantage of not having the system is high, so there is an incentive for market penetration to accelerate. Secondly, the relative advantage of any one driver over the rest becomes smaller. Although there may still be an overall advantage shared among all drivers due to the intelligence of the system in helping to allocate traffic evenly over the network, the advantage of taking any one route over any other diminishes and in the end disappears. Thus the paradox arises that after a certain level of market penetration there is strong pressure for everybody to join in but the advantage of doing so gets less.

An analogy is a crowded football match. When everybody is sitting down, all can see, albeit not perfectly. When a few people stand up, they can see much better but then it is necessary for more and more people to stand up, until everybody is standing, nobody has any particular advantage and the overall visibility is likely to be worse, not better.

During this process it becomes apparent that the actual benefit of using the information function is eroded in two ways: First, the relative advantage disappears with increasing penetration, and secondly, the absolute advantage is reduced because the total flow of vehicles re-approaches the new, slightly higher, effective capacity of the network, reducing any redundancy necessary to recover from breakdown and restoring high congestion on more of the network for longer periods of time. These are precisely the conditions where managing a reduction in total traffic volume to more economically efficient levels, as is achieved by road pricing set at levels, determined by marginal external cost, gives maximum overall benefit. So at the moment when one has finally exhausted all possibility of creating cheap extra capacity—a moment accelerated by the information systems—the latest functionality of the same systems provides the next stage of the solution.

7.5. WAS THE TRANSITION REAL?

It is clear that in the last decade road building 'to meet demand' has not proceeded at the level of some earlier periods and that there have been some changes in associated other policies: little on road pricing, but some important smarter choice initiatives, a greater stated emphasis on public transport, walking and cycling and some shifts in appraisal methodology. There is no observer or agency involved in transport who has concluded that the combined total of these initiatives would be sufficient to make a major change in the dominance of car transport, and some, as noted previously, essentially consider the initiatives to be palliative or exaggerated. It is therefore not necessarily to be expected that there would, yet, be evident empirically observable impacts on the progress of automobility. The question of importance is this: Has there in fact been any evidence of a transition in the actual progress of automobility itself since the late 1980s?

For this, we need to compare the traffic growth (especially since 1989) with the official forecasts of what would be expected from a continuing trend in automobility. This is shown on Figure 7.4, constructed from data in DETR (1997).

What happened after 1989 was that the course of car use at first appeared to be on track for the 'high' version of the 1989 forecasts, but then reduced in speed over a period of about a decade—still, apparently, within the 1989 forecasting envelope—but heading toward the low end. It continued to

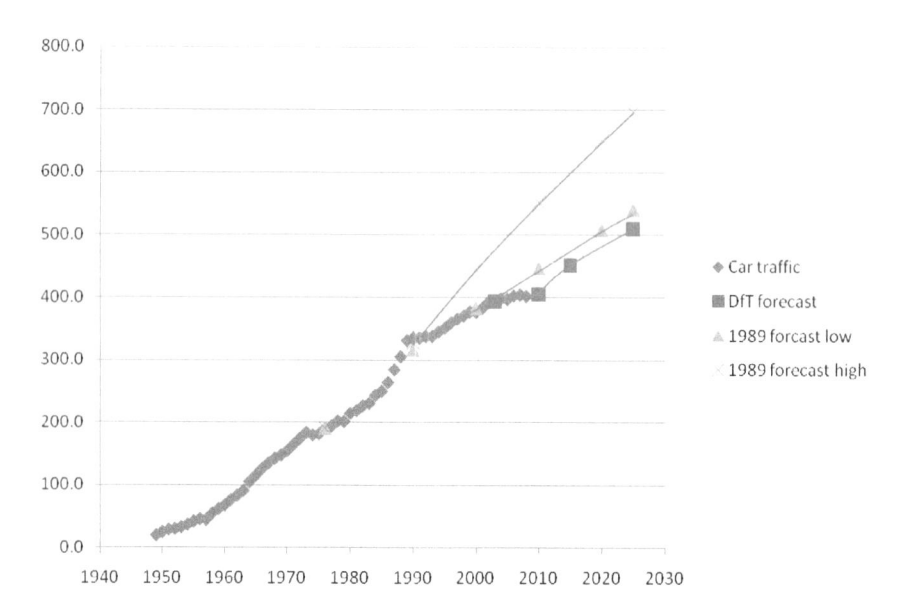

Figure 7.4 Car traffic growth compared with 1989 and current official forecasts.

Providing Road Capacity for Automobility 155

decelerate and has since remained below the lower bound of the low 1989 forecasts. In 2000, revised forecasts were made, rebased on the actual outcomes but evidently showing a growth path expected to be very similar to the low version of the 1989 forecasts. It is clear that this could only happen if there is a substantial acceleration of growth which has been characteristic of the last decade.

An illuminating relationship underpinning this curve is seen in Figure 7.5, adapted from Figure 1.2b, Department for Transportation (2009), which plots the progress of the relationship between traffic and mobility, and economic growth, in the last two decades. There has manifestly been a shift not only in the strength of this relationship but the direction. It occurred somewhere between 1998 and 1994.

The methodological question that arises is this: Suppose, as a hypothesis, that there has indeed been a transition, starting somewhere in the late 1980s in thinking and the early to mid-1990s in practice, which is of a type sufficient to see a major change in the whole course of mobility. In that case, *when* would one be able to test this empirically? There are real policy choices being made all the time, and they are influenced profoundly by interpretations of the past and expectations of the future. It would seem quite unsatisfactory to treat the whole issue of transition as one to be resolved ex post by historical analysis after the whole process has been completed in—say—2050 or 2100. This would be tantamount

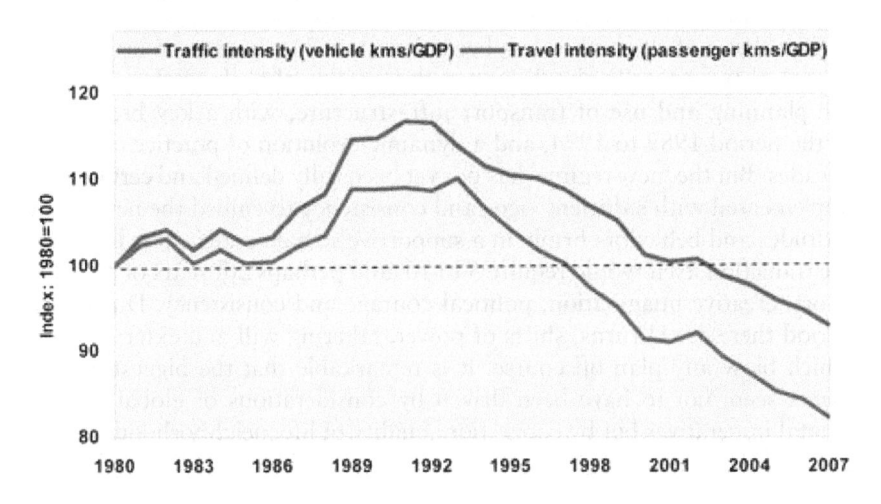

Trend 1.2b – Road traffic and travel intensity: 1980 to 2007, Great Britain

Figure 7.5 A turning point in traffic intensity showing decoupling since the early 1990s.

156 *Phil Goodwin*

to denying the participants in social change any chance of understanding what they are doing.

But a historical growth curve which is subject to a temporary interruption, or one which saturates and levels off to a stable level, or one which grows, reverses and turns down, would *look* identical to the contemporaneous observer. Those three cases are roughly equivalent to the three propositions that transition to an automobility regime is 'still strengthening' or 'largely complete' or 'in the process of transition to something different'. The choice among those three propositions is not yet resolved and may not be agreed among the authors of different chapters in this book, but it is probably the most useful task that a transition approach could set itself.

7.7. REFLECTIONS AND CONCLUSIONS

There are three ways of reading the recent and long term development of road infrastructure during the age of automobility. The first is that we have seen a dominant regime, oriented toward car use and the provision of road capacity for it, with any challenge being subordinate and confined to intellectual and practical niches. From this point of view the current regime remains that of automobility and perhaps with little end in sight. The second view is of a long-lasting, almost permanent battleground of ideas and practices, with no resolution in sight between long-lasting ideas of providing for the car and containing it. The third is that there has been a notable and important transition away from a car-dominated regime, albeit one which is still at a stage where the outcome is unclear.

The thesis of the author has been that there has been a process of fundamental transition in the discourse and analytical tools of thought about the planning and use of transport infrastructure, with a key break point in the period 1989 to 1994, and a dynamic evolution of practice over some decades. But the 'new regime' has not yet been fully defined and certainly not implemented with sufficient vigor and consistency to embed the new trends, attitudes and behaviors firmly in a supportive social context. It is likely that the transition itself would require 5 to 10 (and perhaps 20) years of persistent effort, creative imagination, political courage and consistency. During that period there are U-turns, shifts of power, faltering will and external events which blow any plan off course. It is remarkable that the biggest achievements seem not to have been driven by considerations of global environmental imperatives but by congestion, quality of life, neighborhoods, health, political demonstrations and shortage of money. That is likely to continue: 'Sustainable' transport policies are made politically acceptable and behaviorally effective by aspects other than carbon emissions.

In the context of application of 'the transition approach', the infrastructure discussion raises specific issues. There have been very prolonged periods of contested regimes without resolution of which one is dominant, and at periods there has been no clear proof of even which is the dominant direction of change.

This raises the central question of how to identify 'dominance'. A common measure of the dominance of the car, and hence a core definition of whether it can be described as a regime, is passenger distance traveled. The acceptance of this definition is common, intriguingly, to both the strongest supporters of the continuation of automobility and (some of) its challengers. Thus, in describing the limited effects of a wide range of non-car transport initiatives, Chapter 1 uses distance traveled as the indicator showing that the niches which have failed (yet) to affect the regime:

> While these activities contain seeds for substantial change, they do not yet appear to have substantially influenced the automobility system, which still accommodates for most of passenger miles (around 90%, with some differences between countries).

This is an illuminating example of how a dominant regime successfully imposes its own measures of dominance: Part of the challenge to automobility in transport planning has precisely been the proposition that distance traveled is a biased measure which does *not* accurately reflect individual and social welfare. Alternative measures—the number of trips or the use of that fundamental human resource, time—in most societies substantially reduce the perception that car is overwhelmingly dominant. This may be seen by considering that in the United Kingdom only 3% of the total personal distance traveled is on foot, making walking seem like a very small niche indeed using this measure. But 22% of all trips are made on foot, which is arguably a better measure of the primary function of travel than the distance.

This is an active debate among transport planners, where provision of infrastructure which is suitable for short local movement (within which walking, cycling, local buses and trams, especially in towns, together already command a substantial majority of trips and have done throughout the period in which cars are described as dominant) competes against provision of infrastructure for longer distance movement. Indeed, a major policy issue is whether to favor infrastructure development which has the effect of *increasing* average journey distance (hence car dominance) or reducing it. It is well established that increases in car use lead to more distant patterns of location of workplaces, homes, shops, schools, hospitals and other facilities, thus leading to longer journey distance, more journeys by car and an apparent increase in dominance. This is not always an increase in 'mobility' in any real or useful sense: There is more movement but partly to catch up with destinations which have moved further away.

All this has to be considered in a context of human agency: The future development of automobility is subject to individual and collective choices, not the inexorable result of external forces. To this extent, the idea of a society making its own history at least partly by choice is subject to different dynamic processes than those which determine, for example, whether one technology or another will be favored in a market. It seems most helpful to think of transitions in the field of mobility and the infrastructure for it not as something which just happens but as directions which are chosen or rejected.

NOTES

1. It is useful to note here that in recent discussions two further arguments challenging the provision of roads intended to provide for unrestricted growth in car use would probably tend to be given greater prominence than the three reasons here, namely the effects on health and on global environmental sustainability. They were less prominent in the 1990s and therefore cannot be given as main reasons for what happened at the time, but they do broadly reinforce the same critique.
2. This figure was first published, as far as I know, in the magazine *Local Transport Today* (Goodwin 2006b), but I did not invent it. It was adapted from a hand-drawn sketch shown to me by Alex Macaulay, from Edinburgh, and tidied up by Alan Wenban-Smith. Alex thought it came from a 2004 report edited by Christiane Bielefeldt, of TRI, for a European Commission Consortium called 'Progress'. This is available at www.progress-project.org report 4.3, but the diagram is not there. The 'Gartner Hype Cycle' is almost identical in shape but applies to the economic demand for new technology, not political acceptability, and the mechanisms and dynamics seem to be quite different. Its own precursors include the long-standing idea of a takeoff curve, but that is usually S-shaped and only turns down when new technologies replace old ones, which is an entirely different matter (though some of the insights are similar, especially about the need to get the timing right). So far, I have not found any evidence of cross-fertilization from this to road pricing: They seem to be independent.

REFERENCES

Cairns, S., Sloman, L., Newson, C., Anable, J., Kirkbride, A. & Goodwin, P. (2004) *Smarter Choices: Changing the Way We Travel* (2 vol.), London: Department for Transport. Available at http://eprints.ucl.ac.uk/1224/1/1224.pdf (accessed 30/20/2011).

Department for Transport (2009) *Transport Statistics Great Britain*, London: Department for Transport.

Department for Transport (2009) Transport Trends, 2008 Edition.

Department of the Environment, Transport and the Regions (1997), National Road Traffic Forecasts (Great Britain) 1997 http://webarchive.nationalarchives/gov.uk/+/http://www.dft.gov.uk/pgr/economics/ntm/ntmdatasources/nrtf1997/onalroadtrafficforecasts3014.pdf (accessed 30/20/2011)..

Department of the Environment, Transport and the Regions (1988), A new deal for transport: better for everyone, Cmnd 3950, The Stationary Office, London.

Docherty, I. & Shaw, J. (2010) "The transformation of transport policy in Great Britain? 'New Realism' and New Labour's decade of displacement activity" *Environment and Planning A*4 43(1) 224–251.

ECMT (1998) 'Infrastructure-induced mobility', Report of the 150th Round Table on Transport Economics 1996, ECMT/OECD, Paris.
Freeman Fox and Partners (1996) London Traffic Survey Volume II, Greater London Council, London, July
Glanville, W.H. & Smeed, R.J. (1958) *The Basic Requirements for the Roads of Great Britain, Highway Needs of Great Britain*, London: Institute of Civil Engineering.
Goodwin, P. (1996), Empirical evidence on induced traffic: a review and synthesis, *Transportation*, 23, 1, 35–54.
Goodwin, P. (2006a) 'Conjectures on the dynamic functional transformation of intelligent infrastructure', *Institute of Electrical Engineers Proceedings, Intelligent Transport Systems*, 153(4): 267–275.
Goodwin, P. (2006b, June) 'The gestation process for road pricing schemes', *Local Transport Today*, 444, 17.
Goodwin, P., Hallett, S., Kenny, F. & Stokes, G. (1991) *Transport: The New Realism*. Oxford, England: University of Oxford, Transport Studies Unit, and Rees Jeffreys Road Fund.
HM Treasury (2006), Volume 3 - Meeting the challenge: priortising the most effective policies. http://collections.europarchive.org/tna/20070129122531/http://www.hmtreasury.gov.uk/media/39E/F8/eddingtonreview_vol3.0_011206.pdf (accessed 30/20/2011).
McCulloch, R.J.G. (2001) *Is the Motorway Network Complete? Proc. Universities Transport Study Group* (vol 2), University of Oxford, January.
Ministry of Transport (1963) *Traffic in Towns*, HMSO, London.
Owens, S. (1995) 'From 'predict and provide' to 'predict and prevent'?', *Transport Policy*, 4(1), 43–49.
Standing Advisory Committee on Trunk Road Appraisal (1994) *Trunk Roads and the Generation of Traffic*, London: HMSO.
Sloman, L., Cairns, S., Newson, C., Anable, J., Pridmore, A. & Goodwin, P. (2010) *Effects of Smarter Choices Programmes in the Sustainable Travel Towns*, London: Department for Transport. Available at http://www2.dft.gov.uk/pgr/sustainable/smarterchoices/programmes/index.html (accessed 30/20/2011).
Smeed, R.J. (1964). Road pricing: the economic and technical possiblities, HMSO London.
Tripp, A. (1942). Town Planning and Road Traffic, London, Edward Arnold.
Webb, S. & Webb, B. (1913) *English Local Government: The Story of the King's Highway*, London: Longmans Green.

8 A Socio-Spatial Perspective on the Car Regime

Toon Zijlstra and Flor Avelino

8.1. INTRODUCTION

The 'radical monopoly' (Illich, 1974) of the car in modern societies is inextricably linked to socio-spatial developments, ranging from globalization, privatization and individualization to (sub)urbanization, urban sprawl, zoning, the rise of large scale concentrations, like hypermarkets and edge cities, and the fading of the difference between urban and countryside (Dupuy, 2008a; Harms, 2008; Zijlstra, 2009). Cities that were dense, compact and continuous have become diffuse, loose and discontinuous (Brand, 2002), resulting in car dependency with ecological (Newman & Kenworthy, 1999), economical (Litman & Laube, 2002) and social problems like exclusion, isolation, alienation and the loss of social capital (Dupuy, 1999; see Section 8.3.1).

Although many of the socio-spatial problems of automobility are acknowledged, they often remain underexposed in political discussions on sustainable mobility, which mainly revolve around the alleged tension between economy and environment (Baeten, 2000). This chapter perceives the socio-spatial dimensions of automobile dependency (Dupuy, 1999; Harms, 2008) as a central issue and as the main obstacle in the transition toward sustainable mobility.

In order to unravel this barrier to the transition to sustainable mobility, this chapter provides a multi-level analysis of the car regime from a socio-spatial perspective. This means that the chapter is structured by the multi-level perspective model (see Chapter 3) and that we analyze social and spatial mobility phenomena at the regime, the landscape and the niche level[1].

8.2. CHARACTERIZING THE CAR REGIME

In order to understand the dominance of automobility, the car needs to be 'decentralized' (Dupuy 1999, 2008a; Dennis & Urry, 2009; Peters, 2003). Characteristics assigned to 'the car'—for example, speed, flexibility and comfort (Steg, 2005)—are not necessarily "a product of 'the automobile' but rather [. . .] a consequence of the way its use is promoted and organized,

A Socio-Spatial Perspective on the Car Regime 161

as part of an auto-centered transport *system*" (Freund & Martin, 2000, p. 52–53). This system includes a "vast material infrastructure of roadways, repair facilities, auto supply shops, gas stations and service facilities, motels and tourist destinations, storage spaces, and an extensive social infrastructure of bureaucracies for the control of traffic, the education of drivers, and the regulation of drivers, vehicles, and fuels (among other things)" (ibid., p. 53). The car is a typical network technology; its value depends on spatial, economical, institutional and social configurations that have to be created and maintained. The "power of automobility is the consequence of its system characteristics" (Dennis & Urry, 2009, p. 41).

While talking about an 'autopoietic' system like Urry (2005) does serves quite well as a metaphor, it leaves the agency aspect out of sight. For the "notion of a system tends to underplay collective human agency in the production of automobility and to avoid the political questions about the shaping of the automobile 'system'" (Böhm, Jones, Land & Paterson, 2006, p. 5). Rather than addressing the car regime as a predetermining structure, we discuss how 'regime actors' *reproduce* existing structures. Transport planners, traffic engineers, transport economists, spatial planners, project developers and policymakers occupy a central role in the system. These 'level-one operators' (Dupuy, 2008b), or 'system builders' as we call them, are responsible for the (re)organization of the physical dimension of the system[2], one of their objectives being the configuration of space in order to support movement (of cars).

8.2.1. System Builders and Underlying Principles

The work frame of conventional system builders is mainly influenced by four central principles: (a) efficiency, (b) calculability, (c) predictability and (d) control by non-human techniques, also known as the principles of 'McDonaldization' (Ritzer, 2000). Traditional system builders search to reduce the generalized costs as much as possible (Banister, 2008; Metz, 2008) to make the network more efficient. Generalized costs are the combination of time and money spent on traveling. In cost benefit analysis, projects are primarily seen as beneficial when they promise to reduce these generalized costs. Up to 80% of the benefits are calculated based on 'gaining time' (Banister, 2008). An English manual for traffic engineers emphasizes that the role of a traffic engineer "is increasingly to improve the *efficiency*" of the road network (Slinn, Guest & Matthews, 2005, p. 2, emphasis added). While in the past this was mainly focused on (endless) expansion to facilitate increasing car use ('predict and provide'), this strategy is now perceived as 'inefficient' (see Chapter 7).

There is a strong tendency among 'system builders' to use quantitative discourse. Whether it is the number of cars, average speed, travel time, road capacity, parking spaces or emissions, the language is primarily focused on calculability (Böhm et al., 2006). Compendiums like the one by Slinn et al.

(2005) typically prescribe how to calculate the capacity of a street or how to estimate the needed number of parking lots. Traffic planners, transport economists and road building agencies are often criticized for being (too) technocratic and for focusing on (seemingly) neutral figures presented as 'scientific facts' (Brown, 2006).

Predictability is advocated in order to promote traffic safety and reduce congestion (see CROW, [Center for Regulation and Research in Engineering] 2004). The main ingredient of such predictability is uniformity, fostered by standardized designs that aim for clear, comprehensible and simple traffic situations (see CROW, 2004; Slinn et al., 2005). Predictability is also enforced by the spatial separation of traffic modes; often only similar modes of traffic are allowed in the same space.

Control by non-human techniques is manifested in an automobile landscape filled with traffic lights, speed cameras, toll ports, visual constrictions and physical barriers. "The accumulation of 'street clutter' [. . .] is the most evident visual manifestation of measures aimed to regulate and control traffic" (Hamilton-Baillie, 2008, p. 164). Eye contact is out of the question, since the speed is too fast and drivers are too far apart, leading to 'car only environments'; no humans allowed (Urry, 2007).

8.2.2. Irrational Rationalities and Counterproductive Effects

The four principles described in the previous section demonstrate a certain irrational rationality, also referred to as the 'fifth dimension of McDonaldization' (Ritzer, 2000). The idea of 'irrational rationality' is that there is a single focus on a seemingly rational measure or policy but one which overlooks rebound effects. A good example can be found in the introduction of mandatory safety belts in cars, in itself a logical and demonstrable safety measure. However, the introduction of safety belts came with a rebound effect, as people showed more reckless driving. The average driving speed increased by 3% in Australia (Conybeare, 1980). The overall safety may have been improved, but for those *outside* of the car the environment became less safe and less attractive, and some claim that the casualty rates of non-occupations are higher than they would have been in the absence of seat belt legislation (ibid.).

One of the most telling examples of the irrational rationality is the unilateral focus on the traffic jam as the central mobility problem (Böhm et al., 2006; Harms, 2008; Peters, 2003). The large investments in congestion solutions harbor several 'irrational rationalities' and subsequent counterproductive effects. First, congestion is not a new phenomenon; it has been around for ages. It has been a steady aspect of urban life and therefore probably as old as urban life itself. There are records of congestion in the streets of London and New York in the 17th and 18th century (Jacobs, 1961/1993; Weinstein, 2006).

Second, addressing the 'problem' of congestion makes the car more popular, as driving speeds increase and more resources are spent on the car

A Socio-Spatial Perspective on the Car Regime 163

system at the expense of other modes of transport. This does not only have economic consequences but also socio-psychological effects as alternative modes receive a lower social status (Sheller & Urry, 2000).

Third, the problem of congestion is expressed in terms of 'lost time'. The choice of displacement by car is supported by the idea that it is the fastest mode available. In most cases it will certainly be the fastest choice, even including the 'time lost' due to traffic jams (Bakker & Zwaneveld, 2009). However, the idea of 'gaining time' is an illusion, when increased driving speeds are primarily used in order to travel further away, instead of traveling for a shorter period of time. People now have to travel further to access the same things that were accessed locally a generation ago. Since these distances can, in most cases, only be bridged by the car, it ends up in a 'radical monopoly' for automobiles (Dupuy, 1999; Illich, 1974) and even more 'congestion misery'. Hence the phenomenon of induced travel, where a seemingly new demand for car travel is created by improvements in the car system (see Chapter 7).

8.3. THE AUTOMOBILE LANDSCAPE AND COUNTER-MOVEMENTS

The characteristics discussed in the previous section are not just a transport issue but part of broader societal trends (see also Chapter 9). In this section we discuss landscape trends related to the socio-spatial regime of automobility. First, we discuss dominant trends and paradigms that support and promote car dominance; ranging from private property, time management, individualization to neo-liberalization. Second we discuss undercurrent counter-movements that challenge the automobile landscape.

8.3.1. Dominant Landscape Trends That Support Automobility

One of the foundations of capitalism, namely *private property*, has had a deep impact on the urban form. In 1998 there were approximately 552 million cars in the world, by 2020 this probably will be 720 million (Geffen et al., 2003, as cited in Urry, 2007). Although the car remains unused for 97% of the time (Harms, 2008), car sharing schemes are still marginal. This does not only lead to road expansion, but also to major parking issues (though this is often overlooked; Henderson, 2009b). Even though cars are mostly private property, road building, parking, traffic management and other efforts to support the car system are state investments, not to mention the externalities that are also not privatized. As such only the benefits are privatized, while most of the costs are socialized (Paterson, 2007; Rogers, 1997).

From the beginning of the 19th century there has been an obsession with time, mainly in the advanced capitalist world (Castells, 1996; Gleick, 1999). One could argue that not the steam engine but the clock or metronome was the crucial factor in the industrial revolution (Castells, 1996). There is a focus on time discipline and tuning but even more on time *gaining*. The coercive laws of competition force capitalists to work on their time efficiency and to teach their workers time discipline (Marx, 1867/1971). This of course results in the process of acceleration, which contributes to the ideal of the free market. The theoretical free market is in reality obstructed by physical barriers; trade has to deal with distances and has to literally overcome seas, rivers, deserts and mountains. Getting the products from the factory to the consumer can be understood as a part of the production process, and for that there has been an incessant 'annihilation of space though time' (Harvey, 1990).

Acceleration has a central place in discussions about the car (Peters, 2003). Supporters of the car push for higher speeds and want to get rid of congestion problems, while critics suggest that the alternatives should be made faster. In that respect the high speed trains are promoted in China, the United States and Europe, and in the Netherlands 'bicycle highways' are constructed. Acceleration is not just a principle of automobile system builders but a principle that prevails within modern society at large.

The car can be seen as an icon of the neoliberal era (Henderson, 2009a). The Dutch Liberal party, for instance, suggested car driving should be a 'right' and presented a bill to get this notion legally enforced. The car supports flexibility and offers speed, but most importantly it represents independency. People with a car are able to "take care of themselves," teenagers with a driving license become adults (Dupuy, 1999). This self-reliance is an important aspect that dates back to the emergence of the car, when one was not just a driver, but a mechanic as well (Peters, 2003). Or, like the British Prime Minister Margaret Thatcher stated during a parliamentary debate in 1986: 'A man who, beyond the age of 26, finds himself on a bus can count himself as a failure' (Litman, 2007). Next to these associations on a personal level, there is also an image of the car on a broader societal level. Here the car is a symbol of modernity and economic progress (Paterson, 2007); it is 'the embodiment of progress' (Böhm et al., 2006). Many car brands are iconic for the victory of capitalism in the 20th century; Ford and Toyota can be directly linked to the revolution in the production process (Urry, 2007). When manufacturers suffer, due to mismanagement or declining sales, national authorities jump to aid. The negotiators are not simply staff members but include Ministers, and the bags of money they bring to the table are considerable (Paterson, 2007). This corresponds with the strong focus on gross national product growth within neoliberal societies, where investments made in the automobile system are relatively high (Dennis & Urry, 2009).

A climax of the automobile as a neoliberal icon is the rise of the sports-utility vehicle (SUV). It represents an 'SUV model of citizenship', centered on privatized, unhindered, cocooned movement through public space (Mitchell, 2005). The vehicle showed a steady rise in sales from the end of the 1980s onward (until recently), particularly in the United States, with names that are reminiscent of the world of soldiers and fighters, such as the Shogun, Raider, Trooper and Defender, all in defiance of the 'urban jungle' (Wilkinson & Pickett, 2009). The extreme case of the SUV-hype, the Hummer, actually came from the army. The popularity of the vehicle may be linked to social inequality (ibid.), and greater inequality in turn contributes to more crime, more fear and higher sales for the 'Chelsea-tractor' (ibid.). Levels of individualization rise in a self-reinforcing process.

The difficulty to break away from spatial automobile dependency is not only a question of competent urban designers or spatial planners but also a question of breaking with a lifestyle that is made possible by the automobile (Litman, 2007). During the massive suburbanization in the post World War consumers did not just buy a car; they bought a whole lifestyle, including a spacious single-family house, a kitchen with machines, furniture and a lawnmower. And while doing so, they moved upward on the social ladder. In this typical manner the car creates the preconditions for its own inevitability (Urry, 2007), as acquired comforts and vested interests become too strong to abandon this lifestyle. Meanwhile, the power of spatial planners declines, and urban planning looses ground. The neo-liberal notion that planning is a reduction of freedom and an unnecessary form of government interference left the field of spatial planning as an 'institutional zombie' (Baeten, 2000; Gleeson, 2000).

8.3.2. Counter-Movements That Challenge Automobility

"While the car is everywhere, it is also everywhere contested" (Böhm et al., 2006, p. 6). Critique toward the car might be as old as the car itself. In the early years of the car there were strong restrictions on car use (Mom & Filarski, 2008). Fierce protests arose in reaction to the first known car related death in 1896 in the United Kingdom (Böhm et al., 2006). Several anti-car novels where published in the early 20th century, like *The Wind in the Willows* by Kenneth Grahame and *A la recherche du temps perdu* by Marcel Proust (ibid.; Peters, 2003). Most of the negative attitudes toward the car are not just anti-car movements. These are movements with anti-car aspects, arising from several underlying principles, beliefs, dreams and ideologies. Paterson (2007) distinguishes seven of such movement, arguing that one of the common denominators within these counter-movements is the call to *restrict car movement*. We now discuss these seven counter-movements and how they challenge car dominance.

1. *Technocratic Environmentalism*. This movement is primarily concerned with the environmental consequences of modern times, mainly focusing on pollution and resource use. It gained momentum during the 1970s, when several downsides of the dominant lifestyle in the advanced capitalist world became apparent (e.g., see Club of Rome, Limits to Growth, Greenpeace, etc.). The evident relation between

economic progress, increased auto-mobility and environmental decline, gave rise to a discourse of 'zero-sum balance'. More (auto) mobility corresponds with economic success, while less mobility might be the savior of the environment (Stoep & Kee, 1997). In this discourse, the central challenge for 'sustainable development' is maintaining the perceived positive aspects of the car, while reducing its negative effects (Baeten, 2000).

2. *Automobile Space.* There are many movements concerned with the decline of rural and urban landscape due to modernization. Organizations such as the United Nations Educational, Scientific and Cultural Organization, Monuments preservation and Cultural Heritage foundations, aim to preserve, maintain or mend traditional (urban) landscape, fed by cultural, conservative, esthetic, social or other motives. A popular 'victim' of this movement is 'the car', or more precisely, the damaging effect of the way in which the car is promoted and supported (Jacobs, 1961; Paterson, 2007). The first concerns among 'traditional' architects and urban designers arose in the 1920s and 1930s (Mom & Filarski, 2008). During the 1960s and 1970s this counter-movement was expressed by Jacobs (1961/1993), Gehl (1971) and others (see Paterson, 2007). The devastating effect of the car infrastructure to the city led to a reappraisal of traditional urban architecture. Sub-movements like *Carbusters*[3] and *Reclaim the Streets* and numerous publications (e.g., Kunstler, 1994; Peters, 2003; Rogers, 1997) represent the continuation of this movement.

3. *Atomistic Individualism.* This movement has been an opposition to the ideology of the car as a symbol of atomistic individualism (Mitchell, 2005). For the adherents of this movement, the car symbolizes urban alienation and loss of community. Central line of critique is the way in which the personal freedom of the car denies the dependence on others (Paterson, 2007). Putnam (2000) claims that the car is-next to the television—the main source of the loss of 'social capital', and that social capital is not only a goal in itself, but also contributes to psychological and physical health: "The more integrated we are with our community, the less likely we are to experience colds, heart attacks, strokes, cancer, depression, and premature death of all sorts" (p. 326). These notions of social capital and a lost sense of community also played an eminent role in the critique of 'Automobile Space' in the 1960s and 1970s (Gehl, 1971; Jacobs, 1961/1993).

4. *Social Inequality.* Debates about (in)equality are long standing in politics and at the basis of classical distinctions between left and right-wing policies. Discussions mainly tend to focus on income differences (Wilkinson & Pickett, 2009), but in transport other aspects come into play. Several authors have argued that inequality is inherent to the car (see Freund & Martin, 2000; Gorz, 1973; Wajcman, 1991). The phenomenon of forced car ownership (Banister, 1994) refers to

the phenomenon that households are forced to spend a relatively large share of their income on the car (Coutard, Dupuy & Fol, 2004; Currie & Senbergs, 2007), put aside 'green' principles and/or overcome a fear of driving. The car is no longer a luxurious item but a necessity (Paterson, 2007). The alternative option, using other means of transportation, typically results in longer journey times or a smaller scope (Coutard et al., 2004; Taylor & Ong, 1995). Financial support by the state for poor families in order to wield a car, increases citizens' dependency on the state and does not offer a structural solution (Lyons, 2004).
5. *Against Speed.* This movement is skeptical about the notion of increased speed. A popular argument is presented by Wajcman (1991), who claims that average traffic speed in London, Tokyo and New York did not increase between 1890 and 1990. In a database with traffic data of 32 international cities, the figures of travel speed and hours spent in traffic were compared, and the results indicate that "as the speed of the traffic system increases, so does the actual personal time commitment necessary to maintain participation in the urban system, which would appear to be the reverse of the common assumption about reducing congestion to save time" (Newman & Kenworthy, 1992, p. 163). Illich (1974) calculated that the average American devotes a quarter of his available time per day to the car, by riding it and working to pay for it. The time gained by the use of a car, is lost by the extra hours of work to pay for it.
6. *Consumerist Geopolitics.* Christians, environmentalists, pacifists and others join together in the movements of 'consumerist geopolitics' (Paterson, 2007). They are mainly concerned by the increasing dependency on foreign oil. Oil is a heavily burdened issue, associated with geopolitical conflicts, tensions, wars, totalitarian regimes and environmental degradation arising from 'oil thirst' (Seifert & Werner, 2005). Ninety-eight percent of the car fleet is still using fossil fuels. Cars worldwide use up to 50% of the oil production and almost a quarter of the total available energy (Dennis & Urry, 2009).
7. *Safety.* This movement is mainly concerned with the direct and indirect victims of modernization. Worldwide fatal car accidents include 1.2 to 4 million people, another 20 to 50 million people are hospitalized each year with severe injuries (cf. Paterson, 2007). Indirect victims suffer from car related health concerns, like obesity, cancer and asthma. These indirect victims are not only car drivers; people living along major roads are confronted with exhaustion, lights, noise and vibrations, resulting in all kinds of health issues. Concerns about safety have resulted in a major infringement on children's freedom as parents hope to protect them from the dangers of traffic (Gehl, 1971; Peters, 2003).

8.4. RADICAL SOCIO-SPATIAL MOBILITY NICHES

While the counter-movements described in the previous section were observed at a macro-level, we now turn to discuss niches that deviate from the auto-mobility regime at the micro-level. We focus on 'radical socio-spatial mobility niches', defined as niches that (a) do not only deviate from the current socio-technical automobility regime but also from the *broader automobile landscape*, by pro-actively challenging underlying dominant paradigms and (b) have an explicit socio-spatial dimension that moves beyond technology-oriented solutions and tackles automobile dependency at its 'socio-spatial roots'. These radical niches are intimately related to the 'counter-movements' discussed in the previous section. We distinguish four clusters of niches[4]: modal shift, deceleration, sustainable urban planning and localism. For each of these niches we discuss how they challenge the current automobility regime and to what extent they qualify as a 'radical socio-spatial niche'.

8.4.1. Modal Shift

The idea of 'modal shift' had a prominent place in the political agenda in the Netherlands from the mid 1970s until the mid 1990s (Mom & Filarski, 2008). The car was clearly profiled as 'a problem' that could be overcome by promoting modes like walking, cycling or public transport. The bicycle is a popular 'competitor' to the car, presented as energy efficient, environmentally friendly, healthy, cheap and social. Dutch cities like Groningen, Delft and Amsterdam have high scores in the list of the bicycle capital of the world (mainly competing with Danish, German and Chinese cities). The share of trips by bike varies from 10% up to 42% in the Dutch cities (Ververs & Ziegelaar, 2006), and there are more bicycles then residents in the Netherlands. The success of the bicycle is not just at the expense of the car, as there is clearly a lower share of public transport use in the Netherlands compared to the United Kingdom (Metz, 2008).

Most policies promoting more sustainable transportation tend to take a 'positive' approach; offering infrastructure, services, promotion actions, tax cuts, grants and campaigns. However, research showed that success of promoting bicycle use also depends on a 'negative' approach in terms of discouraging car use (e.g., high parking fees) (Ververs & Ziegelaar, 2006). The same applies to public transport policies: The success of public transport in London, Singapore and Barcelona is supported by car unfriendly policies (Banister, 2008). There is a need for push and pull measures (Metz, 2008), maybe with the emphasis on push (Mackett & Robertson, 2000). However, most urban policymakers fear car unfriendly policies, which are perceived as 'restrictions of freedom' in popular discourse (Bach, 2006; Mom & Filarski, 2008).

Many policies promoting cycling are in themselves not a radical socio-spatial niche in the sense that they do not challenge the prevailing paradigms at the landscape level. The bicycle is, like the car, mainly an individual means of transportation, and many of the improvements focus on speeding up bicycle traffic and providing bike lanes. This niche only qualifies as 'radical' when it is pro-actively combined with the restriction of car movement and the discouragement of car use.

8.4.2. Deceleration

The concept of deceleration directly opposes the idea of acceleration, one of the identified dominant paradigms at the landscape level. Deceleration in traffic engineering and planning is known as 'traffic calming' (see Chapter 7; Newman & Kenworthy, 1999). The major objective of traffic calming is to reduce local air and noise pollution, the barrier effect of heavy traffic and the overall dominance of the car. At the same time traffic calming tries to improve the street environment for non-car users and enhance local economic activity (Newman & Kenworthy, 1999, p. 146). The idea of slowing down cars opens op more possibilities for walking, cycling and public transport. Differences in travel times shrink, the quality of the outdoors is improved and walking and cycling becomes more pleasant, saver and easier. Therefore traffic calming is presented as an important ingredient in order to realize modal shift.

A central principle of traffic calming is the concept that a street is more than just a road. This is most adequately presented by the homezone[5] ('woonerf') as introduced in Delft in the 1960s (Bach, 2006). Here, calming is realized by an integral street design, including the positioning and proportioning of the buildings. 'Woonerf' design techniques also include soft edges, a staggered street axis and a visual narrowing of the space. The residential zone is, firstly, a social space where people can meet and where children can play, a place where exterior activities are pleasant. The traffic function of the 'woonerf' comes second, with the walking pace as 'natural' speed limit.

A contemporary revival of 'woonerf' ideas can be found in the concept of 'shared space' (see www.sharedspace.eu; Hamilton-Baillie, 2008). "Shared Space helps to generate public spaces where traffic, social and all other spatial functions can be in harmony—people can move, meet each other, do things together or get to know somewhere" (Shared Space Institute, 2005, p. 19). This design concept is a critique aimed at the dominant role of traffic experts in arranging public space, where the street is often reduced to just a road (Banister, 2008). The concept of shared space is not restricted to residential areas, as also more intensively used streets can be designed with the shared space principles. Recently three Dutch urban design studios took traffic deceleration one step further by presenting plans for the urban highways A10 and A13[6], in which the highway is transformed into an urban corridor with a maximum speed of 50 km/h and accessible for busses, cyclists and pedestrians. Reliability and socio-spatial integration are preferred over separation, speed and (alleged) efficiency (Boomen, 2010).

8.4.3. Sustainable Urban Planning

Sustainability in urban planning is approached from many different angles, sustainable mobility just being one of them. The goal of sustainable urban planning in relation to mobility is to physically restrict and discourage car movement, while promoting other modes of transport (Newman & Kenworthy, 1999). Visions to enable more public transport mostly focus on the clustering of important destinations, with intensification at stations, terminals and stops. Visions promoting bicycle use or walkable cities aim for an overall reduction of distances between different functions. Examples of urban design and planning movements with strong mobility components are New Urbanism, Compact Cities and Smart Growth (Henderson, 2009a).

The *New Urbanism movement* (NU) was initiated by non-conventional architects and urban designers in the 1980s in the United States (Haas, 2008). The ideas of NU focus on (a) the relation between urban land and hinterland, (b) the relation between urban form and transport, (c) protecting historical and cultural heritage, (d) the importance of public space as the social and cultural backbone of the city and (e) the human being as the measure of all things. A point of critique of NU is the loss of a 'sense of place', or the 'geography of nowhere' as Kunstler (1994) termed it. The movement is often positioned as neo-traditional as many of the NU concepts already existed before the first Congress of NU in 1993. "What New Urbanism did was build a popular movement with global spillover effects, one that would comprehensively advance a coherent theory of urbanism, a solid conceptual framework, and a new, revised set of tools and techniques" (Haas, 2008, p. 9).

Although *Compact Cities* policies have been around in the Netherlands for six decades (Hajer & Zonneveld, 2000), it can still be seen as a niche considering the counter-productive policies and disappointing results (Brand, 2002). The compact city concept promotes high density, mixed use and clear boundaries (Geurts & Wee, 2006). At first, in the 1950s, compact building was promoted to preserve agricultural land. Since the 1970s it was promoted to preserve nature and prevent additional traffic movement. Nowadays high ground prices are the main motive (Geurts & Wee, 2006; Hajer & Zonneveld, 2000). This planning strategy could not prevent the rise of low-density areas due to demographic developments, such as an increased number of divorces, one person households and small families. The singular focus on density could not prevent the rise of mono-functional areas, a growing number of edge cities and other urban field characteristic (Brand, 2002). However, "without compact urban development policies, urban sprawl in the Netherlands is likely to have been greater, car use would have been higher at the cost of alternative modes, emissions and noise levels in residential and natural environments, and the fragmentation of the wildlife habitats would have been higher" (Geurts & Wee, 2006, p. 139).

The *Smart Growth* movement in the United States is spurred by 'demographic shifts, a strong environmental ethic, increased fiscal concerns and more nuanced views of growth'[7]. Smart Growth has several similar principles as New Urbanism but with a more society broad view and approach compared to the urban design focus of NU. Smart Growth advocates the following: (a) creating a range of housing opportunities and choices; (b) creating walkable neighborhoods; (c) encouraging community and Stakeholder Collaboration; (d) fostering distinctive, attractive communities with a strong sense of place; (e) making development decisions predictable, fair and cost effective; (f) mixing land use; (g) preserving open space, farmland, natural beauty and critical environmental areas; (h) providing a variety of transport choices; (i) strengthening and directing development toward existing communities; and (j) taking advantage of compact building design.

Overall, sustainable urban planning opposes the dominant trends in recent decades, like sprawl and edge cities, while at the same time revitalizing planning practice itself. It can be seen as a radical socio-spatial niche that challenges automobility, as long as policies do not only focus on just one aspect, like the policy of compact cities tends to do, but cover different socio-spatial aspects.

8.4.4. Localism

Since the turn of the century there has been a renewed interest in 'localization', especially in alternative food networks (Watts, Ilbery & Maye, 2005)[8]. The idea of localization is not new; all economies where 'local' in pre-modern times. The first wave of revival occurred in 1960s and 1970s with Illich (1974), Schumacher (1973) and Jacobs (1969/1970). Globalization in combination with neo-liberalism is currently often pointed out as the problem (Watts et al., 2005) and going back to a local scale is presented as the solution. The neo-liberal vision on how to promote economic growth is criticized, and instead "the end goal of the economy becomes maximum self-reliance, rather than open markets and international competitiveness" (Hines, 2000, p. 62). Import-substitution is positioned as an important ingredient for a successful economic development. Jacobs (1969/1970) referred to the economic boom of Chicago from the 1840s to the 1850s, when the city became more self-reliant, as an empirical case. Hines (2000) presents seven central rules as guiding principles toward a self-reliant community:

1. Reintroduction of protective safeguards for domestic economies
2. A site-here-to-sell-here policy
3. Localizing money
4. Enforcing a legal competition policy to eliminate monopolies from more protected economies
5. The introduction of resource taxes
6. Increased democratic evolvement, both politically and economically

7. Reorientation of the end goals of aid and trade rules such that they contribute to the rebuilding of local economies

Contemporary empirical examples of localism can be found in the *Transition Towns* and *Slow City* movement. The first *Transition Town* was the English town of Totnes, started by Rob Hopkins a decade ago. Since then the concept spread around the globe[9]. The rebuilding of local resilience is presented as a reaction to climate change and peak oil (Hopkins, 2008, part 1), main ingredients being diversity, modularity and tightness of feedbacks (Hopkins, 2008, p. 55). In 1999 the *Slow City* movement was initiated which—much like the Slow Food movement—challenges multinationals, consumerism and globalization. Central principles are a quiet and less polluted environment, maintaining local and esthetic principles, appreciating handicraft, regional products and local cuisine (Knox, 2005). The actual list of principles of the Slow City movement is 54 points long, all aimed at the traditional lifestyle and local qualities (ibid.). Both the Transition Towns and Slow City movements have a strong transport component, supporting (car) free city areas, the clearance of unnecessary traffic signs and lights and the promotion of public transport, walking and cycling (ibid.). The overall idea of self-reliant local communities reduces distances and decreases the need for global transport flows. As such these empirical cases can be framed as 'radical socio-spatial mobility niches' that explicitly deviate from the automobile landscape by tackling automobile dependency at its 'socio-spatial roots'.

8.5. DISCUSSION AND CONCLUSION

This chapter has provided a multi-level analysis of the car-regime from a socio-spatial perspective. This multi-level analysis is schematically captured in Figure 8.1. The regime is characterized by formal, normative and cognitive rules (Geels, 2004) of the 'system builders' that reproduce the automobility system, mainly based on efficiency, calculability, predictability and non-human control, resulting in a focus on fighting congestion, door-to-door transport, traffic management, edge cities and urban sprawl. At the landscape level, the automobile landscape is characterized by dominant trends and paradigms supporting and promoting car dominance, private property, acceleration, just-in-time management, individualization and neo-liberalization. At the landscape level we also find undercurrent counter-movements that challenge the automobile landscape. At the niche-level we find examples of 'radical socio-spatial mobility niches' that are related to the 'counter-movements' at the landscape level and deviate from the regime and dominant landscape trends.

Obviously, the three different levels interact with one another; dominant trends at the landscape level enforce the automobility regime (and

A Socio-Spatial Perspective on the Car Regime 173

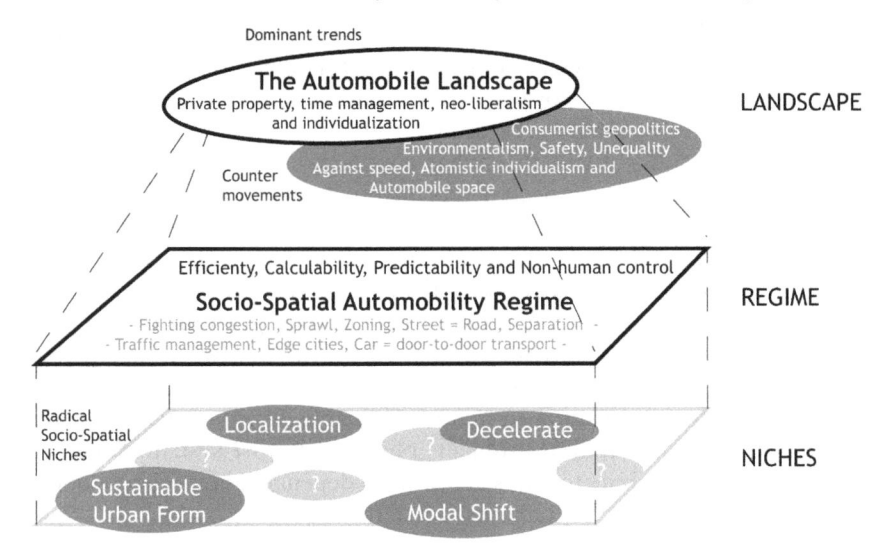

Figure 8.1 Multi-level analysis of automobility from socio-spatial perspective.

vice versa), while counter-movements inspire niche-activities that challenge this regime. The question that fascinates transition researchers is how and when such multi-level dynamics have enabled transitions in the past and to what extent they enable a transition to sustainable mobility now and in the near future. On the one hand, the analysis in this chapter may give rise to skepticism regarding the possibilities for a mobility system that is sustainable also in socio-spatial terms. The counter-movements and 'radical' socio-spatial niches that we discussed have been around for decades, and even though they have deviated from and challenged the automobile landscape and regime, they could not prevent the far reaching and ever-growing dominance of the car in contemporary society.

On the other hand, our analysis can also be interpreted in a more optimistic and hopeful manner. First, the current car regime would probably have been even more widespread if it were not for these counter-movements and niches (Geurts & Wee, 2006). Second, many of the 'old' counter-movements and niches have known new and recent revivals (e.g., Shared Space, Slow City, Transition Towns, as described in Section 8.4). When we consider the growing sustainability discourse in political and business circles, combined with the ongoing sequence of financial and ecological crises, one can reasonably expect that these counter-movements and niches will persist, grow and play into current 'crises' and subsequent 'cracks' in the regime.

The challenge for transition researchers is to systematically analyze the ongoing dynamics between landscape, regime and niches, in order to provide an empirical basis for discussions that are inherently susceptible to speculation and subjective interpretation. The aim of our chapter has been to broaden such empirical transition research in the field of mobility, first and foremost by drawing attention to the socio-spatial dimension of mobility. More specifically, the following insights can be drawn from our multi-level analysis.

First, our analysis shows that what counts as (radical) niche strongly depends on the chosen perspective. For instance, in our socio-spatial perspective 'greening the car' hardly qualifies as a niche. Rather it is seen as a technology that allows the current regime to respond to landscape pressures (i.e., technocratic environmentalism) and thereby strengthens the regime rather than challenging it. In this perspective, 'greening the car' is a continuation of the current regime and not a transition toward sustainable *mobility*. Although 'green cars' might bring about the changes and rebound effects needed for a transition to more sustainable *energy* system, they do not address the *socio-spatial* problems of the *transport* system. Whether green cars will contribute to a less automobile dependent landscape is questionable, and even though it remains an interesting research topic, we argue that transition research should dedicate at least as much attention to the socio-spatial dimension of mobility.

Second, our analysis questions the functionalistic subsystem focus that prevails in transition studies. Many ideas that may seem 'innovative' within the limits of the transport sector, are in fact in line with prevailing neo-liberal trends involving further acceleration, privatization and environmental policies based on market principles. The dominance of the car is not merely a regime characteristic within the boundaries of the transport sector. Rather it dominates and impacts our physical and discursive space across sector-boundaries, which gives rise to countless 'unsustainability issues' in other sectors as well. Many aspects (e.g., spatial and social) are ignored if we discuss transport without taking other sectors or 'functional subsystems' into account. Holding on to sector-boundaries hampers the questioning of paradigms that are *inherent* to these sector-boundaries. Perhaps the most important form of system innovation is questioning and altering the very boundaries of the system. Is 'mobility' or the 'transport sector' the right starting point to understand the drivers of radical change in the way people move through space and time? Can one really change or even understand dominant mobility paradigms without questioning the very concept of mobility? Would the field of transition studies, with its trans-disciplinary and inter-disciplinary ambitions regarding long-term radical change and sustainable development, not be better served with an *issue-based* and *problem-driven* delineation of study objects, rather than a functionalistic sector approach?

Third, we argue that transition research should dedicate more attention to 'radical socio-spatial niches', that is, niches that (a) do not only deviate from the current socio-technical automobility regime but also from the *broader automobile landscape*, by pro-actively challenging underlying dominant paradigms and (b) have an explicit socio-spatial dimension that moves beyond technology-oriented solutions and tackles automobile

A Socio-Spatial Perspective on the Car Regime   175

dependency at its 'socio-spatial roots'. Whether or not one believes that these niches will or might be 'successful' in the future, our point is that the relative success of radical socio-spatial innovation is an empirical question that deserves to be studied accordingly. This can be related to the work of Smith (2006, 2007), who analyzed radical niches in agriculture and housing construction, and the 'niche-regime translation' between radical innovations and mainstream practices. Smith argues that niches need to be flexible in terms of being *both* radical and reforming; while some niche elements are transferred to the mainstream, more radical components of the niches are 'kept alive' by committed actors that 'remain advocates for more radical system innovation' and continue 'radical experimentation'. In the same way that Smith analyzed radical niches in agriculture and housing construction, we need to further research counter-movements and radical niches in the field of mobility and question to what extent they provide concepts that have been, are being or can be 'mainstreamed'.

Last but not least, our chapter has demonstrated how spatial configurations play an eminent role in the reproduction of automobile dependency. We also discussed how the position of spatial planning and urban design is losing ground, as the neo-liberal notion that every form of planning is a reduction of freedom and an unnecessary form of government interference left the field of spatial planning as a 'institutional zombie' (Baeten, 2000; Gleeson, 2000). Many authors are skeptical about the power of spatial planning and urban design to alter mobility patterns (Baeten, 2000; Vilhelmson, 2007). We argue that these arguments are often beside the point. It is not about *changing* mobility patterns, it is about creating the conditions under which mobility behavior *can* change; it is about creating possibilities for now and the near future (Bach, 2006; Banister, 2008; Dupuy, 1999). In this perspective is justified to put "physical planning back into the center of the debate" (Dupuy, 1999, p. 13) and re-enforce planning practices. As a policy recommendation, we would suggest a stronger position for sustainable (urban) planning in governmental policies and argue that such planning and design can make an essential difference for the transition toward sustainable mobility.

DISCLAIMER

This chapter has been written on a personal title not on behalf of any of the employers of the authors.

NOTES

1. A few co-authors on this book advised us to use this structure for our chapter at the second book workshop in March 2010 in Bristol. We thank them for their feedback.

2. There are more system builders than only those who work on the physical side, like builders working on financial, organizational or social aspects.
3. See www.carbusters.org.
4. The selection of niches are loosely derived from earlier work with a target group of Dutch urban designers and planners (Zijlstra, 2009) and are therefore, for a large part, based on Dutch examples and references.
5. See www.homezones.org.
6. See www.langzamestad.nl.
7. See http://www.smartgrowth.org/about/default.asp.
8. See for instance Shuman (1998), Hines (2000), Knox (2005) and Hopkins (2008).
9. See http://www.transitionnetwork.org/initiatives.

REFERENCES

Bach, B. (2006) *Urban Design and Traffic*, Ede, the Netherlands: CROW.
Baeten, G. (2000) 'The tragedy of the highway: Empowerment, disempowerment and the politics of sustainability discourses and practices', *European Planning Studies*, 8(1): 69–86.
Bakker, P. & Zwaneveld, P. (2009) *Het belang van openbaar vervoer, de maatschappelijke effecten op een rij (The importance of public transport, the social effect at a glance)*, The Hague, the Netherlands: KiM & CPB.
Banister, D. (1994) 'Equity and acceptability questions in internalising the social cost of transport', (published in) European Conference of Miniters of Transport, *Internalizing the Social Cost of Transport*, Paris, OECD and ECMT, pp. 153–175.
Banister, D. (2008) 'The sustainable mobility paradigm', *Transport Policy*, 15(2): 73–80.
Böhm, S., Jones, C., Land, C. & Paterson, M. (2006) 'Introduction: Impossibilities of automobility', *Sociological Review*, 54(Supp. 1): 3–16.
Boomen, T. van den (2010) 'Langzaam is sneller, de opwaardering van de snelweg', *De Groene Amsterdammer*, 134(17): 38–39.
Brand, A. (2002) 'Het stedelijke veld in opkomst, de transformatie van de stad in Nederland gedurende de tweede helft van de twintigste eeuw' (The rise of the Urban Field, the transformation of dutch cities in the second half of the 20th century. (a Phd Thesis), doctoral thesis, University of Amsterdam.
Brown, J. (2006) 'From traffic regulation to limited ways: The effort to build a science of transportation planning', *Journal of Planning History*, 5(1): 3–34.
Castells, M. (1996) *The Rise of the Network Society*, Oxford, England: Blackwell.
Conybaere, J.A. (1980) 'Evaluation of automobile safety regulations: the case of compulsory Seat Belt Legislation in Australia', *Policy Sciences*, 12(1): 27–39.
Coutard, O., Dupuy, G. & Fol, S. (2004) 'Mobility of the poor in two European metropolises: Car dependence versus locality dependence', *Built Environment*, 30(2): 138–145.
CROW (Centrum voor Regelgeving en Onderzoek in de Grond-, Water- en Wegenbouw en de Verkeerstechniek—Center for Regulation and Research in Soil-, Water-, Road- and Traffic Engineering) (2004) *ASVV*, Ede, the Netherlands: CROW
Currie, G. & Senbergs, Z. (2007) 'Exploring forced car ownership in metropolitan Melborne', conference paper posted at Social Research in Transport Clearinghouse. Available at http://www.sortclearinghouse.info/research/90/ (accessed 10/11/2009).
Dennis, K. & Urry, J. (2009) *After the Car*, Cambridge, England: Polity

Dupuy, G. (1999) 'From the 'magic circle' to 'automobile dependence': Measurements and political implications', *Transport Policy*, 6(1): 1–17.
Dupuy, G. (2008a) 'The automobile system: A territorial adapter', in J. van Schaick & I.T. Klassen (eds) *Urban Networks, Network Urbanism*. Amsterdam: Techne Press.
Dupuy, G. (2008b) 'Urban planning in the network age: Theoretical Pointers', in J. van Schaick & I.T. Klassen (eds) *Urban Networks, Network Urbanism*. Amsterdam: Techne Press.
Freund, P. & Martin, G. (2000) 'Driving South: The globalization of auto consumption and its social organization of space', *Capitalism Nature Socialism*, 11(4): 51–71.
Geels, F. (2004) 'From sectoral systems of innovation to socio-technical systems: Insights about dynamics and change from sociology and institutional theory', *Research Policy*, 33: 897–920.
Gehl, J. (1971) *Life Between Buildings*, Copenhagen, Denmark: The Danish Architectural Press.
Geurts, K. & Wee, B. van (2006) 'Ex-post evaluation of thirty years of compact urban development in the Netherlands', *Urban Studies*, 43(1): 139–160.
Gleeson, B. (2000) 'Reflexive modernization: The re-enlightment of planning?' *International Planning Studies*, 5(1): 117–135.
Gleick, J. (1999) *Faster, the Acceleration of Just About Everything*, New York: Pantheon.
Gorz, A. (1973, September/October) 'The social ideology of the motor car', *Le Sauvage*.
Haas, T. (2008) 'New urbanism and beyond', in T. Haas (ed) *New Urbanism and Beyond: Designing Cities for the Future*. New York: Rizzoli.
Hajer, M. & Zonneveld, W. (2000) 'Spatial planning in the network society—Rethinking the principles of planning in the Netherlands', *European Planning Studies*, 8(3): 337–355.
Hamilton-Baillie, B. (2008) 'Shared space: Reconciling people, places and traffic', *Built Environment*, 34(2): 161–181.
Harms, L. (2008) *Overwegend onderweg, de leefsituatie en de mobiliteit van Nederlanders (Mainly on the move, socio-economic conditions and mobility of the Dutch)*, The Hague, the Netherlands: SCP (Sociaal en Cultureel Planbureau—The Netherlands institute for social research).
Harvey, D. (1990) *The Condition of Postmodernity*, Oxford, England: Blackwell.
Henderson, J. (2009a) 'The politics of mobility: De-essentializing automobility and contesting urban space', in J. Conley & A. Tigar McLaren (eds) *Car Troubles: Critical Studies of Automobility and Auto-Mobility*, Farnham, England: Ashgate.
Henderson, J. (2009b) 'The spaces of parking: Mapping the politics of mobility in San Francisco', *Antipode*, 41(1): 70–91.
Illich, I. (1974) *Energy and Equity*, London: Calder & Boyars.
Jacobs, J. (1969/1970) *The Economy of Cities* (2nd ed.), New York: Random House.
Jacobs, J. (11961/993) *The Life and Death of Great American Cities* (3rd ed.), New York: The Modern Library.
Kunstler, J.H. (1994) *Geography of Nowhere, the Rise and Decline of America's Man-Made Landscape*, New York: Touchstone.
Litman, T. (2007) *Mobility as a Positional Good*, Victoria, British Columbia, Canada: Victoria Transport Policy Institute. Available at http://www.vtpi.org (accessed 9/11/2009).
Litman, T. & Laube, F. (2002) *Automobile Dependency and Economic Development*, Victoria, British Columbia, Canada: Victoria Transport Policy Institute. Available at http:// www.vtpi.org (accessed 7/10/2009).

Lyons, G. (2004) 'The introduction of social exclusion into the field of travel behavior', *Transport Policy*, 10(2003): 339–342.
Mackett, R. & Robertson, A. (2000) *Potential for Mode Transfer of Sort Trips: Review of Existing Data and Literature*, London: Department of Environment, Transport and the Regions.
Marx, K. (1971) *Het Kapitaal, verkorte uitgave (Capital, short version)* (O. Rühle, Trans.), Amsterdam: Van Gennip. (Original work published 1867)
Metz, D. (2008) *The Limits to Travel, How Far Will You Go?* London: Earthscan.
Mitchell, D. (2005) 'The SUV model of citizenship: Floating bubbles, buffer zones, and the rise of the 'purely atomic' individual', *Political Geography*, 24(1): 77–100.
Mom, G. & Filarski, R. (2008) *Van transport naar mobiliteit: De mobiliteitsexplosie 1895–2005 (The explosuin of mobility 1895–2005)*, Zutphen, the Netherlands: Walburg Press.
Newman, P. & Kenworthy, J. (1992) *Cities and Automobile Dependence: A Sourcebook*, Aldershot, England: Avebury.
Newman, P. & Kenworthy, J. (1999) *Sustainability and Cities: Overcoming Automobile Dependence*, Washington, DC: Island Press.
Paterson, M. (2007) *Automobile Politics: Ecology and Cultural Political Economy*, Cambridge, England: Cambridge University Press.
Peters, P.F. (2003) *De haast van Albertine, reizen in de technologische cultuur: naar een theorie van passages (Albertine's hurry, travelling in the technological culture: towards a theory of passages)*, Amsterdam: De Balie.
Putnam, R.D. (2000) *Bowling Alone: The Collapse and Revival of American Community*, New York: Simon & Schuster.
Ritzer, G. (2000) *The McDonaldization of Society, New Century Edition*, London: Pine Forge.
Rogers, R. (1997) *Cities for a Small Planet*, London: Faber & Faber.
Schumacher, E.F. (1973) *Hou het klein, een economische studie waarbij de mens weer mee telt*, Bilthoven, the Netherlands: AMBO.
Seifert, T. & Werner, K. (2005) *Scharzbuch Öl (Blackbook oil)*, Vienna: Deuticke im Paul Zsolnay Verlag.
Shared Space Institute (2005) *Shared Space: Room for Everyone*, Leeuwarden, the Netherlands: Provincie Friesland. Available at http://www.sharedspace.eu (accessed 23/5/2008).
Sheller, M. & Urry, J. (2000) 'The city and the car', *International Journal of Urban and Regional Research*, 24(4): 737–757.
Slinn, M., Guest, P. & Matthews, M. (2005) *Traffic Engineering: Design, Principles and Practice* (2nd ed), Oxford, England: Elsevier.
Smith, A. (2006) 'Green niches in sustainable development: The case of organic food in United Kingdom', *Environment and Planning C: Government and Policy*, 24(3): 439–458.
Smith, A. (2007) 'Translating sustainabilities between green niches and socio-technical regimes', *Technology Analysis & Strategic Management*, 19(4): 427–450.
Steg, L (2005) 'Car use: Lust and must. Instrumental, symbolic and affective motives for car use', *Transportation Research Part A*, 39(2/3): 147–162.
Stoep, J. van der & Kee, B. (1997) 'Hypermobility as a challenge for systems thinking and governmental policy', *Systems Research and Behavioral Science*, 14(6): 399–408.
Taylor, B. & Ong, P. (1995) 'Spatial mismatch or automobile mismatch? An examination of race, residence and commuting in US metropolitan areas', *Urban Studies*, 32(9):1453–1473.

Urry, J. (2005) 'The 'system' of automobility', in M. Featherstone et al. (ed.) *Automobilities*, London: Sage.
Urry, J. (2007) *Mobilities*, Cambridge, Enlgand: Polity Press.
Ververs, R. & Ziegelaar, A. (2006) *Verklaringsmodel voor fietsgebruik in gemeenten (Explanatory model for bicycle use in municipalities)*, Leiden, the Netherlands: Research voor Beleid.
Vilhelmson, B. (2007) 'The use of the car-mobility dependencies of urban everyday life', in T. Gärling & L. Steg (ed) *Threats From Car Traffic to the Quality of Urban Life*, Amsterdam: Elsevier.
Wajcman, J. (1991) *Feminism Confronts Technology*, Cambridge, England: Polity Press.
Watts, D.C.H., Ilbery, B. & Maye, D. (2005) 'Making reconnections in argo-food geography: Alternative systems of food provision', *Progress in Human Geography*, 29(1): 22–40.
Weinstein, A. (2006) 'Congestion as a cultural construct: the 'congestion evil' in Boston in the 1890s and 1920s', *The Journal of Transport History*, 27(2): 97–115.
Wilkinson, R. & Pickett, K. (2009) *The Spirit Level, Why More Equal Societies Almost Always Do Better*, London: Allen Lane.
Zijlstra, T. (2009) 'Autoafhankelijkheid: over 'auto'-centrisch denken bij ontwerpers en planners' ('Car dependency: on car-centric thinking among urban designers and planners'), master's thesis, Eindhoven: Technical University of Eindhoven, the Netherlands.

9 The Emergence of New Cultures of Mobility
Stability, Openings and Prospects
Mimi Sheller

In recent years there have been significant shifts in the planning, design and funding of major urban infrastructure projects to include 'sustainable mobility' systems. Examples of the shift toward sustainable mobility include improved bike lanes and multi-user vehicle lanes; congestion charging and dynamic road pricing; new investments in energy-efficient public transport systems, light-rail systems and high-speed railway; the emergence of car-sharing and public bike-sharing schemes; and the design of pedestrian-friendly streets and smaller electric vehicles. Yet arguably none of these niche-level changes has seriously challenged the existing system of automobility, which continues to be the dominant mode of transportation, especially in the United States, while rapidly expanding across high-growth developing economies like China and India.

The United States especially trails behind other advanced economies in bringing about a transition in its transportation and mobility systems. Despite rises in the price of gas, worries about peak oil and fears of global warming, climate change and environmental catastrophe, Americans continue to produce and purchase internal-combustion private automobiles (with a slight increase in hybrid engine technologies), to under-invest in public transit and alternative energy systems, and to only marginally inhabit built environments that are bicycle and pedestrian "friendly" (a term that also implies car-friendly). The American transport system remains overwhelmingly dependent on oil whether in the form of gasoline, diesel or aviation fuel, and political leaders continue to endorse more offshore oil drilling even in the face of the BP disaster in the Gulf of Mexico. Moreover, the rate of growth in vehicle ownership, miles traveled per day and time spent traveling continues to increase, aside from a brief dip during the recent recession. Because the United States remains significantly behind other developed countries in making a sustainable mobility transition (and in particular lags behind the massive investments taking place in high-speed rail in China)—despite its technical, political and economic capacities—it may be very instructive to consider the factors contributing to the stability of the mobility regime in the United States, the potential "openings" that may be appearing and the prospects for a future transition away from the existing dominant system of automobility.

The Emergence of New Cultures of Mobility 181

Building on previous work on cultures of automobility, I will argue for a culturally based understanding of the problems of system lock-in and potential transition. A cultural analysis of automobility systems begins from the view that car cultures have social, material and affective dimensions that are overlooked in current strategies to influence car-driving decisions. Car use is never simply about rational economic choices but is as much about aesthetic, emotional and sensory responses to driving, as well as patterns of built environment, political process, sociability, habitation, family and work (Sheller, 2004a). The dominant culture of automobility is implicated in a historically sedimented context of material and embodied relations between people, technologies and everyday ("quotidian") spaces of mobility and dwelling. Any transition process must activate these deep cultural structures if it is to take hold.

This chapter aims to assess the openings and prospects for the emergence of new cultures of mobility in the United States, while also being realistic about the stabilities in the current mobility system. It begins with a model of culture as a combination of practices, networks and discourses, each of which is enacted across multiple levels in the transition process. The next section turns to examples of cultural mechanisms that have stabilized the dominance of automobility in the United States, as well as instances of openings in the regime at both the national and urban scale. Finally, it concludes with some speculative ideas about the emergence of a more far-reaching technological—and cultural—transition arising out of the dynamics of new information and communication technologies (ICT) entering the realm of transportation planning, design and innovation.

9.1. A CULTURAL PERSPECTIVE ON MULTI-LEVEL TRANSITIONS

A relational approach to culture as a complex amalgam of practices, networks and discourses operating at multiple levels can help to explain the limitations of existing transition efforts and the overall stability of the dominant system of automobility. Figure 9.1 depicts the multi-level perspective on interactions between the niche, regime and landscape levels, but in this case gives each one a cultural inflection rather than simply a structural mapping. I aim to demonstrate that systemic transformations such as the transition toward sustainable transportation systems will unfold not only through social, technological and political interactions amongst niches, regimes and larger landscapes but also through dynamic cultural processes at each level, including the following: cultural experimentation, innovation and improvisation at the niche level; cultural alignment, diffusion, contestation and brokerage at the regime level; and large-scale transitions in the material cultures and discursive "master frames" that shape cultures of mobility, legitimacy, spatiality and connectivity at the landscape level.

182 *Mimi Sheller*

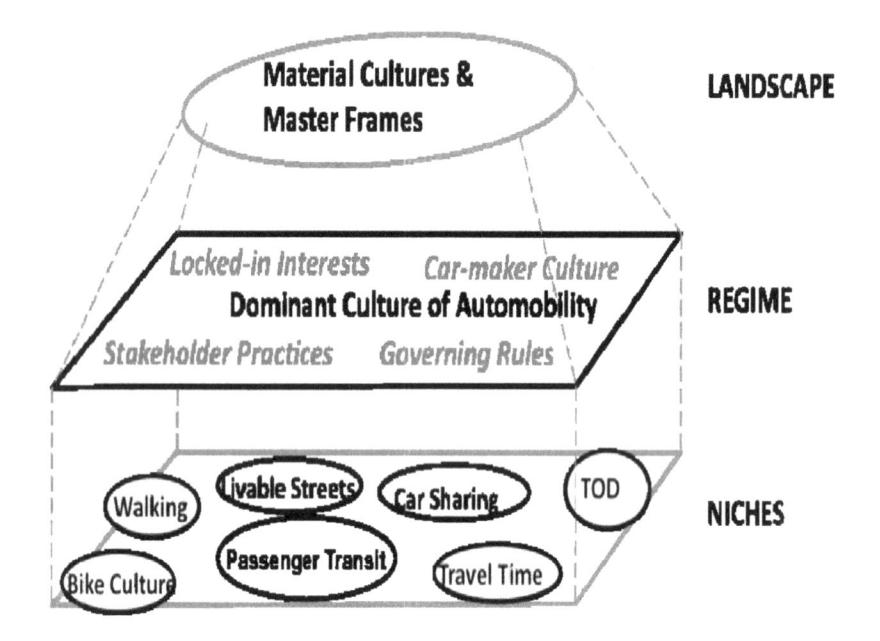

Figure 9.1 Dominant cultural structure for mobility.

Culture is not simply a "resource" or a "tool" used by actors to achieve particular ends or to frame debates but is a crucial *performative* part of the transition process as a whole, which is also *materialized* in environments, technologies and patterns of interaction. I am utilizing a sociological concept of "culture" that draws on relational sociology (Emirbayer, 1997), cultural approaches within network analysis (Mische, 2011) and social movement theory (Benford & Snow, 2000; Tilly, 2003) and cultural dynamics within technological "innovation journeys" (Geels & Verhees, 2011). This approach to culture emphasizes dynamic interaction between cultural discourses, cultural practices and networks of cultural relations at several different levels. It acknowledges complexity and emergent processes, yet maintains a place for human agency and cognition. However, at the same time I also draw on elements of actor-network theory (Law, 2002) and science and technology studies (Suchman, 2006), which acknowledge the role of non-human actants (such as technologies or infrastructures) in enrolling people into action. In contrast to theories of a systems "tipping point" grounded in complexity theory (Dennis & Urry, 2009), this cultural approach emphasizes the agency of particular actors, political discourses and interactional networks in performing—and potentially changing—cultures of mobility at each level in multiple ways (practices, networks, discourses).

The Emergence of New Cultures of Mobility

Table 9.1 presents a tri-partite model of cultures of mobility that encompasses interlocking practices, networks and discourses operating at each of the three levels. There are multivalent cultural interactions occurring within and between each level. In the first column we see how culture operates at the niche level. Niche-level cultural experimentation, innovation and improvisation have in many ways been at the heart of thinking about transitions in mobility systems, since it is here that we already see alternative ways of being mobile in practice. Mobilities research has a strong strand of empirical study and theoretical analysis of cultures of "mobility in daily life" (Freudendal-Pedersen, 2009). In addition to the study of "car cultures" and everyday "cultures of automobility" (Miller, 2001; Sheller, 2004a), a number of cultural and qualitative studies have also investigated how "sustainable mobility" behaviors become embedded in "normalized" routines of daily life (Shove, 2003), how such everyday practices are linked to the emergence of "cultures of alternative mobility" (Vannini, 2009) and how niche mobilities such as bicycling cultures emerge and are sustained by "green lifestyles" (Horton, 2006) and community organization (Furness, 2010). We are also gaining a richer view of the various cultures of

Table 9.1 Cultural Dimensions of the Mobility System

	Niche Level	Regime Level	Landscape Level
Practices	Embodied dispositions or habits and subaltern practices within niche sub-cultures of alternative mobilities (e.g., biking, boating, walking, motorcycling, etc.)	Mainstream legitimized practices (e.g., using cars for shopping, school run, socializing, commute to work, etc.)	Locked-in material cultures stabilized as "background" quotidian practice and infrastructural interactions
Networks	Social movement networks, community organizations and alternative "green" lifestyles that form a counter-public to the dominant culture of automobility	Durable interest groups and governing structures; car-maker and other stakeholder cultures of interaction or brokerage	Family, friendship and work networks of connectivity, social capital and affective economies that shape mobility
Discourses	Counter-discourses that challenge dominant order: sustainability, health, anti-consumerism, ecology, livable streets and complete streets	Standard Discourses used to legitimate existing actors and practices and shape "common sense," rationality and values	Structured Stories: 'Joy of driving'; 'Love affair with the car'; 'Can't do without it'; freedom and individuality (Freudendal-Pedersen, 2009)

public transport use, their forms of sociality (Bissell, 2009; Jain, 2009) and varieties of "traveling times" (Urry, 2006).

Yet supporting these niche developments there is also a growing array of more formally institutionalized political actors, alliances and networks, which in specific locations have promoted alternatives to the dominant culture of automobility. This includes social movements, alternative "free spaces" of non-automobile mobility and the emergence of new political projects such as Transit Oriented Development or Complete Streets/Livable Streets, which lobby for changes in zoning laws, city planning, transport funding and so on, as well as new coalitions of public and private, governmental and non-governmental actors and practices to bring about change. Here one could turn to cultural analyses within the political process and social movements literature to find examples of cultural dynamics of social change such as frame alignment and cultural diffusion in interaction with network structures and network dynamics (Ansell, 1997; Broadbent, 2003; Eliasoph & Lichterman, 2003; Mische, 2010). Within such counter-publics new discourses (and actors) emerge and gather strength, and in some cases (as described next) begin to infiltrate regime practices, networks and discourses.

In the second column I include forms of cultural process at the regime level—defined as "The whole of implicit and explicit rules and associated ways of thinking that guide practical behavior of professional people and which is being reconfirmed by everyday practice." Culture at this "regime" level can be thought of in terms of business as usual, including the legitimatization, normalization and stabilization of "the dominant culture of automobility" (Urry, 2000, 2004), which may be incommensurable with alternative cultural practices and discourses existing at the niche level. As I have argued elsewhere (Sheller, 2004a) this dominant culture of automobility is implicated in a deep context of affective and embodied relations between people, machines and spaces of mobility and dwelling, in which cultures, emotions, bodies and the senses play a key part. Mobility regimes are embedded in historically sedimented and geographically etched patterns of "quotidian [everyday] mobility" (Jain, 2002; Kaufman, 2000; Vannini, 2009), as well as in entrenched repertoires for political interaction and transport planning. Thus it is not only individual level cultural attitudes and mobility choices that matter but also the wider forces of structured interaction that guide practical behavior. There are durable interest groups and stakeholders who dominate the transport-planning process and funding mechanisms (including car makers and related industries such as oil, city and state legislators, the Federal Ways and Means Committee in the US Congress, transport planning agencies, technical experts). But there are also less explicit behind-the-scenes cultural mechanisms of frame alignment and diffusion by which dominant regimes shape practices and discourses that legitimize certain forms of valuation (e.g., economic rationality, utility, growth) or are challenged by counter-frames (e.g., sustainability, ecology, health).

And in the third column, I include cultural processes at the landscape level including the lock-in of material cultures and the complex dynamics of socio-technical evolution and adaptation and the "common sense" understandings that underlie forms of rationality and valuation, including values such as freedom, security, choice and value itself. As Freudendal-Pedersen (2009, 114–15) emphasizes, throughout daily practices of mobility there runs "structured stories" that link mobility to political ideologies such as "freedom" ("more mobility = more freedom"), a kind of master frame (Benford & Snow, 2000) that is often connected to the affordances of automobility (Freudendal-Pedersen, 2009, pp. 114–115; Sheller, 2008). Geels and Verhees (2011) also refer to higher order cultural "discourses, which are shared, sedimented, more general ways of thinking and talking about general topics and technologies," drawing on Benford and Snow (2000):

> Some collective action frames are quite broad in terms of scope, function as a kind of master algorithm that colors and constrains the orientations and activities of other movements. We have referred to such generic frames as "master frames" in contrast to more common movement-specific collective action frames that may be derivative from master frames. (pp. 618–619)

Such master frames shape culture at the landscape level and provide structured stories that explain the existing material culture. The social and the technical are intertwined and happen in a socio-technical landscape of values, settlements, roads and space, but we could also say happen in a *cultural landscape*, which is always contested and performatively enacted through framing struggles and public interactions. Thus technologies are not only "shaped by social, organizational and economic factors," but also by cultural factors. And in turn "technologies shape the social and physical world, through ownership patterns, patterns of use and infrastructures specially created for them," while also shaping the cultural landscape through master frames, forms of legitimacy, and repertoires for public interaction.

"In these struggles actors develop frames and counter-frames in response to each other and to deep-structural repertoires (which tend to change only slowly)" (Geels & Verhees, 2011, pp. 8–9). At the landscape level these slow changing structural repertoires have social, material and affective dimensions that are overlooked in current strategies to influence car-driving decisions—as well as in current understandings of socio-technical transitions. The cultural dimensions of automobility are not simply discursive but also implicate a broader physical/material setting that includes cultural styles, affects and emotions, or what some refer to as "non-representational" geographies (Thrift, 2008). Such "automotive emotions" (Sheller, 2004a)— the embodied dispositions of car users and the visceral and other feelings associated with cultural landscapes of car use—are as central to understanding the stubborn persistence of car-based cultures as are more technical, political and socio-economic factors. Emotional geographies (Bondi, Smith & Davidson, 2005) are not simply located within individual bodies but extend to familial spaces, neighborhoods, regions, national cultures and landscapes, which together support particular kinesthetic dispositions (Edensor, 2002; Rodaway, 1994). Thus a cultural landscape can come to stabilize a regime in part through taken-for-granted practices, and cultural processes at a far wider landscape level may transform mobility regimes.

Therefore it is imperative to understand the contribution of "mobility cultures" to reproducing mobility systems and to appreciate the capacity of cultural performativity to contribute to (and even lead to) sociotechnical change—or to block it. Most importantly, I will argue that the key cultural transitions at each level may come from *outside* the transport arena, instigated by other changes such as cultural shifts in forms of communication, legitimacy, connectivity, sensory perception, embodiment and spatiality. Car use (or the use of alternative means of transport) is never simply about rational economic choices made by individual "consumers," but is as much about irrational aesthetic, emotional and sensory responses to driving (or passengering or walking) at the niche level; locked-in, or at least relatively durable, dominant cultures of automobility, decision-making networks, political discourses and practices at the regime level; and the normalization of wider sets of cultural practices, networks and discourses which act as master frames and shape material cultures at the landscape level.

In order to flesh out this theoretical approach, in the next section I analyze some of the emergent cultural "openings" afforded by mobility niches (alternative practices, networks and discourses) in the United States, alongside the cultural forces that are constraining a more complete transition and stabilizing the current dominant regime. The stability of automobility in the United States offers a good comparator to the Dutch and British cases, demonstrating how slow the process of cultural transformation at the regime level can be and even then how it may ultimately have only modest impact on the landscape-level material culture and master frames that support automobility. The chapter then turns to two potentially more transformative transition processes associated with the development of "intelligent" infrastructure, on the one hand, and "mobility-on-demand" systems, on the other. These are compared in terms of the wide scope of changes they require in practices, networks and discourses, as well as the different actors and technical "actants" they mobilize into the transport planning, design and innovation arenas. And the concluding section engages with a more speculative investigation of the complex cultural forces that might actually elicit a larger shift in mobility regimes and cultures of mobility, including at the slow-moving landscape level.

9.2. OPENINGS VERSUS STABILITY IN THE US CULTURE OF AUTOMOBILITY

The United States has been much slower to implement regime-level changes in transport practices as compared to other industrialized countries. It also has a pattern of taking one step forward and two steps back, such as the controversial failure of General Motor's EV1 electric car in the 1990s and the irrational expansion of markets for gas-guzzling sport utility vehicles, light trucks and Humvees.[1] Another classic example of backward movement was the repeal of the federal 55 mph speed limit in 1995 (first instituted during the oil crisis in the mid-1970s), which was found to be responsible for 12,545 additional deaths and 36,583 additional injuries in fatal crashes between 1995 and 2005, as well as making a significant impact on reducing fuel economy and contributing to greenhouse gases.[2] More recently we have seen proposals for the opening up of the Arctic Wildlife Refuge and formerly protected offshore areas for oil drilling, framed by Governor of Alaska and Republican Vice-Presidential candidate Sarah Palin's political call of "Drill, Baby, Drill" in the 2008 elections. To many environmentalists, a transition toward sustainable transport in the United States might seem to be a lost cause. Nevertheless there are some recent significant developments that merit our attention.

9.2.1. Openings in the National Transport Regime

A recent poll conducted by Transportation for America, a coalition of environmental, smart growth, alternative transportation and public health groups found that two thirds of voters wanted more options besides driving, and that 58% favored spending more on public transportation services.[3] In fact, 51% were willing to have their taxes increased to improve mass transit. More than four out of five voters (82%) say that "the United States would benefit from an expanded and improved transportation system, such as rail and buses," and a solid majority (56%) "strongly agree" with that statement. When asked about reducing traffic congestion, three out of five voters chose improving public transportation and making it easier to walk and bike over building more roads and expanding existing roads (59% to 38%). Two thirds (66%) said that they "would like more transportation options so they have the freedom to choose how to get where they need to go." Along these same lines, 73% currently feel they "have no choice but to drive as much as" they do, and 57% would like to spend less time in the car. In these examples we see a rejection of two major dominant discourses: that road expansion is a solution to congestion and that automobility is a form of freedom. We also see a new willingness to consider public transportation, walking and biking within a discourse of "freedom to choose," which arguably has emerged out of the "smart growth" niche which promotes walking, bicycling and transit-oriented development.

At the federal level we see the recent use of the American Recovery and Reinvestment Act of 2009 to fund the program known as TIGER, or Transportation Investment Generating Economic Recovery. This involves $37.5 billion dollars of investment in the first round of which $26 billion is going to highways, $8.6 billion to public transit projects and about $1.3 billion to rail (including investment in new high-speed lines), with small but unprecedented (and therefore symbolically significant) amounts for bike trails and other alternative infrastructure investment. Still, though, we can see the disappointing continuation of highway funding even here and the paltry $8.6 billion spent on high-speed rail, which is barely enough to get started on a few projects and certainly not enough for any kind of integrated system. Related to this are current proposals to create an "Infrastructure Bank" to fund major investments in multi-modal and multi-state new infrastructure projects (Schulz, 2010). While these moves toward alternative transport funding structures and investment strategies are limited, they have generated not only a good deal of enthusiasm among niche networks but also vocal support from the US Secretary of Transportation, Ray Lahood.

In the current political climate alternative infrastructures for non-car mobility at least are up for discussion and possible funding. These glimmers of change in the framework for public infrastructure investment have come about due to extensive lobbying efforts, including a broad array of non-governmental organizations and coalitions involved in goal alignment, issue framing, political contestation and brokerage (cf. Furness, 2010). The social and political actors involved in promoting changes in the mobility regime in the United States include relatively new national policy and lobbying networks such as the following:

- The National Complete Streets Coalition, which calls for "Instituting a complete streets policy [which] ensures that transportation planners and engineers consistently design and operate the entire roadway with all users in mind—including bicyclists, public transportation vehicles and riders, and pedestrians of all ages and abilities" (www.completestreets.org).
- Transportation for America, "a broad coalition of housing, business, environmental, public health, transportation, equitable development, and other organizations. We're all seeking to align our national, state, and local transportation policies with an array of issues like economic opportunity, climate change, energy security, health, housing and community development" (www.T4america.org).
- Smart Growth America critiques automobile-centered transportation and sprawling land use and calls for "smart transportation" including transit-oriented development, walking as a priority and bicycle-friendly communities (www.smartgrowthamerica.org/transportation.html).

The Emergence of New Cultures of Mobility 189

- Advocacy groups like the Livable Streets Initiative (www.openplans. org) and Reconnecting America (www.reconnectingamerica.org).
- The Streetsblog, an online coalition that seeks to transform cities by improving conditions for cyclists, pedestrians and transit riders, through social media like StreetFilms.

The use of the Internet and social media have greatly enhanced the organizational capacity to build alliances that span the regional and national levels and thus enhance not only their lobbying capacity but also their ability to foster new frame alignments and shifts in the dominant discourse. Major national and international conferences are also taking place, like Next American City, or Walk21, which promotes walkability and shares examples of good practice. Thus in a number of cities across the United States there have been highly visible efforts to put into daily practice improved infrastructures for alternative mobilities, including new public bike-sharing schemes, new bike lanes, investment in light rail and car-sharing companies.

Are such networks finally contributing to bringing about a regime change in the taken-for-granted planning practices and the cultural discourses of automobility? While the programs advocated by these groups might appear to be niche-level actions, I would argue that as they interact with exogenous landscape factors such as climate change and the predicted scarcity of oil in the future, they have the potential to elicit new master frames. Thus, Robert Paaswell (2009), a US expert on transport planning, argues that:

> The current period—2010 and beyond—represents a radical shift in planning and managing our transportation infrastructure. [...] Energy has become expensive and will become scarcer. And the debate over policy approaches and responses to global warming has opened the question of whether any auto oriented urban form—even "green autos"—should still be the preferred form of lifestyle for most Americans. And this debate includes the role of the government in policy approaches that will be sufficient and appropriate to address this complex, but major 21st C problem. To cope with this shift, changes in the transportation planning process, and the regional planning process must be made (p. 3).

Paaswell implies that both policy debates (a niche level strategy to influence decision making) and global warming (a landscape level shift that impacts on government policy) have "opened" the regime to new questions, new processes and new actors. He suggests that new funding structures, new institutional structures and new uses of information and communication technology will all transform transport planning in coming years.

9.2.2. Openings in the Urban Transport Regime

Some of these regime changes are already taking hold in urban transport planning networks. The National Association of City Transportation Officials connects major city transportation leaders and promotes urban transportation investment that includes public transit, bicycling and pedestrian infrastructure.[4] New York City, for example, has implemented new pedestrian plazas on existing roadways, new bike and bus-only lanes and a far-reaching planning vision known as PlaNYC. There has been growing federal support for many of these urban initiatives, including from the Department of Transportation under the relatively progressive leadership of the US Secretary of Transportation Ray Lahood (http://fastlane.dot.gov/). Mainstream cable news networks like CNN now carry reports such as this one, suggesting a shift in the dominant discourse:

> Americans are rebuilding their cities and communities to make people, not cars, the center of a more environmentally friendly lifestyle, urban planners and transportation experts say. "We're creating infrastructure for human beings, rather than automobiles," says Michael Smith, CEO of Center City Partners in Charlotte, a group of business leaders that has helped lead a revival of the city's downtown. Creating a new infrastructure means new rules, experts say. What's on the way out: sprawling interstates, suburban living, long car commutes. What's now in: light rail, green space and vibrant downtown districts.[5]

Here it seems that certain regime actors may be looking at niche discourses with new eyes, as the cultural framing of "environmentally friendly lifestyles" comes into the mainstream and begins to impact on "new rules" for urban development which for the first time are recognizing not only light rail but even walking and bicycling as legitimate forms of transport.

To take just one example of niche-level discourses impacting on the regime level, the city of Philadelphia has made major progress under the administration of Mayor Michael Nutter, who implemented a plan called Greenworks Philadelphia that departs from business as usual. This includes plans to develop a "Pedestrian and Bicycle Master Plan" for the city, to explore implementation of a city-wide bike sharing program and to reduce vehicle miles per capita:

> Although Philadelphia already has one of the country's lowest rates of vehicle miles traveled per capita, Greenworks Philadelphia calls for reducing miles driven by 10 percent by 2015. Philadelphia has the type of transit system that many progressive cities are now trying to build and gives the city a competitive advantage over other cities. Greenworks Philadelphia applauds SEPTA's[6] efforts to increase transit ridership through further services and capital improvements and the adoption

The Emergence of New Cultures of Mobility 191

of new fare technologies. Transit Oriented Development investments supported by the City will also help induce more people to take transit. Finally, the appointment of Philadelphia's first Bicycle and Pedestrian Coordinator in 2008 will help grow the city's thriving bicycle culture as it seeks to create a city-wide trail network.[7]

This reflects a significant shift in regime-level practices, networks and discourses, which in the past focused almost exclusively on road and bridge maintenance and expansion. The existence of this new discourse of sustainable mobility, Transit Oriented Development and "bicycle culture" inside government networks is in and of itself a significant shift in the cultural discourse at the regime level.

The city of Philadelphia also has a very successful non-profit car-sharing organization, PhillyCarShare (www.phillycarshare.org), as well as for-profit Zipcar, while groups such as the Bicycle Coalition of Greater Philadelphia (www.bicyclecoalition.org) and Bike Share Philadelphia have long promoted bicycling, including advocacy for new bike lanes (which the city has recently been testing) and public bike sharing. While these appear as niche-level "alternatives," such goals are now also shared by the Delaware Valley Regional Planning Commission, a government agency which handles transportation planning in a three-state area including parts of Pennsylvania, New Jersey and Delaware. It was in part because of the synergy between these various networks of both governmental and non-governmental organizations that Philadelphia (in partnership with Camden, NJ) was one of the few regions to be awarded TIGER funds of $23 million for a bicycling infrastructure project to build 10 different segments of the Schuylkill River Trail, the East Coast Greenway and the Camden GreenWay. Thus the coming together of new governmental agencies and non-governmental advocacy groups is slowly pushing transport planning in new directions, linking together the urban and the federal scale and creating bridges between various actors and sites of action.

Yet it still remains questionable to what extent even these cultural shifts will impact on the overwhelmingly automobile-centered pattern of majority mobility. More far-reaching policies such as New York Mayor Michael Bloomberg's 2007 effort to institute a congestion-charging scheme in Manhattan are unachievable, in his case stymied by conflict with the State Legislature despite support for the plan among all relevant planning agencies. Philadelphia's public transit systems such as SEPTA suffer from perpetual budget shortfalls, and the development of high speed rail on AMTRAK's Northeast Corridor seems unlikely any time in the near future, despite being the most important connection between the major metropolitan conurbations of New York, Philadelphia and Washington, D.C.

In the case of car cultures, not just the built environment, but also the cultural landscape is deeply enlaced with practices of driving and dispositions toward automobility. Car cultures build on patterns of kinship, sociability, habitation and workforce participation, all of which are realized in particular built environments *and* institutional environments that are not easy to change. Philadelphia remains deeply divided into inner city neighborhoods and surrounding suburbs that are wealthier and whiter; these white suburban inhabitants remain wedded to their cars, even as some gentrifying districts of the inner city take to their bikes. These geographies are not just physically stable but also emotionally embedded in forms of familial life and racial coding, such as fear of the public transit system that goes through poor, black neighborhoods. Even within poor neighborhoods there are subaltern counter-publics for whom automobility represents a form of freedom (Gilroy, 2001). This goes hand in hand with a disenabling, disavowal or disparagement of other less desirable means of mobility (like the public bus system) which do not fit the cultural imaginaries of the dominant system—what Sarah L. Jain (2006) describes as the violent "taking" of space by cars, as celebrated in television and film car chases—or games like Grand Theft Auto 4. Transport choices are part of embodied sensibilities that are socially and culturally embedded in familial and sociable practices of car use, bike use or transit use and the circulations and displacements performed by vehicles, roads and drivers—including the creation of ghettoes with limited means of mobility versus privileged areas only accessible by car.

Indeed, as the introduction to this volume notes, many of the developments in "green" transportation have in fact tended to shore up the existing system of automobility by offering cleaner vehicles for the private consumer market. At the same time, the new niche developments such as bicycling infrastructure or improvements in public transit remain marginal to the overall built environment of most American cities, even if a higher percentage of people make use of them. Roads and highways dominate the built landscape, and the over-arching culture at the "landscape" level remains one in which automobility is normalized as freedom and associated with wealth and privilege. Simply inserting more public transit, a few bike lanes and some electric vehicles into existing patterns of automobility actually resists transformative change because it leaves unchallenged the underlying culture of autonomous mobility, the spatial and social relations that go along with automobility and the landscape of cultural discourses that equate personal mobility with freedom.

9.3. INTELLIGENT INFRASTRUCTURE AND MOBILITY-ON-DEMAND

This section turns to another set of emergent cultural transitions making use of new ICT via both top–down planning and bottom–up user innovation (see Chapter 13). The growing integration of ICTs into automobiles and other vehicles is leading to a lacing of technologies of transportation with capacities for conversation, entertainment, information access, navigation, automation, tracking and surveillance (Sheller, 2007; Sheller & Urry,

The Emergence of New Cultures of Mobility   193

2000). But the transformation goes far beyond vehicles to the planning process itself and to the wider cultural landscape of mobility. As Paaswell (2009) suggests, "The impacts of the combination of modern wireless communications and ubiquitous computing power are having profound impacts on the supply, demand for transportation and on the culture of those using transport" (p. 9). It is crucial that we pay attention to this changing "culture of those using transport," as well as of those planning the transport systems of the future.

Technologies do not necessarily produce expected effects (or produce unintended effects) and indeed new transport technologies are often very slow in their uptake. As the late Bill Mitchell, Director of the Smart Cities Group at MIT Media Lab and one of the key innovators in urban mobility systems, put it:

> The really big fundamental transformations are fundamentally unpredictable. And that's the very definition of a creative disruptive transition in fact; and that is where the big payoffs are [. . .] Incremental advances are not going to get us where we need to be . . . we need in parallel to have this imaginative search for the big transformative changes that are going to be real game changers. (see http://bigthink.com/billmitchell2)

Are those big transformative changes taking place? And if so, where should we look for them? Large complex transitions cannot take place in a single sub-system, such as the transportation system, without realizing the necessary parallel transitions in a whole range of other systems (including power generation, electricity storage, electronic information, smart infrastructures, real time information flows, urban design, etc.). Thus, projects such as the Smart Cities Group at MIT or IBM's "A Smarter Planet" envision not simply new concept cars but how new transport technologies such as mobility-on-demand systems will interface with other smart infrastructure, such as smart electric grids and pervasive computing.

As noted previously, regional and urban transportation planning agencies in the United States are beginning to adopt forward-looking plans that incorporate bicycling, pedestrians, light rail and expanded public transit. However, another set of constituents and stakeholders is also investigating the use of new "intelligent infrastructures" to improve existing road systems, in some cases with federal Department of Transportation funding. I will consider two different trajectories of change. First, the development of "intelligent transport systems" (ITS) is potentially reconfiguring forms of traffic management, transport planning and the overall cultural landscape of mobility. Second, the design of new "mobility-on-demand" systems has the potential not only to transform everyday cultures of mobility but also to bring ICT designers, systems and users into the core of how we think about and conceptualize transport infrastructure planning, management, pricing and use within a changing cultural landscape.

.3.1. IntelliDrive

The idea of an "electronic highway of the future" had already been envisioned by General Motors in the Futurama exhibition of the 1939 New York World's Fair, designed by Normal Bel Geddes to carry viewers into a world of futuristic skyscrapers, seven-lane highways and raised walkways, a vision later extended into the traveling Motorama exhibits of the 1950s and the futuristic Space Age design of the 1960s (Packer, 2008, pp. 277–279). That imagined future has now arrived in many ways. One significant development is the re-making of entire road, freight and transport systems to support embedded electronic infrastructures such as ITS and automated highways. The trend toward "cybercars" (vehicles enhanced with data processing, information transmission and mobile communications capacities; see Sheller, 2007) and digitally "enacted environments" is inseparable from the growing prevalence of "code-written spaces" (Thrift & French, 2002) and the "software sorting" of all mobility systems (Wood & Graham, 2006). This offers an exogenous landscape-level change in the socio-technical practices and actor networks that enter into transport planning.

The US Department of Transportation's ITS Strategic Plan for 2010 to 2014 includes an extensive research program known as "IntelliDrive," which has certain limitations but does suggest a far more high-tech approach to transition management and a somewhat different vision of the future of mobility and how we might get there (http://mtc.ca.gov/planning/intellidrive/). The Department of Transportation-funded national program incorporates transit systems with features such as driver-condition monitoring and real-time information delivered to passengers.[8] ITS allow for "real-time" information, with adjustments by transport managers or individual drivers made based on concurrent events, and hence the shift to "real-time, dynamic scheduling" (Paaswell, 2004, pp. 22–24). ITS solutions to congestion problems "now include advanced traffic control, electronic toll collection, incident management, incorporating computer video, and advanced vehicle systems, including route guidance and automatic vehicle separation" (ibid., p. 27). These socio-technical capabilities introduce a whole new set of practices and discourses into the transport-planning arena, shifting the cultural landscape.

Already at the discursive level, journalists and technology commentators have latched onto the idea, promoting it as a way to improve the driving experience:

> The promise of vehicles communicating with each other and with the road, coupled with advancements in transportation infrastructure, has planners, technocrats and futurists creating an Intelligent Transportation System. Unlike the future once envisioned in Disney's Magic Highway, which predicted a world of fog-eliminating machines and nuclear robots building highways, the ITS will work within a proven

technological infrastructure that already exists. This future of transportation will be based around smart phones, mobile navigation systems and other common gadgets and will drastically change how we navigate and interact with cities. The idea of ITS, at its most basic, is to connect every vehicle in a network of transportation users that instantly tracks and shares information. Ideally, everyone will be able to quickly determine where the accidents and tie-ups are and what routes can be taken to avoid them. What this means for the average commuter is quicker drive times by the way of more efficient traffic patterns and planned out routes created for you in real time (Kambitsis, 2010).[9]

Here we see discourses relating to speed, efficiency and instantaneous information harnessed to ideas of new forms of "intelligent" navigation. The U.S. Department of Transportation (DOT) has supported the "IntelliDrive" initiative, which "seeks to improve transportation safety and mobility while reducing the environmental impact of surface transportation through the use of networked wireless communication among vehicles, infrastructure, and travelers' personal communication devices." (http://www.fhwa.dot.gov/publications/research/safety/10073/001.cfm, accessed 24/10/11) But insofar as these are built into the existing transport system, this is a very limited vision of "intelligence." It maintains the idea of a car-based transport system (not surprisingly, since the automotive industry remains a major partner and stakeholder) and applies itself mainly to traffic management, crash avoidance and congestion mitigation. Such systems also carry further negative implications for mobility cultures.

Cultural theorist Jeremy Packer (2008) points out that Automated Highway Systems and Intelligent Vehicle Highway Systems are being designed as part of larger networked systems known as Integrated Communications and Navigation in the industry, which is closely linked to military research into automated vehicles. The capability to network "already existing technologies, including mobile telephony, GPS, light emitting diodes, satellite radio, black boxes, digital video recorders, computer processors" and fleet tracking systems, according to Packer, "has made and will continue to make the expansion of surveillance and ultimately control possible" (p. 281). Automated vehicles are of interest not only to new manufacturing companies who make the new communication and surveillance technologies, but also to new modes of state securitization and governance of mobile populations and new cultures of scientific expertise interested in pervasive computing. Given these changes in transport technologies, plus the growing concerns over distractions of driver attention caused by phone calls and text messaging, it seems likely that the emerging "intelligent" mobility system will incorporate automatic vehicle control, changing the cultural meanings of driving and what it means to drive. This has the potential, crucially, *to reconfigure the landscape level discourses* around freedom, choice and ultimately the meaning of mobility by subsuming these values under the banner of security, control and surveillance. In effect, the incorporation of new technologies into mobility systems empowers new networks to legitimate their projects in ways that reshape cultural discourses and forms of valuation of mobility systems.

9.3.2. Mobility-on-Demand

On the other hand, a potentially more creative-disruptive transformation is offered by what are called Mobility-on-Demand systems, which incorporate public shared vehicles, real-time transit information and dynamic pricing. The Smart Cities Group at the MIT Media Lab has been studying Sustainable Personal Mobility and Mobility-on-Demand Systems through a collaborative team project (awarded the 2009 Buckminster Fuller Challenge Prize). Their project relies on mutual adaptation amongst multiple newly designed vehicle types (light-weight electric cars, bikes, motorcycles), abundant access nodes for re-charging and parking the vehicles and continuous feedback loops from GPS tracking and real-time information flow between users and mobility managers (including dynamic pricing systems). In this way the entire system is able to adjust in an evolving process of dynamic equilibrium. The system is flexible, scalable and responsive, as well as incorporating new energy sources and cradle-to-cradle recycling of materials. It encourages user-innovation and is suggestive of the kind of comprehensive integrated "anticipatory" design that will be needed for major systems innovation and transition from the bottom up. But in what kind of cultural contexts will such anticipatory and comprehensive designs take hold?

Advanced human sensor systems already have been developed especially by the automotive industry, and cities are beginning to use Radio-frequency identification (RFID), vehicle-to-vehicle communication and parking sensors for real-time traffic information. The widespread deployment of ubiquitous computing, sensor networks and mobile media into the urban environment leads to emerging technologies that can sense and react to what is happening around them. The various systems throughout a modern city are beginning to maintain persistent memories of their own use, communicate with each other about their status and even reconfigure themselves based on dynamic needs. High-density broadband will make Open Data Cities increasingly possible. Within a dynamic urban infrastructure, city-scale services like power (smart grids), data (ubiquitous computing) and transportation (ITS) will soon begin to adapt in real time to the changing needs of the public, according to proponents. Such systems are anticipatory rather than reactive. Pervasive data-surveillance and forms of continuous real-time calculation—referred to by Nigel Thrift as "qualculation" (Thrift, 2008)—create an artificial world that is increasingly sentient and potentially adaptive. Again, this suggests a fundamental change in the everyday practice of mobility, in the networks involved in transport planning and in the discourses that are used to legitimate projects.

The Emergence of New Cultures of Mobility 197

Above all, we should consider how such systems might transform the culture of *relationality* between human beings, vehicles and environments. The phenomenology of car use highlights the driving body and the passenger as assemblages of social practices, embodied dispositions, emotional orientations and physical affordances which produce a "tacit automobilized embodiment" (Dant, 2004; Sheller, 2004a; Sheller & Urry, 2000; Thrift, 2004). Developments in digital control of the car and in mobile information technologies are transforming the very ways in which drivers sense and make sense of the world, such that cars are (according to Nigel Thrift) "one of the key moments in the re-design of modern urban environments" and in the production of "new bodily horizons and orientations" (Thrift, 2004, pp. 48–49). This relational approach to human agency suggests the possibility of more fundamental changes taking place in the practices and rules governing what we understand mobility to be and how we understand transportation to function. In other words, the taken-for-granted cultural background may be shifting at the landscape level, with profound implications for both niche actors and dominant regimes.

More broadly, one could argue that new socio-technical systems are already re-shaping corporeal existence, material environments and urban informatics in diverse and complex ways, with direct and indirect impacts on all people, spaces and temporalities. The emergent convergences between transport systems, mobile communications, locative media and "smart" infrastructures are generating new kinds of social coordination, real-time scheduling and hybrid private-in-public life (Sheller, 2004b), while also posing problems of governance, social exclusion, public fragmentation and automated control. It is these combined processes, and the new cultural dynamics that arise out of them, I argue, that will lead to a cultural transition in the system of automobility. Crucially, we need to explore the ways in which all vehicles are becoming *more than* technologies of movement—they are being hybridized with the rapidly converging technologies of the mobile telephone, the personal entertainment system and the computer, all linked together by GPS, geographic information systems (GIS), RFID and mobile broadband networks (Sheller, 2007; Sheller & Urry, 2000). And they are also increasingly going to be linked to ubiquitous computing, pervasive sensing and the "internet of things" in new forms of "hybrid urbanism" or "everyware" (Greenfield, 2006). These future scenarios offer forecasts with both utopian and dystopian elements, suggestive of the complex transition pathway ahead.

9.4. SPECULATIONS ON FUTURE CULTURAL TRANSITIONS IN MOBILITY

In the absence of complete ecological disaster (which of course remains a possibility and for some a reality already occurring due to the Deepwater Horizon blowout still pouring oil into the Gulf of Mexico as I write), what are the cultural changes that could conceivably push us toward a post-car form of mobility? And where can we detect these changes starting to occur? I suggest, in conclusion, that in spite of extensive niche-level changes in the quotidian practices of mobility of some actors, and some emergent shifts in regime-level discourses and networks even in the United States, truly transformative mobility cultures may not arise directly out of the search for sustainable transportation; they might instead come from tangential socio-technical and cultural realms that are leading broader landscape-level changes in the cultural practices, networks and discourses of mobility.

I have argued that for a real socio-technical transition to occur a fundamental cultural transformation must take place at all three levels. A full transformation in the culture of mobility will not simply be about bikes or buses or light rail being built alongside existing car and road systems (we have all those possibilities now), or even about pricing and tax incentives encouraging a shift away from private cars. No matter how "livable" streets become, they will remain subsumed to private cars so long as urban and suburban built environments are still designed to support cultures of automobility. More radical change will require new cultural articulations of the person with some as yet unimagined "cyborg complex" of vehicles, infrastructures and communicative mobile devices. It will require an entirely re-designed modern urban environment, as well as new bodily orientations. As human "hybridization" with the car incorporates intelligent software agents that can assist drivers and carry out certain vehicle operations automatically (Sheller, 2007), newly emergent (and somewhat unpredictable) cultural practices of "cybermobility" will play a crucial part in any future transitions—leading toward "cybercities" (Graham, 2004).

Cities of cybermobility are simultaneously mobile and intensely communicating. Connected mobility will be crucial to the forging of real-time, 24-hour, cellular participation, interactive media and new forms of individuation, all of which will contribute to the transition in passenger transport. The transition to sustainable passenger transport will require far more than niche developments within the transport sector. The entire culture that has been built up and materialized around automobility will have to be displaced and replaced with a different material culture as well as a transformed political culture surrounding transport planning processes, and it is the incorporation of new ICT that is contributing most to this cultural shift.

My discussion here is not meant to detract from the efforts of individuals, organizations or transition managers to promote or foster more sustainable passenger transport at the niche and regime level, but it is meant to question the targeting and feasibility of any such projects if they focus only on individual choice and local transport planning processes. At the niche level, it may not be advocates of sustainability who drive the transition in passenger transport but more pervasive market forces assembled around personal entertainment, ICT and niche developments within cultures of urban entertainment and infotainment. Accompanying these changes in urban culture there is also a counter-movement toward user innovation, participatory urbanism and cultures of "hacking," do-it-yourself fabrication and living "off the grid." All of these forces suggest a diversity of initiatives that are not necessarily grounded in the transport realm at all but may well have a cultural impact on it.

At the regime level, the forces leading toward transition might not come out of the policy networks and governing agencies that appear to be involved with passenger transport but might instead come out of the regimes of new communications technologies and intelligent infrastructures, where there is a market-led convergence of technologies taking place. As John Urry (2008) points out, "Just as the internet and the mobile phone came from 'nowhere', so the tipping point here [in systems of automobility] will emerge unpredictably, probably from a set of technologies or firms or governments not currently a centre of the mobility industry and culture" (p. 272). Yet I would suggest that this requires not a "tipping point" emerging out of complexity (Dennis & Urry, 2009) but a landscape-level cultural shift that re-arranges the key practices, networks and discourses that inform transport planning and counter-publics.

There may be other logics than sustainability guiding the technological transition toward automated vehicle control, which may ultimately displace the private automobile of today's mobility system. Research and development of automated transport infrastructures may be promoted by those elements within the state and the scientific community who are also committed to new modes of automated warfare, pervasive sensing and anticipatory surveillance. In short, those for whom security is more important than freedom. This injects into the conversation on transitions a rather more dark vision of the forces that might be driving the wide-scale transition in mobility systems. Certainly the weakening of the US automobile industry, the opening of new debates in the transport planning field and the ecological pressures on the oil industry already point toward a regime opening that may allow other corporate and governmental actors more space to maneuver. The regimes of military technology and state surveillance policy may have a strong impact on the direction transport systems move in, even as wider cultural forms of freedom and control are culturally reconfigured.

Finally, at the landscape level, ironically, it may not be peak oil or the collapse of the automotive industry that consolidates a transition but far more diffuse cultural forces that accompany the emergence of new kinds of urban space, new forms of communicative individuality and orientation toward others and new modes of governance and state control. My point, in conclusion, is that once we introduce culture into our models—which I strongly believe we need to do—we need to broaden the scope of investigation to a much wider range of arenas. Transition prediction, management or advocacy then becomes far more difficult to locate, because the necessary cultural transition is not centered on individual choices, local niche practices or even on transport policy regimes. Yet at the same time a broad cultural lens gives us a far more realistic assessment of what forces and actors might actually be in play. When a transition to sustainable mobility finally does take place it may bring the hoped for "green" transition away from the polluting technologies of the past, yet at the same time lock us unpredictably into a new path dependence on socio-technologies that are culturally problematic in other, perhaps even more troubling, ways.

NOTES

1. See "Who Killed the Electric Car?" (dir. Chris Paine, Sony Pictures, 2006).
2. While the difference between 55 mph and 65 mph may not seem so large, the relationship between speed and fuel economy is highly non-linear due to engine design and the physics of wind resistance. A car that gets 30 mpg at 55 mph gets about 27.5 mpg at 65 mph and 23.1 mpg at 75 mph (Friedman et al., 2009).
3. The telephone survey contacted 800 registered voters between February 27 and March 2, 2010 and has a margin of error of plus or minus 3.5%. See http://t4america.org/resources/2010survey/.
4. For some examples from across the country see http://www.nacto.org/videos.html.
5. See http://www.cnn.com/2010/LIVING/04/01/infrastructure.rebuild/?hpt=Sbin.
6. SEPTA is the South Eastern Pennsylvania Transit Authority; the national AMTRAK rail system also passes through Philadelphia.
7. See http://www.phila.gov/green/greenworks/economy_Vehicles.html.
8. See http://www.intellidriveusa.org/whoweare/program.php.
9. See WIRED Magazine http://www.wired.com/autopia/2010/02/intelligent-transportation-systems/.

REFERENCES

Ansell, C.K. (1997) 'Symbolic networks: The realignment of the French working class, 1887–1894', *American Journal of Sociology* 103(2): 359–390.
Benford, R.D. & Snow, D.A. (2000) 'Framing processes and social movements: An overview and assessment', *Annual Review of Sociology*, 26(1): 611–639.
Bissell, D. (2009) 'Moving with others: The sociality of the railway journey', in P. Vannini (ed) *The Cultures of Alternative Mobilities: The Routes Less Travelled*, Farnham, England: Ashgate.
Bondi, L., Smith, M. & Davidson, J. (eds) (2005) *Emotional Geographies*, Aldershot, England: Ashgate.
Broadbent, J. (2003) 'Movement in context: Thick networks and Japanese environmental networks', in M. Diani & D. McAdam (eds) *Social Movements and Networks: Relational Approaches to Collective Action*, Oxford, England: Oxford University Press.
Dant, T. (2004) 'The driver-car', *Theory, Culture and Society*, 21(4/5): 61–79.
Dennis, K. & Urry, J. (2009) *After the Car*, Cambridge, England: Polity.
Edensor, T. (2002) 'Material culture and national identity', in T. Edensor (ed.) *National Identities in Popular Culture*, Oxford, England: Berg.
Eliasoph, N. & Lichterman, P. (2003) 'Culture in interaction', *American Journal of Sociology*, 108(4): 735–794.
Emirbayer, M. (1997) 'Manifesto for a relational sociology', *American Journal of Sociology*, 103(2): 281–317.

Freudendal-Pedersen, M. (2009) *Mobility in Daily Life: Between Freedom and Unfreedom*, Farnham, England: Ashgate.
Friedman, L.S., Hedeker, D. & Richter, E. (2009) 'Long-term Effects of repealing the national maximum speed limit in the United States', *American Journal of Public Health*, 99(9): 1626–1631.
Furness, Z. (2010) *One Less Car: Bicycling and the Politics of Automobility*, Philadelphia: Temple University Press.
Geels, F.W. & Verhees, B. (2011) 'Cultural legitimacy and framing struggles in innovation journeys: A cultural-performative perspective and a case study of Dutch nuclear energy (1945–1986)', *Technological Forecasting & Social Change*, 708:(6) 910–930.
Gilroy, P. (2001) 'Driving while black', in D. Miller (ed) *Car Cultures*, Oxford, England: Berg.
Graham, S. (ed.) (2004) *The Cybercities Reader*, New York: Routledge.
Greenfield, A. (2006) *Everyware: The Dawning Age of Ubiquitous Computing*, Indianapolis, IN: New Riders Publishing.
Horton, D. (2006) 'Environmentalism and the bicycle', *Environmental Politics*, 15(1): 41–58.
Jain, S.L. (2002) 'Urban errands: The means of mobility', *Journal of Consumer Culture*, 2(3): 385–404.
Jain, S.L. (2006) 'Urban violence: Luxury in made space', in M. Sheller & J. Urry (eds) *Mobile Technologies of the City*, London: Routledge.
Jain, J. (2009) 'The making of mundane bus journeys', in P. Vannini (ed) *The Cultures of Alternative Mobilities: The Routes Less Travelled*. Farnham, England: Ashgate.
Kambitsis, J. (2010, 19 February) 'Glimpse the wireless future of transportation', *Wired Magazine*. Available at http://www.wired.com/autopia/2010/02/intelligent-transportation-systems/ (accessed 30 April 2010)
Kaufman, V. (2000) *Mobilité Quotidienne et Dynamiques Urbaines: La question du report modal*, Lausanne, Switzerland: Presses polytechniques et universitaires romandes.
Law, J. (2002) *Aircraft Stories: Decentering the Object in Technoscience*, Durham, NC: Duke University Press.
Miller, D. (ed.) (2001) *Car Cultures*, Oxford, England: Berg.
Mische, A. (2011) 'Relational sociology, culture and agency', in J. Scott & P. Carrington (eds) *Sage Handbook of Social Network Analysis*, London: Sage.
Paaswell, R. (2004) 'Intelligent transportation systems: Creating operational, institutional, and labor force changes in the United States', in R. Hanley (ed) *Moving People, Goods and Information in the 21st Century: The Cutting-Edge Infrastructures of Networked Cities*, London: Routledge.
Paaswell. R. (2009, April) 'A new paradigm for transportation planning', unpublished paper presented at the American Planning Association National Planning Conference, Minneapolis, Minnesota.
Packer, J. (2008) *Mobility Without Mayhem: Safety, Cars and Citizenship*, Durham, NC: Duke University Press.
Rodaway, P. (1994) *Sensuous Geographies: Body, Sense and Place*, London: Routledge.
Schulz, J. (2010) 'Transportation infrastructure: Is a U.S. Infrastructure Bank an idea whose time has come?' *Logistics Management*. Available at http://www.logisticsmgmt.com/article/455228 (accessed 2 April 2010).
Sheller, M. (2004a) 'Automotive emotions: Feeling the car', *Theory, Culture and Society*, 21(4/5): 221–242.
Sheller, M. (2004b) 'Mobile publics: Beyond the network perspective', *Environment and Planning D: Society and Space*, 22(1): 39–52.

Sheller, M. (2007) 'Bodies, cybercars and the production of automated-mobilities', *Social and Cultural Geography*, 8(2): 175–197.
Sheller, M. (2008) 'Mobility, freedom and place', in S. Bergmann & T. Sager (eds) *The Ethics of Mobilities: Rethinking Place, Exclusion, Freedom and Environment*, Aldershot, England: Ashgate.
Sheller, M. & Urry, J. (2000) 'The city and the car', *International Journal of Urban and Regional Research*, 24(4): 737–757.
Shove, E. (2003) *Comfort, Cleanliness and Convenience: The Social Organization of Normality*, Oxford, England: Berg.
Suchman, L. (2006) *Human-Machine Reconfigurations: Plans and Situated Actions* (2nd ed), Cambridge, England: Cambridge University Press.
Thrift, N. (2004) 'Driving in the city', *Theory, Culture and Society*, 21(4/5): 41–60.
Thrift, N. (2008) *Non-Representational Theory: Space, Politics, affect*, New York: Routledge.
Thrift, N. & French, S. (2002) 'The automatic production of space', *Transactions of the Institute of British Geographers*, 27(4): 309–335.
Tilly, C. (2003) *Stories, Identities and Political Change*, New York: Rowman and Littlefield.
Urry, J. (2000) *Sociology Beyond Societies: Mobilities for the Twenty-First Century*, London: Routledge.
Urry, J. (2004) 'The 'system' of automobility', *Theory, Culture and Society*, 21(4/5): 25–39.
Urry J. (2006) 'Travelling times', *European Journal of Communication*, 21(3): 357–372.
Urry, J. (2008) 'Climate change, travel and complex futures', *British Journal of Sociology*, 59: 61–79.
Vannini P. (ed) (2009) *The Cultures of Alternative Mobilities: The Routes Less Travelled*, Farnham, England: Ashgate.
Wood, D. & Graham, S. (2006) 'Permeable boundaries in the software-sorted society: Surveillance and the differentiation of mobility', in M. Sheller & J. Urry (eds) *Mobile Technologies of the City*, London: Routledge.

Part III
Dynamics of Change

10 The Electrification of Automobility
The Bumpy Ride of Electric Vehicles Toward Regime Transition

Renato J. Orsato, Marc Dijk, René Kemp and Masaru Yarime

10.1. INTRODUCTION

Cars powered solely by electricity—or *pure battery* electric vehicles (EVs)—can help to reduce carbon from transport. When electricity is generated via hydro, solar, wind or atomic energy, EVs present the lowest carbon emissions of all power train technologies, including plug-in hybrid-electric vehicles (PHEVs; Japan Automobile Research Institute, 2006). EVs have a long track record of being regarded as the ultimate green car and thus as a logical component of more sustainable mobility patterns. But their history is one of twists and turns. After the announcement of the zero-emissions vehicle (ZEV) regulation in California in 1990, for instance, most large car manufacturers launched EVs via various urban demonstration projects in Europe, Japan and the United States. By the end of the decade, however, the number of EVs on the roads remained marginal. Today, EVs are back: Almost all major car manufacturers are working on prototypes to be commercialized soon, and some companies such as Renault-Nissan and Mitsubishi Motors are already offering them to customers. Nissan will start commercial production of its EV Leaf at the scale of 50,000 per year in the end of 2010 and has plans to increase production to 150,000 per year by 2012. In 2003 BYD, a Chinese manufacturer of batteries for cell phones, entered the EV market, and since 2008 Better Place, a Californian start-up, has been working with governments, businesses and energy producers to provide electric mobility services. Car manufacturers are teaming up with battery providers, municipalities start deploying recharging infrastructure and several countries have electric mobility programs.

In this chapter, we analyze the rationale behind this new momentum for EVs. Why have EVs, once more, gained attention from policymakers, car manufacturers and investors? Is the new momentum for EVs another hype to be followed by frustration? In this chapter we examine the past, present and future of EVs. The past consists of two parts: the episode of disappointing sales and disillusionment (until 2005) and an episode of renewed interest (2005–2010). Departing from the work of Cowan and Hulten (1996) and Dijk and Yarime (2010), Section 10.2 uses the notion of path-dependence to examine EV niche developments before 2005. Section 10.3 analyzes EV developments in the 2005 to 2010 period, using ecological modernization as a framework to discuss barriers and facilitators toward the electrification of *automobility*. The section builds on the previous work of Orsato (2004) on the automobile industry and, in broader terms, on the work of Huber (1985), Spaargaren and Mol (1992), Mol and Spaargaren (1993), Mol (1995) and Spaargaren (1997). Overall, ecological modernization theory is about the modernization of society or subsectors (organizational fields) thereof through a modernization of the institutions of society (politics, markets and civil society) to assist technical change with which it evolves. Section 10.4 reflects on those developments in *transition* terms and examines prospects for EVs. We examine two possible trajectories: a substitution trajectory in which electric cars are simply another car and a transition trajectory, based on various innovation patterns and interactions. Section 10.5 summarizes the factors behind EV activity in recent years. In our mind, EV development has passed a critical threshold but it is unclear whether they will come to dominate the market.

10.2. EV NICHE DEVELOPMENTS UNTIL 2005

After the early appearance and decline in the late 19th and early 20th centuries, interest in EVs re-emerged in the 1960s and 1970s in the United States mainly due to the negative effects of air pollution and rising oil prices. The 1965 Clean Air Act triggered several research institutes and firms to develop electric cars, but results were poor in terms of both technological performance and price compared to their gasoline counterparts (Mom, 1997). At the end of the 1970s, less than 4,000 EVs had been sold worldwide. After a period of stillness, public interest on EVs revived once again in the second half of the 1980s and the early 1990s, bringing renewed hopes to environmentalists that EVs would finally become a mass market reality. This was mainly due to the new regulatory push done by the US state of California and, to a lesser extent, to the environmental policies and programs promoted in Europe.

10.2.1. Regulatory Push and Bottom–up Developments of EV Enthusiasts

Following a tradition of being in the vanguard of emission legislation, in the early 1990s the US state of California led a technology forcing approach for the introduction of ZEVs. The California Air Resources Board (CARB) had the ambition to set strict emission standards to curb health problems in the Los Angeles area provoked by motor vehicles' toxic emissions. Coincidentally, in January of 1990, General Motors presented an EV concept car (later marketed as the EV1) in the Los Angeles Auto Show, which greatly

The Electrification of Automobility 207

impressed the public and sent signals to CARB that EVs were ready for mass commercialization. Though GM did not intend the car to be mass-produced, it encouraged CARB to include EVs in the mandate,[1] which was adopted in September of that year (see Hoogma, Kemp, Schot & Truffer, 2002). With the standards, CARB intended to trigger further development and sales of EVs. Since California represented about 4% of the world market for cars and about 12% of the US car market at that time, the ZEV Mandate was quite important for automakers (Kemp, 2005). By 1994, four additional states (New York, Massachusetts, Vermont and Maine) had adopted the California ZEV mandate and eight more joined the National Low Emissions Vehicle Program, approving stricter requirements than the federal ones from the Environmental Protection Agency.

The organization of the European Union with its system of environmental Directives made it difficult to adopt a regulatory framework similar to the American ZEV mandate. Although national or local authorities could impose a ZEV regulation, there was an apparent consensus among policymakers that the use of incentives, rather than disincentives, was a more desirable and potentially more effective way of promoting cleaner vehicles (Nieuwenhuis & Wells 1997). In Europe, interest in EV technology had its main origins in engineering schools—in Germany, Denmark and Switzerland, in particular. Ecologically conscious students and technicians in small enterprises were able to move from developing solar vehicles to the artisanal manufacturing of lightweight EVs. After being showcased to the public, these vehicles motivated politicians to promote their mass production and commercialization. This led to the support of research and development programs in several Western European countries, involving the sponsorship of demonstration projects, subsidies and tax reductions for such vehicles.

10.2.2. Pilot and Demonstration Projects

In the early 1990s, a few small companies outside the (high volume) car industry were dominating EV developments. These niche players adopted a different design for the car body, which depended less on economies of scale and allowed them to be profitable by selling only a few hundred vehicles (for a discussion of car body technologies, see Chapter 6). Forced by the Californian ZEV mandate, high volume car manufacturers showed increasing commitments to the EV technology, and after presenting prototypes in auto shows, some started to sell a small number of EVs. Different from the dedicated EV producers, automakers opted for a low-risk, low-cost strategy of converting existing models into EVs (the Renault Clio and Peugeot 106 are good examples).

Hoogma et al. (2002) studied the European demonstration experiments with EVs in Germany (Rugen Island, 1992–1996), Switzerland (in the town of Mendriso, after 1995) and Norway (via the development of an EV called Th!nk, after 1991), among others. Possibly the most remarkable project was the one led by EDF, the French electric utility, which ordered 2,000 EVs for the experiment in the city of La Rochelle on the west coast of France. The experiment initially seemed a small miracle, since users loved EVs. Public attention was high, and much was learned about user acceptance and the conditions needed to support EVs. As it turned out, however, only a few consumers were willing to buy the new car outside the experiment. People's willingness to pay for an EV was not really tested by the experiment.

More positive results were achieved in the large-scale pilot and demonstration project for lightweight EVs in the Swiss town of Mendriso between 1995 and 2001. The aim of the project, initiated by the Swiss Federal Office of Energy, was to demonstrate and evaluate the usefulness of lightweight EVs, to identify measures of promotion and to demonstrate the electric mobility concept. By 2001 the project had helped to bring 396 EVs onto the roads (174 cars, 20 light duty vehicles, 97 scooters and 96 electric bikes); two thirds of the vehicles were owned by individuals and one third by companies (Hoogma et al., 2002, p. 102). The program and its follow-up heavily relied on subsidies (50% to 60% of the purchase price). When the subsidies ceased, the enthusiasm for EVs also faded away.

10.2.3. Doubts and Disappointments

The interest in the early 1990s developed rapidly and rose substantially above the public interest registered in previous decades. Nevertheless, around the year 2000 attention shifted to hydrogen fuel cell vehicles (FCVs), indicated by the number of prototypes for EVs and FCV (see Figure 10.1). EVs prototypes emerged in the 1990s, fell after 1999, rising again after 2005. FCV prototypes fell sharply after 2007, which shows that both types of vehicles are prone to cycles of optimism and pessimism.

All along the entire period, the interest of the car industry remained mostly on internal combustion engine (ICE) technology. The number of patents and new product launches in the period 1990 to 2005 clearly indicates the focus of European automakers on further developments of ICEs, such as the variable valve timing and direct fuel injection systems. On average, around 80% of the patents were awarded to ICE-related technology, against only about 20% for technologies associated with pure battery EVs and hybrid-electric vehicles (HEVs; Oltra & Saint Jean, 2009). In Japan, the number of patent applications on EVs began to rise sharply in the early 1990s, stabilized in 1995 and declined rapidly afterwards (Yarime, Shiroyama & Kuroki, 2008). Overall, most firms did not regard electric propulsion as a profitable strategy, and strong competition on gasoline and diesel engines triggered a great refinement of ICE performance in the 1990s. Although the efforts of regulators to make EVs a commercial success motivated the formation of design and production networks, the few large-scale demonstration projects in Europe and in the United States did not seem to appeal to consumers, discouraging carmakers to scale them up.

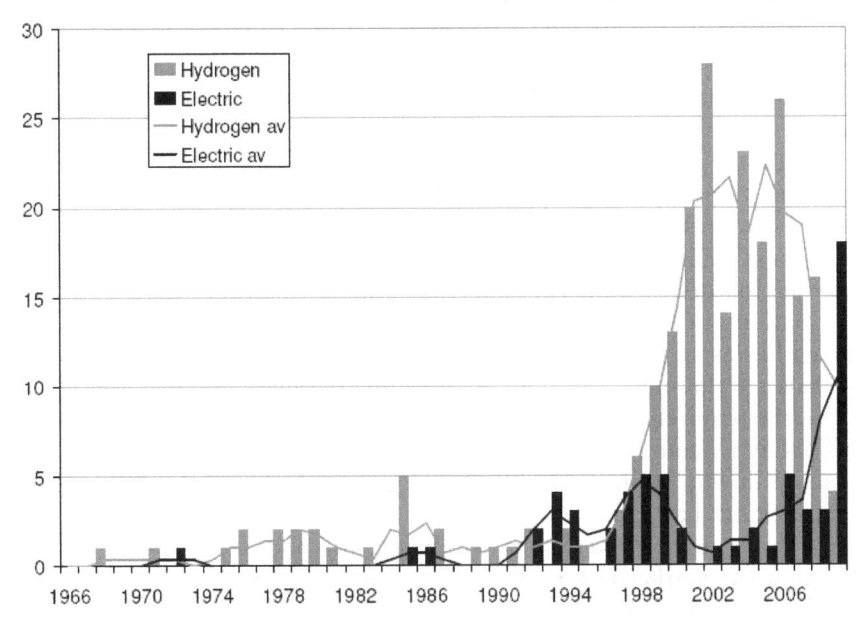

Figure 10.1 Number of prototypes for hydrogen fuel cell vehicles and electric vehicles.

An important reason for the disappointment in mass-commercializing EVs is the limited technological progress achieved during the 1990s, particularly in batteries. In that period, EVs were mainly equipped with lead-acid batteries, resulting in very limited lifetime and range. In the late part of the decade, the focus of research and development shifted to nickel metal-hydrate and lithium-ion batteries, which were expensive at low production volumes. The two-seat EV of the company Think Nordic is a good example. At €25,000 at that time, it was simply too costly to have a chance to succeed in the marketplace. Even the efforts made by Ford to rump up production, who acquired Think Nordic in 1999, were not sufficient to make the business profitable. Ford sold Think Nordic in 2001. In Japan, several EVs were released to the market in the middle of the 1990s by major automakers (such as the Toyota RAV4 and Honda EV Plus). By the early 2000s, however, commercial production of EVs had almost stopped (Yarime et al., 2008).

Another reason for the failure in scaling up experiments and, broadly, for the commercial failure of EVs in the 1990s, was consumer preference. Dijk (2011) identified the main attributes consumers were looking for when buying cars, back in 1996. The research was limited to the Netherlands, but the results are indicative of behavior in industrialized countries. Range and price were the most important attributes mentioned by potential consumers of EVs (75% and 55%, respectively). They were dissatisfied with both the functionality and price of EVs. Environmental impact (35%) was positively appraised, although to a lesser extent than by 1990. By 2000, however, there was only one salient negative attribute: range (71%). Overall, appreciation for EVs in the 1990s was low. The available vehicle during that period cost twice as much as a conventional car and would take several hours to refuel.

In addition, the lobbying efforts made by the auto industry to loosen up regulations certainly contributed to depress the commercial success of EVs. From the early 1990s, car firms voiced their dissatisfaction with California's ZEV regulation and put pressure on US federal and European legislators to limit emission regulations. Although there were EV associations in Europe, the United States and Asia (mostly created by EV enthusiasts), public support for EVs was not strong enough to counter-balance the industry lobby. As a result, the mandate was relaxed in 1996 (the requirements for 1998–2002 were abolished) and again in 1998 (ZEV credits could be earned through partial EVs). With the mandate watered down, by the early 2000s the political support for EVs in the United States had faded away. As a result, between 1995 and 2000 only a few thousands EVs were sold worldwide, and poor sales records clearly reflected the market failure of EVs in California and elsewhere, closing another EV hype-disappointment cycle.

10.3. EV NICHE DEVELOPMENTS AFTER 2005

From 2005 onwards, there is a new momentum for EVs. This time, climate change concerns rather than urban pollution are driving the efforts toward the electrification of mobility, with *peak oil* also playing a role. The aftermath of Hurricane Katrina in 2005 sensitized public opinion about the negative effects of climate change, and Al Gore's *Inconvenient Truth* documentary (May 2006) raised global awareness.[2] Altogether, such events influenced policymakers to develop regulatory frameworks and market instruments to curb carbon emissions (Orsato, 2009). Macroeconomic and political trends were also pointing toward a new era for the auto industry and, perhaps, for mobility patterns.

In the light of these new developments, this section discusses factors fostering and inhibiting the transition toward low carbon mobility, which have been previously identified in studies about the eco-modernization of the automobile industry (Orsato, 2001, 2004). Combined, these factors indicate the process of ecological modernization occurring in a specific *organizational field* (DiMaggio & Powell, 1983). Increasingly stringent emission regulations, consumer demand for low carbon vehicles, the influence of related businesses, interest groups, competition, among other factors, all may influence the developments of EV technology and the overall commercial success of EVs. Within the framework, the factors summarized in Figure 10.2 and described in the following sections provide an overview about the possibilities of EVs becoming a commercial success in the coming decades.

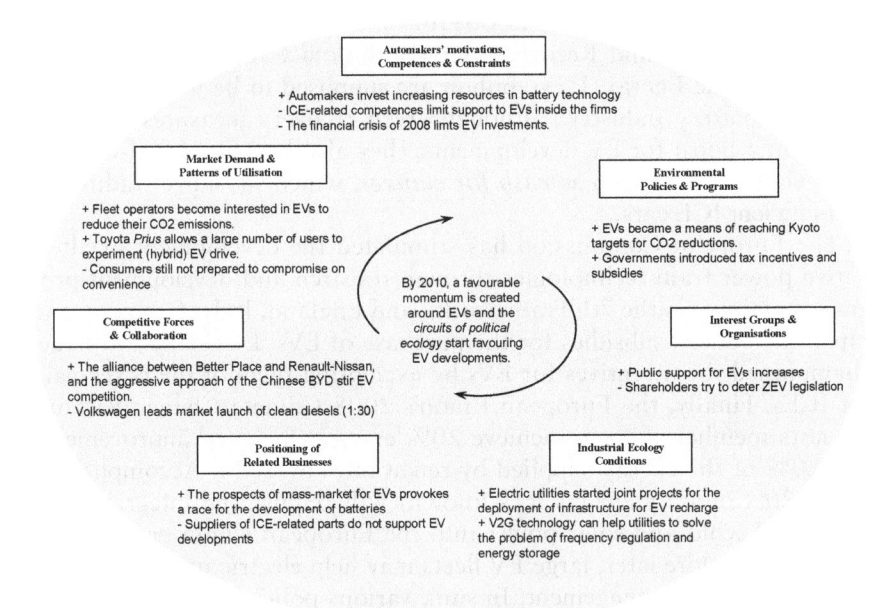

Figure 10.2 Influences on the electrification of automobility after 2005 (adapted from Orsato, 2004).

10.3.1. Environmental Policies and Government Programs

At the turn of the millennium, new cars emitted only around 5% of the pollutants their counterparts did in 1975 (Graedel & Allenby, 1998). Such improvements, however, have not alleviated the regulatory pressure faced by carmakers. Regulatory measures have continuously intensified in Europe and other industrialized countries such as Japan and some states of the United States.[3] From 2005 onwards, concerns about climate change motivated governments worldwide to demand the car industry to decrease vehicle CO_2 emissions even further. In particular, the emissions targets of the Kyoto protocol gained momentum in this period. Regulatory measures were introduced after Annex 1 countries realized that they would not meet the Kyoto targets. With the looming threat of having to purchase emission allowances, many countries started regarding EVs as a means of reducing CO_2 emissions.

Politicians and policymakers also used climate concerns and policies promoting the diffusion of EVs as a means to profile their green credentials, which were high in the public agenda in that period. Green, more fuel-efficient vehicles also featured well in the packages for economic recovery policies after the financial crisis of 2007. In Europe, regulations will require the average CO_2 emissions of vehicles to be reduced to 130g/km by 2012, while plans to lower to 95g/km by 2020 are underway.

In the United States, from the US$16.8 billion provided in the American Reinvestment and Recovery Act for the office of Energy Efficiency and Renewable Energy, US$2 billion are supposed to be used to build a domestic battery industry. Although green recovery measures certainly represent a boost for EV developments, they also benefitted fuel-efficient ICEs via programs such as *cash for clunker*, which subsidize traditional fuel-efficient ICE cars.

The European Commission has stimulated the development of alternative power train technologies through research and development programs (mainly via the 7th Framework), and England, Italy, Germany and Japan introduced subsidies for the purchase of EVs. Denmark and Israel championed the incentives for EVs by exempting them of the taxes paid for ICEs. Finally, the European Union's 2008 Climate Change Package requests member states to achieve 20% energy efficiency improvements and 20% of the energy supplied by renewables by 2020. Accomplishing the *2020 Commitment* (as it is known) will require the integration of additional wind and solar power into the European Union grid system, and as we explore later, large EV fleets may help electric utilities to optimize power-grid management. In sum, various policies and programs in the largest economies in the world provided signals and incentives for carmakers to invest increasing amounts of money in research and development and acquisitions in order to build competences in pure and hybrid EV technologies.

A good example of the role played by government programs is the one of the United Kingdom. More than half of the £420 million of investment needed to start production of the Nissan EV Leaf will come from the UK government and the European Investment Bank. For Nissan, the United Kingdom's commitment to a low-carbon future (by supplying basic EV infrastructure, customer incentives and educational promotions) was an important factor behind the decision to invest in the Leaf. The commitment included a pledge from One North East, a regional development agency, to provide a network of 1,300 charging points in the region during the period from 2010 to 2013, parking spaces dedicated to EVs, priority road lanes for EVs, as well as the pledge of a government subsidy of up to £5,000 (from 2011 onwards) for the purchase of ZEVs, which obviously includes the Leaf.[4] Figure 10.3 gives an overview of announced national EV and PHEV sales targets.

The message that these plans and forecasts convey, however, should be carefully interpreted; they may indicate a climate of a structural change, or they may themselves be a product of the EV hype. Over the next years the stability of environmental plans and programs will be tested, as with the ZEV mandate in the 1990s, also in the light of possible new disappointment around the technology. In either case, many national governments are at least momentarily providing stimuli for electric mobility. Initiatives in the niche for EV are not limited to just five or six countries (as in the 1990s), but it is a world-wide phenomenon.

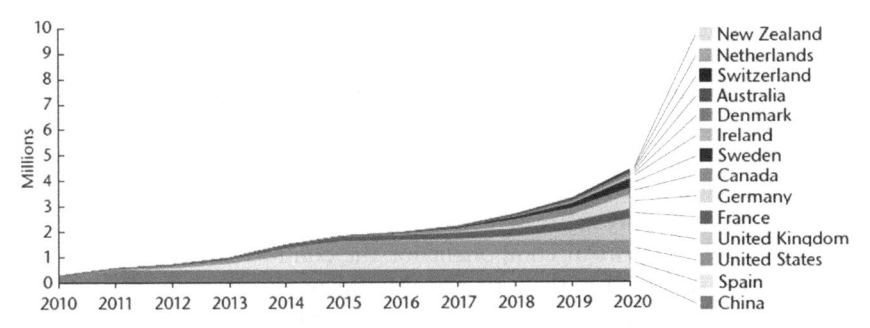

Figure 10.3 Announced national electric vehicle and plug-in hybrid-electric vehicle sales targets (see http://www.iea.org/papers/2009/EV_PHEV_Roadmap.pdf, p. 19).

10.3.2. Automakers' Motivations, Competences and Constraints

In the 1990s, Toyota and Honda were the first carmakers to move toward the mass commercialization of low-carbon vehicles via alternative power train technology. While automakers were relieved by the relaxation of the ZEV mandate in California in 1996, these two Japanese firms saw a business opportunity for the hybrid-electric power train technology, independently from regulatory measures. The Prius I was launched in 1997 and targeted the green market niche in Japan. In part because acceleration and maximum speed were compromised, the sedan had the lowest consumption in its category (3.6 liters/100 km). After capturing the Japanese niche, the Prius II was launched in California in 2000. The new version had increased acceleration (leading consumption to grow to 5.1 liters/100 km), which was well received by American consumers, ramping up sales quickly and motivating Toyota to go one step further and launch the third generation of the hybrid technology (Prius III) worldwide in 2004. Toyota rolled out the Prius vigorously, which appealed to a broad set of consumers, such as the tech-savvy, paving the way for wider applications of hybrid-electric technology in other models. Overall, the car has been a huge success for Toyota, resulting in the reputation of the greenest volume carmaker in the world. In the period 1997 to 2007, Toyota sold more than one million Prius worldwide. In 2004 a production level of 100,000 vehicles per year was reached, which represents the break-even for most cars (for details on break-even point of vehicles, see Wells & Orsato, 2005, and Chapter 6 of this volume).

In sharp contrast to Toyota and Honda, who launched their hybrids Prius and Insight in 1997 and 1998, respectively, all other carmakers were reluctant to invest in the hybrid technology. Disappointing experiences with fuel-efficient cars (the unsuccessful introduction of Volkswagen's Lupo 3 litre in 2000 is a good example) was a factor, as this led them to believe there was no market for more expensive fuel-efficient cars. After 2005, however, there was a shift in perception, with most car manufacturers investing considerable resources in research and development to catch up. Nevertheless, these investments need to be viewed against car manufacturers' strategies and investments to improve ICE. All firms have invested heavily in refining ICEs, and most firms have marketed 'eco' versions of their ICE models, such as Volkswagen's Bluemotion line.

Nissan, on the other hand, remained faithful to its investments in *pure* EVs—even though it was not until the second half of the 2000s that the company became more aggressive in trying to commercialize them. Because infrastructure is a critical component for the commercial success of EVs, Nissan also partnered with both private and public sectors. Nissan and its French partner (and main shareholder) Renault became the main supporters of the technology of battery swapping, proposed by Better Place, a new entrant in the industry. As described in the next section, the battery swapping technology is seen by some as a key solution for the problem of limited range of EVs (today at around 160 km). Carlos Ghosn, the CEO of Renault-Nissan has been the main supporter of both pure EVs and the battery exchange technology. By 2011 consumers will be able to choose between the four EV models produced by Renault, and the EV Leaf rolled out by Nissan in late 2010. The partnership between Renault-Nissan and Better Place—in particular, the adaptation of EVs to the battery exchange technology—provided the legitimacy of new approaches and business models for the mass deployment of EVs. Moreover, it triggered a level of competition around the EV technology unique in the history of the car industry.

10.3.3. Competitive Forces and Collaboration

Somehow, from 2007 onwards, new entrants have influenced the development of EVs in a unique manner. Better Place, a *mobility operator* backed by venture capital, is the most remarkable example. Better Place was founded in 2007 via US$200 million in venture capital ($350 million were added by HSBC in late 2009). Under the inspirational leadership of Shai Agassi, the start-up established a partnership with the State of Israel and Renault-Nissan for the mass deployment of EVs. Israel committed itself to implementing an appropriate tax policy, serving as a test-bed for applications elsewhere. There, more than 90% of the population drives less than 70 km/day, and major urban centers are less than 150 km apart. By deploying more than 500,000 charge spots and 100 battery-swapping stations (to be in place by 2015), Better Place expects Israel to become the leading EV country in the world. Denmark, a country with similarly short driving ranges and densely populated urban areas, will also offer tax advantages, as well as Australia and the states of California and Hawaii.[5]

The business model adopted by Better Place addresses the problems associated with refueling time, infrastructure and battery costs—the most critical problems for the diffusion of EVs. By blanketing major urban areas with recharging spots and battery exchange stations, the company will guarantee that electric power will be available to recharge EVs. Owners of

EVs, who sign up for different types of subscription packages, will be able to recharge their vehicles at home or at parking lots. This should cover 90% of the needs (i.e, for trips shorter than 160 km). For the remaining 10% of journeys (above 160 km), users will be able to swap discharged batteries for fully charged ones in locations similar to the petrol stations (in fact, most swap services will be installed at the existing petrol stations). Better Place will equip EVs with on-board computers that will help the driver to check the battery load, identify free parking spots with charging capacity, connect to swap stations, as well as order battery exchanges.

For users the electric mobility leasing model of Better Place also addresses the problem of upfront costs of batteries (around US$11,000[6]), the uncertainties associated with their lifetime and the residual value at the end of their lifetime. The company includes the battery in the infrastructure, so the cost of an EV becomes the cost of the empty car body. In other words, Better Place will bear both the initial cost of the battery pack and its residual risk value. In order to minimize its own risks, it established partnerships with battery manufacturers, such as Automotive Energy Supply Corp and A123. Together, they are expected to address the challenges of eventual shortages of supply and other issues related to the lithium-ion technology, such as safety and performance, for instance, thermal runway, over-charging, difficulties to operate in extremes weather conditions, as well as battery durability and deterioration (Deutsche Bank, 2008).[7]

The entry in 2003 of the Chinese battery company BYD in the car segment is another important development in the EV domain. BYD is the world's greatest battery producer for cell phones, a position it achieved in 2000, just five years after its foundation. By 2008 the company employed 130,000 people, had revenues close to US$4 billion and a net profit of US$200 million. These are certainly impressive numbers, but what makes the company particularly interesting is the fact that it started making batteries *before* it entered the car business. In this respect, BYD has certainly a competitive advantage when compared to traditional automakers, who are rushing to develop or buy battery technology. Together, Better Place and BYD seem to have shaken up the sector.

Indeed, the competition toward the electrification of cars could be seen in the 2009 edition of the Frankfurt Motor Show, with almost every carmaker displaying EVs prototypes (or *concept cars*, as they are known in the industry). Besides the aggressive marketing campaign around the launch of four models of EVs by Renault-Nissan, other European volume producers, such as Mercedes and Fiat presented EV models with clear plans to be launched before 2015. General Motors, following its fall from grace in 2008, put great emphasis in its plug-in hybrid Chevy Volt as a potential savior of its financial problems—even though most analysts think this hope is unfounded (Fortune Magazine, 2009). In the Japanese auto industry, Mitsubishi Motors started the mass production of its electric vehicle (called i-MiEV) in mid-2009 at the scale of 1,400 vehicles per year, almost at the same time when Fuji Heavy Industry introduced the plug-in Stella. Nissan plans to start commercial production of its EVs Leaf, at the scale of 50,000 per year by early 2011 and to increase its production level to 150,000 per year by 2012. Even the Think Nordic, the Norwegian company mentioned in Section 10.2, succeeded to convince new investors that the time for EVs has finally arrived, avoiding bankruptcy once again.[8] In sum, competition and collaboration for the development and mass commercialization of EVs built up in the second half of the 2000s, at the same time the race to build a minimum recharging infrastructure was also underway.

10.3.4. Industrial Ecology Conditions

To enable widespread use of EVs, there are *industrial ecology conditions* to be satisfied, mostly in terms of recharging infrastructure, skills and collaborations across industries. In this respect, power utilities play an important role. Whereas battery technology and costs are crucial for the market success of EVs, commercial success also depends on the infrastructure for recharging. In a similar attempt to make a transition from diesel engine to compressed natural gas vehicles in Tokyo, it was of critical importance that gas infrastructure providers were involved in cooperation with carmakers and users (Yarime, 2009). In the same fashion, mainly after 2005, national and local governments were deeper involved in the market preparation and the provision of infrastructure for EV recharging, and the level of research and development funds is substantially larger than in the 1990s.[9] In the period of 2010 to 2011 there were thousands of projects with a much larger budget, compared to a few dozen projects in the 1990s.[10] Moreover, many local governments started providing refueling infrastructure, though still in the order of dozens of refuel points, which is too little for a widespread use of EVs.

Electric utilities have been increasingly involved in EV partnerships. Whereas in the 1990s only the French EDF regarded EVs as a business opportunity, the list of utilities engaging in infrastructure developments was much larger in 2010, including the Swiss Energie Ouest Suisse, Oregon's Portland General Electric, San Diego's Gas and Electric, Ireland's ESB, Tokyo Electric Power Company, among others. These large organizations are important enablers of recharging infrastructure, and their involvement in automobility seem to be gaining momentum in the early stages of the decade 2010 to 2020.

For electric utilities, EVs can represent a new source of revenues, but only if the problems associated with their widespread diffusion are addressed via the upgrading of the energy grid. Recharging large amounts of EVs clustered in specific areas may require additional power supply, substations and rewiring, resulting in substantial added costs. In order to curb such problems and associated costs, major EV deployments will require the (re)fitting of hardware and software so the control of the electric grid can be done more efficiently. One way of dealing with peak loads is through a system innovation named *smartgrid*, a concept that has been widely debated in political and academic circles. Although the meaning of a *smartgrid* can vary widely, its main rationale is the optimization of the grid via information technology (IT)-related equipment and software, which are capable of measuring and controlling end-points (such as appliances in a house), resulting in the optimization of energy production, transmission, storage and use.

There is a synergetic relationship between the *smartgrid* and EVs. When equipped with smart equipment that allows both upload and download of electricity, batteries can be used to store energy and serve as a reserve of power during the time EVs are dormant. Banks of used batteries can also provide up to 40% of their (remaining) power capacity to other uses (backup stations for instance), which means extending their economic value beyond their life to power EVs. Under vehicle-to-grid operations, batteries connected for recharging can also provide backup power for a number of purposes, such as: (a) contingent load, in the event of failure of a generating unit connected to the grid; (b) frequency regulation, fine tuning the necessary instantaneous balance between supply and demand on the grid; (c) reactive power, providing local phase-angle corrections important in alternate current (AC) networks; and (d) load-following reserves, necessary in particular to backup and absorb variations in power provided by renewables such as wind and solar. Thus, for electricity suppliers, the electrification of mobility offers off-peak demand, as well as uncertainties on the extent of benefits and complexities of involving vehicle batteries in grid storage. In sum, EVs may become the link between the energy and the transportation sectors (which together represent 75% of CO_2 emissions). Besides new clients, EVs can help utilities to reduce system inefficiencies embedded in today's grid (Bauknecht 2009).

10.3.5. Market Demand and Patterns of Utilization

Although individual consumers have always been seen as the main buyers/users of EVs, a much more powerful player emerged in the 2000s as the best candidate to speed up the uptake of EVs: the fleet operator. Differently from the 1990s, when local emissions were the main driver for the interest in EVs, from 2005 onwards, the motivation for large fleet operators to move toward alternative power train technologies was the goal of reduced carbon footprint. Fleet operators emerged as a key force influencing the directions of EV development and commercialization. In France, for instance, a consortium formed by large fleet operators is expected to create the demand for at least 100,000 EVs by 2015 (French Ministry of Energy and Environment, 2010). This should not be a surprise. In the early days of the automobile, EVs were the preferred method for delivery in the postal service and other daily-route sectors, such as dairy delivery. In particular, EVs are the most appropriate vehicles for postal delivery services. The average length of routes in France, for instance, is 33 km, well below the 100 km range that is easily achievable with current EV technologies.

At current electricity and fuel prices, the cost per kilometer is already lower for EVs than for ICEs.[11] Expected increases in gasoline and diesel prices in the coming decades reported by the International Energy Agency will help to augment the cost difference. Fleet operators are sensitive to such prospects, more than consumers are. EVs also cost less to operate and maintain and, as it was mentioned in the previous section, when the cars are parked, there is the possibility of using the batteries for reserve power and grid buffering. For forward looking fleet owners such things are important. Although vehicle-to-grid entails costs in connecting and controlling batteries for bidirectional flows, fleet operators could profit from both transportation services and battery reserves, when EVs are not in use. EVs also help to reduce carbon emissions. Even when the eventual carbon credits resulting from the move from ICEs to EVs are almost negligible,[12] fleet operators can expect to reap up reputational benefits from the decarburization strategies (Orsato, 2009).

10.3.6. Positioning of Related Businesses

Within the car industry, component suppliers are important actors. The high costs of research for alternative power trains have motivated car firms to establish alliances with some of these suppliers, often sponsored by governments, such as California's Fuel Cell Partnership. Similar to diesel and gasoline innovations, which have been developed mainly by first-tier suppliers (e.g., Bosch, Denso, Valeo and Delphi), EV research occurred mainly within the supplier network—Japanese ones in particular (Pilkington & Dyerson, 2006). For instance, Toyota and Matsushita (Toyota's battery supplier at the time), formed a joint venture for battery development in March 1995, allowing them to share research and development costs and risks associated with battery technology (Magnusson & Berggren, 2001). In April 2007 Nissan established a joint venture (called Automotive Energy Supply) to produce lithium-ion batteries, with NEC Corporation and NEC Tokin in the electronic industry. In March 2009 Honda entered into collaboration for developing batteries through a joint venture (called Blue Energy) with the specialized battery maker GS Yuasa—the same company Mitsubishi Motors created a joint venture (called Lithium Energy Japan)—with the particular focus on lithium-ion batteries, in December 2007. While Japanese automakers basically chose to work with battery makers through joint ventures, US auto manufactures preferred to maintain arms-length relationships with battery suppliers (Yarime et al., 2008). All these developments indicate the increasing centrality of battery producers in the near future of the auto industry. The level of partnerships for the development of battery technology is significantly higher in the 2000s than it was previously. Because battery technology is the key to improving the performance of EVs, automakers started to collaborate closely with battery producers to generate or strengthen competencies. Such collaborative efforts will affect battery prices, although expert studies differ in how significant the probable price decrease will be.

Many traditional suppliers also realized that a move away from the ICE vehicle will not necessarily hurt their business. For manufacturers of glass and lights, for instance, such a move would have very little impact; these components are necessary in any car, independent of the type of power train. Since alternative power train technology may entail a more important role for electronics, suppliers of electronic components may even benefit from hybrid and pure EVs. For battery manufacturers, on the other hand, the potential demand for EVs represents a huge growth potential for existing business or new entrants in the industry. According to expert studies, by 2015 the market value of car batteries is expected to reach €15 billion, rising to €30 billion in 2030 (Lache et al., 2000). In the longer term, suppliers of steel (used in car bodies and structure) will be negatively affected by a shift to lightweight materials such as aluminium and carbon fiber. At the downstream side, millions of people and organizations survive on the basis of fixing engines and body parts. These people will certainly resist change.

In the past few years, EVs have been identified as an investment opportunity by investors in the world. In 2009 Warren Buffett, the world's most renowned financial investor, bought 10% of the Chinese automaker BYD, a decision which sent a signal to the world of investors that EVs are a genuine business opportunity, even when EVs still have to prove themselves in the market.

10.3.7. Interest Groups and Organizations

Environmentalist and green parties at the national and local level openly supported EVs and developed action plans for it. In the Netherlands the green environmental nongovernmental organization Foundation of Nature and Environment put forward an action plan for the introduction of EVs in the Netherlands, aimed at 1 million EVs in 2020. The action plan was supported by a broad coalition of private companies which included Rabobank, the Eneco power company, Athlon car lease, Prorail and the mobility association ANWB. The city of Amsterdam supported the action plan and formulated an action plan of its own for EVs. This is showing that various interest groups are putting themselves behind EVs.

For some car manufacturers, however, EVs offer a threat to past investments, and they will do everything they can to protect their investments. In so doing they are often supported by policy. Indeed, governments have been careful not to weaken the position of their national (and traditional) car industry. In the aftermath of the financial crisis in 2008 to 2009, most governments supported their established car industry. In many countries, the substantive support schemes went together with the requirement to develop more fuel efficient and greener cars, but it was not transparent to what extent it would be spent on ICE or EV technology. At present, national governments have not put their enthusiasm for EVs above the interest of their national industries.

In sum, the developments described in the previous pages indicate the factors fostering and limiting the ecological modernization of the car industry around the electric power train technology. As Figure 10.2 summarizes, there is a (mostly positive) tendency toward EV developments in the organizational field in which the car industry is embedded. Starting with government incentives and developments in battery storage technology, the new EV momentum is also powered by the enthusiasm of investors in cleantech, new market entrants such as BYD and Better Place, as well as the new power train diversification strategies of some large automakers. Overall, the ecological modernization of the automobile industry around electric power train technologies centers around a few distinctive electric drives, not just EVs, but also HEVs, and electrified ICEs, and the interactions between these trajectories and what this implies for a possible transition is explored in the next section.

10.4. THE TRANSITION PERSPECTIVE OF EV DEVELOPMENTS

Throughout the history of the car industry, EVs have gone through cycles of hype and disillusionment—particularly during the 1970s, 1980s and in the period of 1990 to 1997. It is difficult to be precise but in 1990 a hype cycle started with the ZEV mandate within the low-emission vehicle program of CARB, which peaked somewhere in the mid-1990s. In 1997, for instance, the *Financial Times* published an article with the heading 'EVs are a turn-off'. From 1998 onwards, the number of EV prototypes started to fall. Since 2005, expectations have been rising steadily, and according to some we are at the peak of a new hype.

In transition terms, as it was the case in the early days of the car industry in the 1900s, in the last 40 years EVs have not succeeded in breaking out of their small niche. Neither traditional cars converted to EVs (produced by regular manufacturers) nor especially dedicated EVs manufactured by market entrants were able to compete with regular vehicles. EVs have mostly been sold in unconventional markets: demonstration projects, fleet users committed to green issues, with the help of subsidies.

The reality of EVs is in sharp contrast with the sales of the Toyota Prius, a hybrid vehicle, which by Spring 2010 had sold more than 1.6 million units. Indeed, it seems that the most successful electric car configuration today is the hybrid power train, based on two propulsion systems: an ICE and an electric motor powered by batteries. Hybrid electric cars use small batteries that are charged by an ICE, adding two benefits to the vehicle: electric drive in urban centers and higher fuel economy. In other words, incumbent car manufacturers have incorporated unconventional electric engines and other electronic components into the dominant design of ICEs.

The Electrification of Automobility 221

In terms of the transition approach, hybrid vehicles can be seen as another example of a well-known hybridization pattern, which has been observed earlier in steamships (and therefore it has been referred to as "the sailing ship effect"; Ward, 1967). The pattern results, primarily, from defensive strategies of incumbent firms (discussed in Section 10.3), who find hybrid solutions a less risky and, therefore, more attractive strategy. Apart from hybrid-electric models, the eco-friendly versions of ICEs, which include electronic start-stop systems and transmission software, could also be considered examples of hybridization. The fast diffusion of these systems shows how prominent this pattern is.

Policymakers have played an important role in facilitating hybridizations, mainly through tax benefits and emissions requirements that could be met by improved ICEs. Regulations were based on the technological feasibility of ICEs, with industry being the source of information on that. Also, with the support schemes during the financial crisis in 2008 to 2009, governments have favored national industries with sunk investments and competences around the ICE more than they had supported EV developments. This inconsistency is quite typical in transition processes. Governments tend to offer support to struggling established business alongside the support for new business. Such policies are part of transition processes and cannot singularly support radical change that can potentially disrupt existing business. Politicians are sensitive to job losses far more than they are sensitive to new types of jobs. People whose job is at stake appear on television and in newspapers. The media report about new product developments, and when a new development is under way, national lead market concerns may encourage government to support a new (eco-innovation) development quite forcefully, which is exactly what we are seeing now for pure EVs.

In addition to patterns of hybridization, earlier transition studies have observed *"fit-stretch"* patterns (see Chapter 3), and these can also be applied to the case of EVs. They are sketched in a two dimensional graph, with the first dimension (horizontal axis) being the fit or stretch of an innovation in terms of *technical form and design*; the second dimension (vertical axis) represents the fit or stretch in terms of *user context and functionality*. The combination of the two dimensions allows us to place pure electric or HEVs relative to each other and delivers some insights in the requirements for the success of these power train technologies in the near future (see Figure 10.4). Earlier studies suggest that new technologies are successful only when the technological and behavioral discontinuity—between the old and the new—is not too significant (Geels, 2005). During a transition, the niche stretches in both form and function.

When plotting EVs in such a fit-stretch scheme, we find that in the 1990s battery EVs involved a significant stretch in the user environment in comparison to established gasoline and diesel car mobility. Although the new type of engine (electric motor) was fitted into conventional car bodies (technological fit), the vehicle autonomy (shorter) and refueling time (longer)

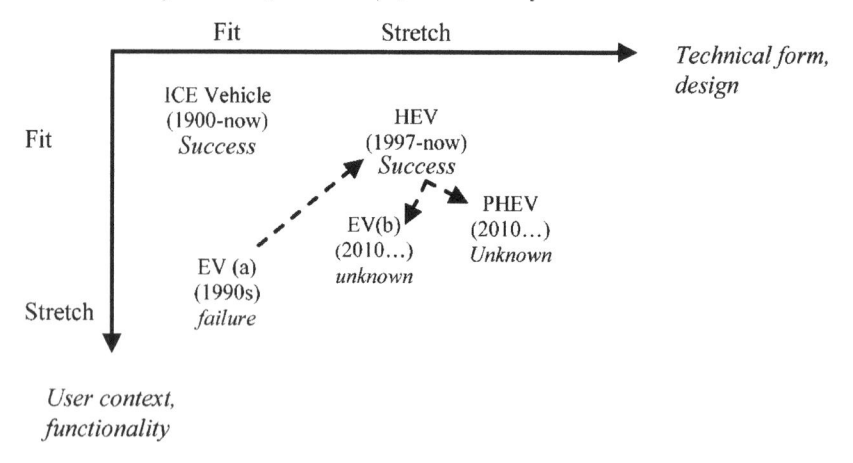

```
                    Fit              Stretch
                                                              →  Technical form,
              ICE Vehicle                                        design
              (1900-now)         HEV
    Fit       Success            (1997-now)
                                 Success
                                            PHEV
                         EV(b)   (2010...)
                         (2010...) Unknown
                EV (a)   unknown
                (1990s)
   Stretch      failure
    ↓

    User context,
    functionality
```

Figure 10.4 Fit-stretch pattern for power train technologies (adapted from Hoogma, 2000, and Geels, 2005).

were significantly different than the cars powered by ICEs. The hybrid vehicles (notably the Toyota Prius), which came to the market in the late 1990s, fit well in the existing user context but involved a partial stretch in terms of technology. As we have described in previous sections, this technological option has been the more successful strategy in market terms.

In the scheme, the Better Place business model is represented by EV(b), which has a better fit in the user environment in comparison to the EVs of the 1990s, because the range and refueling problem is solved by an extensive battery swapping infrastructure. A future PHEV, on the other hand, would further stretch the electric propulsion system. As to the user aspects, the greater range for electric drive can be considered a positive functionality. Home charging could be valued positively or negatively, depending on whether people find this a hassle or not.

The resulting pattern, sketched in Figure 10.4, suggests that at present stretching technical form and design have been more successful than stretching functionality. In this respect, the success or failure of the new generation of EVs and PHEVs depends more on their ability to satisfy consumer requirements than the technology per se. Although technical form and design will certainly play an important role in the future of these vehicles, a successful stretch of the user context depends mostly on the alignment of consumer demands (user context and functionality) with clever value propositions.

From a fit-stretch point of view it is interesting to compare EVs with PHEVs—which most resemble pure EVs and compete with them. Table 10.1 highlights the key differences between the two technologies. Fundamentally, PHEVs retain the entire ICE system but add battery capacity compared to HEVs, which allows for greater electric range (International Energy Agency,

The Electrification of Automobility 223

2009, p. 10). We will probably see a trajectory in which the importance of the battery pack in a PHEV will grow as a function of lower battery costs. PHEVs better fit the current regime than EVs, for reasons having to do with infrastructure, economies of scale and current consumer preferences and practices. At one level PHEVs compete with EVs, but the success of PHEVs may stimulate all kinds of changes (infrastructure, supplier industry, acceptance of electric drive) from which pure EVs can benefit.

Table 10.1 Key Differences Between PHEVs and EVs

PHEVs	EVs
Infrastructure:	Infrastructure:
• Home recharging will be a prerequisite for most consumers; public recharge infrastructure may be relatively unimportant, at least to ensure adequate driving range, though some consumers may place a high value on daytime recharge opportunities.	• Greater need for public infrastructure to increase daily driving range; quick recharge for longer trips and short stops; such infrastructure is likely to be sparse in early years and will need to be carefully coordinated.
Economies of scale:	Economies of scale
• Mass production levels needed to achieve economies of scale may be lower than those needed for EVs, for example if the same model is already mass-marketed as a non-PHEV hybrid; however, high-volume battery production (across models) will be needed.	• Mass production level of 50 000 to 100 000 vehicles per year, per model will be needed to achieve reasonable scale economies; possibly higher for batteries (though similar batteries will likely serve more than one model).
Vehicle range:	Vehicle range:
• PHEV optimal battery capacity (and range on grid-derived electricity) may vary by market and consumer group. Willingness to pay for additional batteries (and additional range) will be a key determinant.	• Minimum necessary range may vary by region – possibly significantly lower in Europe and Japan than in North America, given lower average daily driving levels. 100 km (62 miles) to 150 km (93 miles) may be a typical target range in the near term.
Consumer adoption:	Consumer adoption:
• Many consumers may be willing to pay some level of price premium because it is a dual-fuel vehicle. This needs further research.	• Early adopters may be those with specific needs, such as primarily urban driving, or having more than one car, allowing the EV to serve for specific (shorter) trips. More research is needed to better understand driving behaviour and likely EV purchase and use patterns.
• People interested in PHEVs may focus more on the liquid fuel efficiency (MPG) benefits rather than the overall (liquid fuel plus electricity) energy efficiency. Metrics should encourage looking at both.	
• Electric range should be set to allow best price that matches the daily travel of an individual or allow individuals to set their own range (e.g., providing variable battery capacity as a purchase option).	• With involvement from battery manufacturers and utilities, consumers may have a wider range of financing options for EVs than they have for conventional vehicles (e.g., battery costs could be bundled into monthly electric bill).
	• EVs will perform differently in different situations (e.g., weather) and locations (e.g., Colorado versus California); therefore utility and operating costs may vary significantly.

Note: From International Energy Agency (2009, p. 10). PHEVs = plug-in hybrid-electric vehicles; EVs = electric vehicles.

From a transition point of view, it is interesting to speculate whether EVs and PHEVs will change *automobility* as we know it. They may bring about a process of transformation of the car regime, or alternatively a process of transition toward a new regime (Geels & Kemp, 2007). In the first case, established regime actors, who bring in new engine technologies but sustain car user practices, infrastructures and so on, do the transformation. In the latter case, actors outside the regime have a prominent role, initiating new actor practices and networks. Many industry experts believe that EV and PHEV will not cause major changes in the current mobility behavior, since EVs are expected to be used by urban fleets (taxis, postal delivery, city distribution) and PHEV by individual consumers and car rental companies, for instance.

The rise of car-sharing organizations may also lead more people to have access to individual motorization without having to own cars, but this does not mean that car use will significantly decrease, since space freed from a more selective use may be filled by other cars. The majority of these cars are expected to be powered by ICE vehicles. Although user experiences with EVs have been positive, sales have been low because vehicles powered by ICEs offer more range and power (i.e., more convenience) for a lower price. Besides, we have witnessed a broad range of "green car" offers, such as cars that are fuel efficient and cars fueled by biofuels, with very competitive prices. The transition perspective helps to consider all these developments in an integrated way. In this respect, the transition approach is not a truth machine; its value is on drawing attention to patterns and interactions between multiple developments in and outside the car industry.

10.5. CONCLUSION

The regime around the ICE car has reigned for more than 100 years. One source of lock-in comes from the production side: It has not been economically attractive to invest in a new technology that has been considered non-competitive in terms of costs. Competition has indeed been fierce in the past decades with many large car manufacturers struggling to survive. For these companies, it has been both more attractive and safer to invest in incremental innovation around the existing ICE technology than in technological options that carry the risk of low consumer acceptance. This yields a pattern in which car manufacturers continuously refine the dominant design in order to improve environmental performance of ICEs. In this respect, the development of hybrid technology can also be seen as an attempt by car assemblers to innovate without having to move away from their core competencies.

In the last 5 years (2005 to 2010), however, there has been a spell of EV activity, which has to do with the following developments:

1. Climate protection policies and targets that included EVs as a source of reduction of CO_2
2. EVs becoming an icon for zero-carbon vehicles (mainly after 2005)
3. The peak oil expectation and the unpredictability of future prices which brought attention to vehicles that do not depend on oil
4. The success of the Toyota HEV Prius in the past decade, showcasing electric drive
5. Progress in battery technology spurred by consumer electronic sector, helping to lower the costs of EVs
6. New offers of EVs based on battery leasing and mobility packages such as the one of Better Place, which aroused consumer curiosity and widened consumer choice
7. The realization by fleet operators and, to a certain extent, by individual consumers that EVs may have lower overall driving costs than ICEs
8. Doubts that (hydrogen) fuel cell technology will be ready for commercial use any time soon
9. The economic recovery program in the United States and Europe which favored clean technologies, including EVs
10. Car manufacturers adopting a diversification strategy, including hybrid and pure EVs in the portfolio of car, which remained dominated by ICE vehicles

Altogether, these developments suggest that a trajectory of electrification of cars is underway, led mainly by progress in batteries, carbon reduction policies, new value propositions by business, as well as an increasing positive image of electric drive among consumers and policymakers. It remains to be seen, however, whether these developments will lead to a transformation of the established regime, with a more prominent place for hybrids (HEV and PHEV), or whether it will entail a transition to a new regime dominated by pure electrics.

NOTES

1. The ZEV mandate in 1990 required that 2% of all new cars sold in California should be "zero emission" by 1998. In the year 2000 all new cars sold had to be either "low emission," "ultra low emission" or "zero emission." Moreover, by 2003, 75% had to be low emission vehicles, 15% ultra low emission vehicles and 10% ZEV. For a more detailed analysis on how this regulation came about, see Kemp (2005).
2. The movie received an Oscar award in 2007 and Gore also received the Nobel prize in 2007. In July 2007, an internet survey conducted by The Nielsen Company and Oxford University, found that 66% of viewers who claimed to have seen *An Inconvenient Truth* said the film had "changed their mind" about global warming, and 89% said watching the movie made them more aware of the problem. Three out of four (74%) viewers said they changed some of their habits as a result of seeing the film.

3. In Europe, for instance, a sequence of emissions requirements (Euro 1 to 6) has triggered a significant stepwise reduction of nitrogen oxides, carbon monoxide, particulates and others over the last 25 years.
4. See http://www.telegraph.co.uk/finance/newsbysector/transport/7470370/Nissan-Leaf-to-be-built-in-Sunderland-after-carmaker-gets-Government-support.html.
5. For details about Better Place, see Orsato et al. (2009) and Chapter 7 of Orsato (2009). See also www.betterplace.com.
6. Deutsche Bank (2008) estimates that lithium-ion batteries, depending on which type, will cost around US$500 to 600/kWh, which comes to US$11,000 for a full EV 22kWh.
7. By the unbundling of batteries and chassis, automakers may be able to transfer some of the risk associated with battery development to the reinsurance or capital markets. Securitized financial instruments have been in use for some time to finance the initial sale of commercial fleets and to insure the resale value of such fleets. The risks in the case of EV batteries could then be mitigated by the possibility of Battery-to-Grid applications after their useful life in an EV. Large numbers of batteries grouped in convenient locations could be used for reserve power by a wide number of organizations, as well as frequency regulation, by electric utilities. This type of use would not face the problems associated with distributed locations of EVs.
8. For a broader explanation of the Think trajectory, see Orsato, Wassenove and Wells (2008).
9. For instance, Great Britain reserved 400 million pounds for the introduction of electric vehicles, Germany 500 million euro, France 400 million for electric and hybrid vehicles over the next 4 years.
10. For instance, a 46 million European demonstration project started in 2010, involving 19 cities, 17 vehicle manufacturers or suppliers and 12 electricity providers, planning to install 14.000 charging points and 9,500 vehicles (see www.avere.com).
11. The cost is around €5.00/100 km for EVs and €8.30/100 km for ICEs, assuming an average of 25 kWh per 100 km for an EV and 8.3 liter/100 km for an ICE car, and €0.19/kWh (average between day and night tariffs) and €1.00/liter of gasoline.
12. For a car that does 40 km a day 350 days a year, emitting an average of 150g/km, at €30/ton of carbon, a shift to EVs would generate €63/year. This does not include transaction costs, which could easily nullify the gain.

REFERENCES

Bakker, S. & van Lente, H. (2009, May) 'Fuelling or charging expectations?—A historical analysis of hydrogen and battery-electric vehicle prototypes', unpublished paper presented at the EVS24 International Battery, Hybrid and Fuel Cell Electric Vehicle Symposium, Stavanger, Norway.
Bauknecht, D. (2009) personal communication, 1 June 2009.
Cowan, R. & Hulten, S. (1996) 'Escaping lock-in: The case of the electric vehicle', *Technology Forecasting and Social Change*, 53(1): 61–80.
Dijk, M. (2011) 'Technological frames of car engines', *Technology in Society* 33, 165–180.
Dijk, M. & Yarime, M. (2010) 'The emergence of hybrid-electric cars: Innovation path creation through co-evolution of supply and demand', *Technological Forecasting and Social Change*, 77(8): 1371–1390.

DiMaggio, P. & Powell, W. (1983) 'The iron cage revisited: Institutional isomorphism and collective rationality in organizational fields', *American Sociological Review*, 48(2): 147–160.
French Ministry of Energy and Environment (2010). Press release, 13 April.
Fortune Magazine (2009) 'The great electric car race, 27 April: 28–31.
Geels, F. (2005) 'Processes and patterns in transitions and system innovations: Refining the co-evolutionary multi-level perspective', *Technological Forecasting & Social Change*, 72(6): 681–696.
Geels, F. (2010) 'Ontologies, socio-technical transitions (to sustainability), and the multi-level perspective', *Research Policy*, 39: 495–510.
Geels, F. & Kemp, R. (2007) 'Dynamics in socio-technical systems: Typology of change processes and contrasting case studies', *Technology in Society*, 29(4): 441–455.
Graedel, T.E. & Allenby, B.R. (1998) *Industrial Ecology and the Automobile*, Upper Saddle River, NJ: Prentice-Hall.
Hoogma, R. (2000) 'Exploiting technological niches', doctoral thesis, Twente University, Enschede, the Netherlands.
Hoogma, R., Kemp, R., Schot, J. & Truffer, B. (2002) *Experimenting for Sustainable Transport. The Approach of Strategic Niche Management*, London: Spon Press.
Huber, J. (1985) *Die Regenbogengesellschaft. Okologie und Sozialpolitik*, Frankfurt, Germany: Fisher Verlag.
International Energy Agency (2009) *Technology Roadmap. Electric and Plug-In Hybrid Electric Vehicles*, Paris: International Energy Agency.
Japan Automobile Research Institute (2006) 'Report on JHFC study on total efficiency', Japan Hydrogen and Fuel Cell (JHFC) Special Committee on Total Efficiency, Tokyo, March.
Kemp, R. (2005) 'Zero emission vehicle mandate in California. Misguided policy or example of enlightened leadership?' in C. Sartorius & S. Zundel (eds) *Time Strategies, Innovation and Environmental Policy*, Cheltenham, England: Edward Elgar.
Lache, R., Nolan, P., & Crane, J. (2008) 'Electric cars: Plugged in batteries must be included,' *Global Market Research*, June. Retrieved from: www.d-incert.nl
Magnusson, T. & Berggren, C. (2001) 'Environmental innovation in auto development—Managing technological uncertainty within strict time limits', *International Journal of Vehicle Design*, 26(2/3): 101–115.
Mitchell, M., Borroni-Bird, C. & Burns, L. (2010) *Reinventing the Automobile: Personal Urban Mobility for the 21st Century*, Cambridge, MA: MIT Press.
Mol, A.P.J. (1995) *The Refinement of Production*, Utrecht, the Netherlands: Van Arkel.
Mol, A.P.J. & Spaargaren, G. (1993) 'Environment, modernity and the risk society: The apocalyptic horizon of environmental reform', *International Sociology*, 8(4): 431–459.
Mom, G. (1997) 'Geschiedenis van de auto van morgen. Cultuur en Techniek van de elektrische auto', doctoral thesis, Eindhoven University of Technology, the Netherlands.
Nieuwenhuis, P. & Wells, P. (1997) *The Death of Motoring? Car Making and Automobility in the 21st Century*, Chichester, England: Wiley.
Oltra, V. & Saint Jean, M. (2009) 'Variety of technological trajectories in low emission vehicles (LEVs): A patent data analysis', *Journal of Cleaner Production*, 17(2): 201–213.
Orsato, R.J. (2001) 'The Ecological Modernization of industries: developing a multi-disciplinary research on organization and environment', Sydney: University of Technology (PhD-thesis).

Orsato, R.J. (2004) 'The Ecological Modernization of Organizational Fields: a framework for analysis', in S. Sharma and M. Starik (eds.) *Stakeholders, Environment and Society*, London: Edward Elgar, pp. 270–306.
Orsato, R.J. (2009) *Sustainability Strategies*, Basingstoke, England: Pelgrave Macmillan.
Orsato, R.J., Wassenove, L. & Wells, P. (2008) 'Eco-entrepreneurship: The bumpy Ride of TH!NK', INSEAD Teaching Case 04/2008-5485, France: INSEAD Business School.
Orsato, R.J., Hemne, S. & Van Wassenhove, L. (2009) 'Mobility Innovation for a Better Place (Part A)', INSEAD Teaching Case 09/2009–5630, France: INSEAD Business School.
Pilkington, A. & Dyerson, R. (2006) 'Innovation in disruptive regulatory environments: A patent study of electric vehicle technology development', *European Journal of Innovation Management*, 9(1): 79–91.
Spaargaren, G. (1997) *The Ecological Modernization of Production and Consumption: Essays in Environmental Sociology*, Wageningen, the Netherlands: Wageningen University.
Spaargaren, G. & Mol, T. (1992) 'Sociology, environment and modernity. Towards a theory of ecological modernization', *Society and Natural Resources*, 5(4): 323–344.
Ward, W.H. (1967) 'The sailing ship effect', *Bulletin of Institute of Physics and The Physical Society*, 18: 169.
Wells, P. & Orsato, R.J. (2005) 'Redefining the industrial ecology of the automobile', *Journal of industrial Ecology*, 9(3): 15–30.
Yarime, M. (2009) 'Public coordination for escaping from technological lock-in: Its possibilities and limits in replacing diesel vehicles with compressed natural gas vehicles in Tokyo', *Journal of Cleaner Production*, 17(14): 1281–1288.
Yarime, M., Shiroyama, H. & Kuroki, Y. (2008) 'The strategies of the Japanese auto industry in developing hybrid and fuel cell vehicles', in L.K. Mytelka & G. Boyle (eds) *Making Choices About Hydrogen: Transport Issues for Developing Countries*, Tokyo: UNU Press.

11 Introducing Hydrogen and Fuel Cell Vehicles in Germany

Oliver Ehret and Marloes Dignum

11.1. INTRODUCTION

Since the late 1990s, fuel cell vehicles (FCVs) using hydrogen (H_2) fuel have been promoted as highly promising alternatives to conventional cars and buses burning fossil fuels. DaimlerChrysler was the first large motor manufacturer championing FCVs as *the* vehicle concept of the future and investing heavily in the technology. Most motor manufacturers developed prototype hydrogen vehicles. Mineral oil companies and other energy companies invested in hydrogen production and distribution facilities. Fuel cells and hydrogen received international support from both companies and governmental bodies, for example in the United States, Europe and Japan. Large programs, such as the US *FreedomCAR and Fuel Partnership*, stressed the advantages hydrogen FCVs offer with regard to achieving more sustainable road transport.

Radical cuts in greenhouse gas and pollutant emissions, greater independence from fossil fuel imports and the economic gains of innovation have been cited as the main benefits (Mytelka & Boyle, 2008). Hydrogen can be produced from a range of primary energies including wind power and biomass. Hydrogen produced from renewable energy facilitates resource conservation and very low-emission transport (Geitmann, 2002). Fuel cells emit only water and are twice as efficient as internal combustion engines (ICEs). The promise of FCVs and clean hydrogen guide public expectations and stakeholder activities.

Some early expectations of FCVs anticipated hydrogen FCV commercialization in the early 2000s. These expectations could not be met; vehicles and infrastructure are still in the demonstration phase. Critics pointed to issues hampering market-introduction: high costs, lack of refueling infrastructure, limited performance and life-expectancy of fuel cells and hydrogen storage (Flotow & Steger, 2000). This criticism contributed to a reorientation of public opinion toward other environmentally friendly propulsion concepts. Battery electric vehicles (BEVs) especially have recently gained popularity. FCVs seem to have lost in a perceived competition between propulsion concepts.

Nevertheless, hydrogen maintains its prospective benefits, and car manufacturers continue to invest substantially in FCVs (Ball & Wietschel, 2009).

In September 2009, a consortium of motor manufacturers, representing approximately 50% of the global market share, signed a *Letter of Understanding on the Development and Market Introduction of Fuel Cell Vehicles*. Daimler, Ford, GM/Opel, Honda, Hyundai/Kia, the Alliance Renault/Nissan and Toyota expressed their confidence that from 2015 onwards a few hundred thousand FCVs would be market-introduced worldwide. Germany was identified as the lead-market for Europe. The firms also called for the build-up of large-scale hydrogen-fueling infrastructure (Daimler, 2009a). It was the first time that several large companies jointly made such ambitious statements. This calls for a deeper enquiry into the status of hydrogen vehicle and infrastructure development in Germany.

11.2. RESEARCH QUESTIONS, THEORY AND METHOD

This chapter contributes to answering the central question of the book on whether the automotive regime has reached a stage of transition, focusing on the role of the hydrogen and FCVs niche therein. The chapter analyzes the German *Clean Energy Partnership* (CEP) and related initiatives. Germany is selected because it is one of the most important motor manufacturing countries worldwide, and German companies are among the most active champions of hydrogen and fuel cell technology (Ehret, 2004). Germany has also been identified as the European lead-market for the introduction of FCVs and hydrogen infrastructure. The German *National Innovation Programme Hydrogen and Fuel Cell Technology* (NIP) is one of the largest hydrogen and fuel cell funding schemes worldwide (Garche, Bonhoff, Ehret & Tillmetz, 2009). CEP is one of the biggest hydrogen vehicle and infrastructure demonstration projects worldwide and the largest in Europe.

Strategic Niche Management (SNM) is used to guide analysis. SNM is a research model aimed at understanding and shaping the process of technological development. The focus of SNM is on the development of technological niches that promise benefits in terms of sustainability (see Hoogma, Kemp, Schot & Truffer, 2002, for further explanation). Five propositions identified within SNM are explored (adapted from Geels & Raven, 2006; Schot & Geels, 2008).

1. Expectations contribute to successful niche-building once they become more robust (i.e., shared by more actors), more specific (i.e., offer clearer guidance) and are of higher quality (i.e., are substantiated in projects).
2. Social networks contribute more to niche development if networks are broad (i.e., involve multiple kinds of actors and relative outsiders) and networks are deep (i.e., members can mobilize commitment and resources).
3. Learning processes contribute more to niche development if they change cognitive frames and assumptions beyond the accumulation of facts.

4. Positive learning outcomes give rise to further and often scaled-up development cycles backed up by new resources, expansion of social networks and more refined learning along emerging trajectories.
5. Alignment of different local niche projects facilitates the formulation of generic lessons and rules for the general niche.

The chapter is structured as follows. Section 11.3 explores the emergence of the hydrogen transportation vision. Section 11.4 investigates CEP and other key hydrogen and fuel cell initiatives. Section 11.5 discusses CEP from a niche development perspective. Section 11.6 concludes whether a hydrogen niche is emerging and how it relates to a possible transition of the automotive regime.

11.3. THE EMERGENCE OF THE HYDROGEN TRANSPORTATION VISION

Since the 1990s there has been increased pressure on car manufacturers to reduce tailpipe emissions. This incentive for developing cleaner cars largely came from California. After decades of severe air quality problems, California adopted the *Low Emission Vehicle Mandate* in 1990. The Low Emission Vehicle Mandate included the *Zero Emission Vehicle* (ZEV) *Mandate*, which prescribed the production of vehicles without any tailpipe emissions starting from 1998. Since only BEVs and FCVs could meet this requirement, the ZEV mandate stimulated the development of these two electric drive options (e.g., Sperling, 1994; Van Den Hoed, 2004).

In 1990, BEV development was ahead of hydrogen FCV development (Chapman, 1998). Expectations surrounding BEVs were boosted earlier that year when General Motors presented the Impact, a two-seated BEV. This vehicle was perceived as a big step toward realizing a functional BEV, and further progress was expected (Collantes & Sperling, 2008). At the time, the ZEV mandate was perceived as a way to create a market for BEVs in the near future (Hoogma et al., 2002).

In the second half of the 1990s, the development of the BEV had disappointing results (technological difficulties combined with limited market), and car manufacturers started to focus on hydrogen and fuel cell propulsion as zero emission concepts (see Chapter 10). This revised focus was partly due to breakthroughs in the field of *Proton Exchange Membrane* fuel cells, making fuel cells more suitable for mobile applications. In 1993 the first hydrogen fuel cell bus was built. This was followed by other FCV demonstration projects. This created more visibility for hydrogen as a future fuel option. The hydrogen and fuel cell vision became more widely accepted at the end of the 1990s, and many hydrogen initiatives were built upon this vision. One of the first commitments was made by the Icelandic government. In 1999, it announced its ambition to become the world's first hydrogen economy by 2050 based on hydroelectric and geothermal energy (Sigfússon, 2003).

In the early years of the new millennium, hydrogen enthusiasm and government commitment increased further. In June 2003, Romano Prodi, President of the *European Commission*, called for a hydrogen revolution and announced the *European Hydrogen and Fuel Cells Partnership* (Prodi, 2003). The same year, US President George Bush announced a $1.2 billion investment in hydrogen and fuel cell research and development. In 2004, Arnold Schwarzenegger, Governor of California, announced the *California Hydrogen Highway Network*. This initiative aimed to make hydrogen accessible for every Californian by 2010 (Schwarzenegger, 2004).

This period of great enthusiasm and investment accelerated hydrogen and fuel cell development. Although many of the technical goals were reached (Gronich, 2010), the enthusiasm was combined with a demand for instant visible results. When a delay occurred, important expectations were not met. The Californian ambition was not matched with sufficient funding, resulting in a lag in hydrogen infrastructure development. Due to the financial crisis, little is expected regarding future state funding. On the federal level, much of the hydrogen enthusiasm left the White House together with President Bush. In 2009, Secretary of Energy Chu stated that it would take miracles for hydrogen to come about (Bullis, 2009). This statement was combined with a proposed cut in the hydrogen budget of $100 million (Service, 2009) together with an increase in budget for BEVs and plug-in hybrids. Eventually, the 2009 hydrogen budget got largely restored.

Although public hydrogen investments have largely been maintained in the United States, concern has risen within the hydrogen community. Currently, international attention for hydrogen research and development has shifted more toward Germany, due to Germany's long-term planning and consistent investments (Levin, 2010).

11.4. HYDROGEN AND FCVS IN GERMANY

The *Fuel Strategy* of the German government sets the direction for the envisaged evolution of transport fuels and propulsion concepts. It is part of the *Strategy for Sustainable Development* adopted in 2004 (see www.bmvbs.de).The Fuel Strategy was inspired by the activities of the *Transport Energy Strategy* (TES) and the CEP. Yet, the Fuel Strategy is a self-contained document and the most comprehensive and authoritative outline of public policies toward cleaner road transport (see Figure 11.1). It sets the context in which the TES (focusing on alternative fuels) and CEP (focusing on hydrogen vehicles and infrastructure) are placed.

The Fuel Strategy, drafted by the *Federal Ministry of Transport, Building and Urban Affairs* (BMVBS) after extensive stakeholder consultation, suggests that future vehicle propulsion will be more diversified than today. ICEs continue to play a central role in the medium term, though alternative fuels become more important. At the same time, ICE-hybrids and second-

National 'Fuel Strategy': Evolution of Fuels and Vehicle Technology

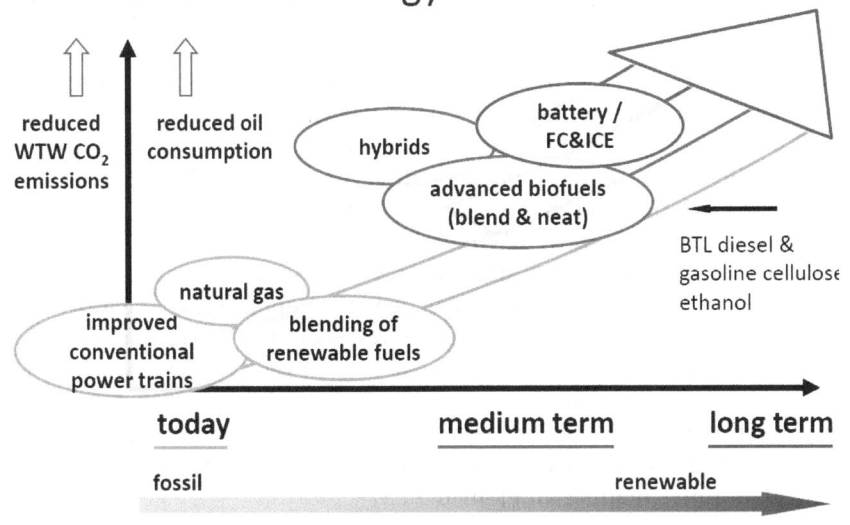

Figure 11.1 National organization hydrogen and fuel cell technology (Ehret, 2009).

generation biofuels play a bigger role. In the medium to longer term, FCVs and BEVs become most important. They are regarded complementary technologies bringing about an *electrification* of road transport seen as the answer to sustainability problems.

Next to TES and CEP, the National Innovation Programme Hydrogen and Fuel Cell Technology, and related initiatives such as H_2 *Mobility*, are central to the German hydrogen and FCV activities. Hydrogen and FCVs are among the propulsion options regarded most promising by the Fuel Strategy and thus receive the strongest governmental support. The initiatives discussed next focus on specific elements captured by the Fuel Strategy, without denying the potential of alternatives.

11.4.1. The TES

The TES was initiated in May 1998 by seven companies and the German *Federal Ministry of Transport, Building and Housing* (BMVBW), the precursor to BMVBS. The motor manufacturers DaimlerChrysler, BMW, MAN and Volkswagen were involved, as well as the mineral oil companies Aral, RWE and Shell. TES aimed at identifying the *fuel of the future* to facilitate more sustainable road transport in the longer term. It was to be renewable and to drastically reduce carbon dioxide emissions and mineral oil dependency (Heuer & Scheuerer, 2002).

DaimlerChrysler and BMW were the main champions of TES. This was motivated by concerns about the supply security of mineral oil, the growing environmental problems of road transport and increasingly stringent emission legislation and the hope for competitive advantages. Also, the United States was an important export market, and ZEV requirements had to be catered to (Lloyd, 2000). This was less important for car manufacturers such as PSA Peugeot Citroën that mainly served European markets. DaimlerChrysler and BMW were convinced that the introduction of fuel cell and hydrogen vehicles required the consensus and support of all major stakeholders.

DaimlerChrysler and BMW approached the other parties that later became TES members. BMVBW had goals similar to DaimlerChrysler and BMW and was keen to encourage industry-wide solutions rather than scattered innovation activities and thus pledged to support TES. Generally, the prospective corporate partners had interests similar to DaimlerChrysler and BMW but were less supportive of hydrogen vehicles. This was due to different appraisals of market positions, technological prospects and oil supplies (Ehret, 2004). Nevertheless, MAN, Volkswagen, Aral, RWE and Shell agreed that alternatives to fossil-based propulsion had to be examined in a collaborative manner and joined TES.

The companies committed expertise and funding to TES; BMVBW ensured legitimacy and alignment with wider governmental policies (Fabri et al., 2000). TES offered a forum for public-private sector exchange on future propulsion. It focused on technology evaluation but also facilitated some reconciliation of differences. Collaboration was helped by a political tradition of corporatism, as German policy-making places faith in public-private sector collaboration. The industry sought to ensure that future investments would not be devaluated by incompatible political decisions. Government wanted to harness the industrial innovation potential to meet wider societal goals.

The TES partners believed that a transition to more sustainable transport would be inevitable in the longer term. The companies and BMVBW agreed upon a *vision of sustainable energy supply and mobility* aimed at a "long-term contribution to the reduction of CO_2 emissions, the reduction of the transport industry's dependence on crude oil, [and] the conservation of finite resources" (BMVBW, 2001). This suggested an increasing reliance on domestically produced renewable fuels.

The industry participants took over all costs and operational work, and the ministry committed to participate in strategic discussions relating TES to wider political agendas (Nierhauve, 2002). The partners selected 10 possible future fuels for closer scrutiny. In conjunction with the fuel assessments, fuel cells and ICEs were evaluated. The consultancy *L-B-Systemtechnik GmbH* (LBST) and the *Institute for Energy and Environmental Research* (IFEU) were entrusted with the analysis to ensure transparency and credibility. Whereas LBST tends to be affirmative of fuel cells and hydrogen, IFEU is known for its rather skeptical stance. By January 2000,

Introducing Hydrogen and Fuel Cell Vehicles in German 235

hydrogen, methanol and natural gas were preliminarily identified as future fuels. The fuels performed similarly on balance but differently on individual evaluation criteria. By June 2001, following high-level discussions between industry and BMVBW, hydrogen produced with renewable or low-carbon energy was identified as the most promising fuel. This was mainly due to the expected long-term environmental benefits (BMVBW, 2001). Later, high-level discussions recommended the formation of CEP to demonstrate the capability of hydrogen vehicles and infrastructure to facilitate a more sustainable road transport.

11.4.2 The CEP I

In mid-2002, the plans for realizing CEP took shape with the signing of a *Memorandum of Understanding* toward demonstrating hydrogen vehicles, filling stations and production facilities in Berlin. Corporate membership partly overlapped with TES and grew by one. All of the companies held stakes in hydrogen and/or fuel cell technology, had produced or operated hardware and accepted the conclusions of TES. The public presentation of the Memorandum of Understanding was hosted by the BMVBW, which lent political support to the initiative. Filling station operator Aral was to build a hydrogen fueling station (HFS), and a small fleet of vehicles was to be tested for up to 5 years. Aral expected a hydrogen fuel market to emerge in the mid-term and wanted to prepare for it. The government acknowledged the CEP as an element of the Fuel Strategy, and BMVBW continued its involvement in decision making (Ehret, 2004).

Technical learning about performance and required improvements to vehicles and infrastructure were the central tasks of CEP. Raising public awareness and exploring public acceptance represented further goals (CEP, n.d. c). The parties involved understood that customer expectations as well as regulative and political requirements had to be met. However, it was regarded as necessary to ensure that the technology worked before exploring the context of innovation more deeply. The basic design rationale of hydrogen vehicles is to offer a full-fledged alternative to conventional vehicles. Hence, the need to learn about customer behavior was regarded less pronounced than with innovations involving greater changes in user behavior.

In October 2003, the vehicle producers DaimlerChrysler, BMW, GM/Opel and Ford; the energy companies Aral, Linde and Vattenfall; an electrolyzer manufacturer today owned by Statoil; and the bus operator Berliner Verkehrsbetriebe signed the CEP contract (CEP, n.d. a). CEP members were expected to accept the vision of sustainable and emission-free mobility carried over from TES and to commit substantial financial resources. As with TES, CEP was based on public private cooperation. Industry was to undertake all operational tasks and bear most of the costs. The government was to participate in key decisions and contribute limited financial resources. One important goal of CEP was to use hydrogen produced from no or low carbon sources, although full provision of clean hydrogen was only regarded feasible in the longer run.

In November 2004, CEP celebrated a milestone by opening its first filling station. This Aral HFS dispensed both liquid hydrogen (LH_2) and compressed gaseous hydrogen (CGH_2) at 350 bar. Much of the hydrogen was produced by Linde from natural gas, liquefied using renewable energy, shipped to Berlin and dispensed as LH_2. Additional hydrogen was generated on site by electrolysis and dispensed as CGH_2. The electrolyzer was supplied by Statoil and fed with renewable electricity from Vattenfall. Situated at the Aral HFS, a technical service and repair station for hydrogen vehicles was jointly operated by the motor companies. The facilities were necessary to accommodate the higher service and repair requirements of demonstration vehicles and showed the unusual willingness of competitors to engage in joint learning experiences.

The mineral oil company Total became a CEP partner in April 2005 for reasons similar to Aral. In March 2006, Total opened the second HFS of CEP, which could fuel both cars and buses. Hydrogen was partly supplied as LH_2 and partly produced on site with a steam reformer. Both 350 bar CGH_2 and LH_2 were dispensed (CEP, n.d. b, n.d. c). Both HFSs were made publicly accessible. This reflects the principal decision to strengthen public involvement by only allowing public HFSs in CEP.

The technological learning experiences with hydrogen infrastructure were generally positive but mixed in regard to specific components. Supply and storage of LH_2 did not pose substantial problems. Still, based on the operating experience, the initial LH_2 dispensing technologies at the Aral site were improved. A second generation tank coupling performed much better than the original. Thus, straightforward technological learning took place with positive results. The CGH_2 production via electrolysis at Aral exceeded expectations. LH_2 refueling with Total benefited from the experiences at Aral and used the latest tank coupling technology without substantial problems. The on-site reformer technology at Total was not satisfactory in terms of reliability and cost. CGH_2 refueling proved unproblematic at both HFSs (CEP, n.d. a, n.d. b). Technological learning thus largely confirmed expectations or led to technological improvements.

All of the car manufacturers committed hydrogen vehicles to CEP: DaimlerChrysler committed 10 A-Class f-cell FCVs, Ford committed 3 Focus FVCs, BMW committed 2 Hydrogen 7 sedans with H_2 combustion engines and GM/Opel committed 1 HydroGen3 FCV. Volkswagen joined CEP in July 2006 and provided one Touran HyMotion FCV. All FCVs were hybrid cars, using a battery supporting the fuel cell. On average, 17 hydrogen cars were operated between mid-2005 and mid-2007; the highest number operating at one time was 24. The driving range of the FCVs was between 170 and 400 km; the BMW sedans offered greater ranges. Depending on the vehicle, hydrogen compressed at 350 bar or LH_2 was used. Close to 400,000 km were driven by all the cars with an overall performance that

Introducing Hydrogen and Fuel Cell Vehicles in German

was better than expected (CEP, n.d. a, n.d. b, n.d. c). Most of the vehicles used GH_2 storage without energy losses, and boil-off only occurred in the LH_2 tanks used in some cars. This encouraged research and development activities on developing innovative storage solutions. Starting in June 2006, a fleet of 14 buses with H_2-ICEs produced by MAN was introduced and demonstrated as part of the EU *HyFLEET:CUTE* project. The buses used the Total HFS even though the project was not part of CEP (see www.global-hydrogen-bus-platform.com).

The cars were leased to a range of governmental organizations and companies and tested in everyday use. A total of 180 drivers were instructed on vehicle use, and the overall feedback was very positive. Most drivers appreciated the strong torque of the electric engines and the environmental benefits the cars offered. Only some critical comments were made regarding the limited driving range and occasional problems with the HFSs. The vehicles had a high profile and were well received by the public. Similar results appeared in surveys of the HyFLEET:CUTE project. Various public relations events made CEP more widely known and increased technology acceptance. Ministers, corporate executives and senior scientists from Germany and abroad were involved. A convention centre was set up and hosted many information events (CEP, n.d. c).

Learning and decision making occurred on various dimensions. There was learning within individual companies, mainly regarding technological performance. Learning between firms occurred through everyday contacts and working group meetings. The operation of the Aral HFS, jointly run by several partners, was especially conducive to collaboration and learning about infrastructure. Common service and repair facilities at the Aral HFS represented a focal point for exchange regarding vehicle operations (CEP, n.d. a). Senior staff regularly met in steering committee meetings. BMVBW, which was renamed to BMVBS during CEP I, always participated. The steering committee was responsible for strategic decisions and the overall running of CEP. Technological matters were of central interest; but market and policy issues were also discussed.

Often, corporate interests partially diverged, and strategy was negotiated in a collective manner. Individual companies never sought to forcefully impose their views on other partners. The BMVBW reserved the right of veto regarding key policy issues. This veto right had been required for incorporating CEP in the Sustainability Strategy. The primacy of political authority was never questioned by the companies, and the ministry assumed a *primus inter pares* position (Ehret, 2004). Government participation ensured that the progress of CEP was embedded in the wider considerations on alternative propulsion and future road transport outlined in the Fuel Strategy.

The overall project costs of CEP I were €40 million. Industry invested €35 million, largely for vehicle operations. Government contributed €5 million, essentially for infrastructure (CEP, n.d. a, n.d. c). The financial contribution of the public sector and the resource of political legitimacy it provided through inclusion of CEP in the Fuel Strategy helped to ensure that political imperatives were respected. After CEP I, the general perception was that FCVs worked well enough to warrant further consideration as a promising future option, although improvements regarding performance, durability and costs were required.

11.4.3. Upgrading of CEP

Initially, CEP was designed as a 5-year self-contained project. Toward the end of what became known as CEP I, the project was extended by two phases. CEP II covers the period 2008 to 2010 and aims at upgrading and expanding hydrogen vehicles and infrastructure, based on CEP I experiences. It also emphasizes greater public involvement and an increasing share of renewable hydrogen production. CEP III is scheduled for 2011 to 2016 and focuses on commercialization. At the time CEP I was extended, further German hydrogen and fuel cell initiatives materialized. The activities are described and related to CEP next.

11.4.4 CEP II (2008–2010): Expansion and Modernization

The city of Hamburg and the state of North Rhine-Westphalia joined Berlin as regions of CEP II (www.cleanenergypartnership.de). With the powerful newcomers Shell, Toyota and Hamburger Hochbahn, corporate membership grew to 13 (compared to 9 when the contract was signed) and is expected to grow further (CEP, n.d. d).

The extension of CEP and the accession of new members represented milestones within CEP development. However, CEP II also experienced a setback when Aral left due to a strategic reorientation. This resulted in the closure of the HFS operated on the premises of the company. CEP II also involves enhanced international cooperation. CEP entered collaborative agreements with the *California Fuel Cell Partnership* (www.cafcp.org) and with *HyNor*, the Norwegian project for the introduction of hydrogen as fuel for transportation (www.hynor.no).

The Berlin car fleet has been expanded to 47 and continues to grow. The bus fleet formerly run within HyFLEET:CUTE became part of CEP and continues its operation until 2010. In Hamburg a fleet of at least 20 new-generation FCVs is being introduced. Up to 6 Daimler fuel cell buses previously operated within HyFLEET:CUTE by Hamburger Hochbahn joined CEP but are being replaced by 10 new-generation Daimler fuel cell buses. New-generation FCVs, operating at 700 bar CGH_2, have been introduced by GM/Opel and Daimler. Both have a range of up to 400 km, which far exceeds previous models. Ten GM/Opel HydroGen4 are being leased to a diverse set of customers. These and other vehicles were shown to wider audiences in public events such as a rally between Berlin and Hamburg

Introducing Hydrogen and Fuel Cell Vehicles in German 239

(CEP, 2010). The rally was designed to celebrate two new HFSs and demonstrate the increased range of the new models (www.cleanenergypartnership.de). Such events, along with the involvement of more diverse users, indicates increasing user involvement. Marketing studies were not undertaken in CEP, as market research was regarded a company task.

The presentation of the new Daimler B-Class f-cell cars was accompanied by a rare public announcement regarding the lifetime of fuel cells. Daimler's CEO Zetsche expects a stack-lifetime of 110,000 km for today's FCVs and of 250,000 km for FCVs built 2020 (Zetsche, 2010). This is in comparison to a life of 160,000 km for conventional ICE-vehicles today. In the new generation Daimler buses, comparable improvements have also been achieved (www.now-gmbh.de). According to Zetsche, it will be possible to sell FCVs at the price of ICE-hybrid vehicles in the medium term. GM/Opel went public with similar statements. Toyota is the first Asian company participating in CEP and brings in new-generation FCVs with ranges close to 800 km (CEP, 2010). This shows both the growing international reach of CEP and that ranges comparable to conventional vehicles can be achieved.

All of the new generation FCVs in CEP can operate at very low temperatures and have thus overcome a significant problem faced by their predecessors. Costs have significantly decreased while driving performance and reliability have much improved (www.now-gmbh.de). FCVs are produced in growing numbers but are not yet ready for market introduction. The state of the art is roughly in line with the announcement of motor manufacturers anticipating market-introduction in 2015 (see introduction). Due to decisions to shift some research and discovery resources, not all of the manufacturers involved in CEP have recently produced new-generation vehicles. For instance, BMW decided to develop innovative hydrogen storage solutions and to improve H_2-ICE technology on the component level before producing a new demonstration fleet (www.cleanenergypartnership.de). Moreover, all manufacturers started to allocate financial resources to BEVs, competing with research and discovery resources for hydrogen vehicles. Economic problems generally limited research and discovery funds available in the industry.

Within CEP, developments continue with the construction of additional and larger filling stations in Berlin and Hamburg. In addition to the 350 bar CGH_2 technology used to fuel buses and older FCVs, 700 bar CGH_2 technology for latest-generation FCVs is being introduced across CEP. This allows for greater driving ranges and reduced fueling times (~3 minutes), approximating the standards of conventional vehicles (www.cleanenergypartnership.de). In Berlin, Total opened a second and larger HFS and announced the construction of a third (CEP, n.d. d). Both HFSs dispense hydrogen from *wind-hydrogen-systems* that use wind electricity for hydrogen production (www.enertrag.com, www.total.de). In Hamburg, construction work started for the largest European HFS, which will partly use hydrogen made from renewable sources. Additional mobile HFSs are in use. Pursuing a goal set by BMVBS and readily accepted by the companies, the share of renewable energy in hydrogen production is being increased to at least 20% by the end of 2010 (CEP, 2010).

11.4.5. CEP III (2011–2016): commercialization

CEP III is scheduled to last from 2011 to at least 2016 and aims at technology commercialization (www.cleanenergypartnership.de). Berlin and Hamburg are to be connected by an *infrastructure corridor* with HFSs along a motorway. In North Rhine-Westphalia, a new player, gas company *Air Liquide*, will operate an HFS together with Total. Further hydrogen corridors linking regions are envisaged. CEP III shares the goal of CEP II to expand and modernize hydrogen vehicles and infrastructure, benefitting from prior experiences. CEP expects increases in scale and alignment with related initiatives (CEP, n.d. a, n.d. c, n.d. d). Since Germany is the European lead market for the announced roll-out of a few hundred thousand FCVs starting 2015 (see introduction), CEP expects to attract a significant number of the vehicles. In 2009, Hamburg announced that 500 to 1,000 FCVs might be introduced before 2015 (www.hamburg.de/senat/). As for hydrogen production, the declared aim is to rely on 50% renewable energies by 2015 and an increasing share thereafter.

11.4.6. Increasing dynamics beyond CEP

In parallel to the preparations for CEP II, the National Innovation Programme Hydrogen and Fuel Cell Technology was adopted by the German government in May 2006 (BMVBS et al., 2006). The overall task is to support preparations for the market introduction of hydrogen and fuel cell technologies in the mobile and stationary sectors, as well as in special markets such as marine power supply. NIP aims at improving technology performance, reliability and costs. A *National Development Plan* (NEP) specifies the agenda for technology development (Strategierat Wasserstoff und Brennstoffzellen, 2007). BMVBS has contributed €500 million for demonstration projects to the NIP, and the *Federal Ministry of Economics and Technology* committed €200 million for research and development projects. The combined public funds worth €700 million are to be matched by roughly the same amount contributed by industry and other bodies running the projects. The total NIP budget comprises €1.4 billion over the program duration (2007–2016). Over half of the budget is allocated to mobile applications such as fuel cell cars and buses, including hydrogen production and distribution. In March 2009, €15 million from the *Economic Stimulus Package* were added to the NIP budget and allocated to the build-up of additional HFSs (www.now-gmbh.de).

NIP was adopted in response to increasing environmental, energy resource and economic problems, especially those of the transport sector. The program

Introducing Hydrogen and Fuel Cell Vehicles in German

design, with public-private collaboration and its focus on hydrogen and fuel cells, was derived from the CEP learning experiences. To avoid the limited effects that isolated and discontinuous policies often have, the 10-year NIP takes a strategic and long-term perspective, which facilitates the exploitation of learning experiences over time. NIP was conceived as a public-private initiative, with a wide range of stakeholders from government, industry and science influencing implementation (Garche et al., 2009).

The *National Organization Hydrogen and Fuel Cell Technology* (NOW) was established in February 2008 to coordinate and implement NIP, together with the *Project Management Organization Jülich* (www.now-gmbh.de). Since March 2009, NOW has also managed about a quarter of the €500 million made available for BEVs and HFSs within the Economic Stimulus Package. NOW is a company fully owned by the German federal government, mainly represented by BMVBS. One of NOW's main tasks is to align the many activities in the technology areas that NIP supports in order to give strategic direction to technology development. Another core task is to evaluate project funding applications. In considering applications, evidence of substantial environmental benefits is a key evaluation criterion. The organizational setup of NOW ensures close interaction with stakeholders. An *Advisory Board* with 18 members from politics, science and industry supervises the implementation of NIP and, by updating NEP, sets the framework within which NOW operates (Strategierat Wasserstoff und Brennstoffzellen, 2007).

Since its founding, NOW has replaced BMVBS as the representative of governmental interests in CEP (www.now-gmbh.de). Some 50% of the overall costs of CEP II (as compared to a much smaller contribution during CEP I) are borne by NOW and the same funding regime is anticipated for CEP III. Once it was operational in 2008, NOW transferred all activities of CEP II to a *Lighthouse Project* (i.e., a large-scale and highly visible integrated project composed of various individual projects run by different partners). There is a module concerned with project coordination, knowledge management and external communication. Similar to CEP I, a steering committee takes strategic decisions on technological, market and policy-related matters. Joint working groups facilitate learning. NOW participates in all steering committee meetings and ensures that political imperatives are respected. NOW also makes sure CEP is aligned with related research and development and demonstration activities in Germany. As a quasi-governmental organization with additional resources, NOW strengthens the alignment of CEP with wider policy agendas. Further alignment is achieved through international contacts with related organizations such as the EU *Fuel Cells and Hydrogen – Joint Undertaking*.

NOW supports the provision of low and no carbon hydrogen in the wider NIP. A study called *GermanHy* investigated how the expected future demand for hydrogen fuel in Germany could be met (Dena FZK, ISI, L-B-Systemtechnik GmbH & Wuppertal Institut, 2009). A political requirement was that at least 50% of the energy used had to be renewable. The study showed that up to 70% of cars and light-duty vehicles could run on fuel cells by 2050 and that enough clean hydrogen to fuel the vehicles could be made available. GermanHy concluded that hydrogen-based mobility would be possible at costs comparable with today's vehicles if the development targets for FCVs are met. The study also found that CO_2 emissions from cars and light duty vehicles could be reduced to 20g CO_2/km (tank to wheel) by 2050, compared to some 160g CO_2/km in 2010 (car fleet average). In late 2009, NOW presented a draft *Strategy Paper Hydrogen Production* to the Advisory Council (Ehret & Bonhoff, in press). The paper informs the ongoing revision of NEP and strongly emphasizes the importance of wind energy and biomass in hydrogen production. If the Council accepts the propositions, hydrogen production in NIP will largely rely on renewable energy and involve very low emissions. Several projects regarding renewable hydrogen are running in 2010 and may supply vehicles operated in CEP and elsewhere with clean fuel in the future.

A key event was the signing of a *Memorandum of Understanding* in September 2009 on the build-up of hydrogen infrastructure in Germany by the large energy companies EnBW, Linde, OMV, Shell, Total and Vattenfall, as well as Daimler and NOW (Daimler, 2009b). The memorandum introduced the initiative H_2 *Mobility*, which aims at building up hydrogen infrastructure, to facilitate the market introduction of FCVs expected to commence in 2015. H_2 Mobility takes a two-stage approach. Between 2009 and 2011, technological concepts and business models are being developed. Several HFS are constructed in the first phase as well, adding to the total of 30 HFS in place in Germany today. Providing that an agreement between the parties involved can be found, the plans devised in phase one are to be implemented in phase two starting in 2011. While the precise content of the roll-out strategy will only emerge after completion of negotiations, the aim is to achieve nationwide HFS coverage after 2015. Germany is to be the nucleus for the anticipated build-up of a European-wide infrastructure. H_2 Mobility represents the most ambitious push toward building up hydrogen infrastructure thus far. Several companies involved in CEP extended their commitment to H_2 Mobility. Moreover, with EnBW and OMV, two large companies joined the champions of hydrogen fuel in Germany. Apart from the overlap in membership between CEP and H_2 Mobility, the involvement of NOW in both initiatives strengthens alignment.

11.5. DISCUSSION

This section discusses CEP developments in relation to SNM. SNM emphasizes the importance of non-regime actor involvement for successful niche development processes (e.g., Hoogma et al., 2002). In this case study, the role of regime actors in niche development is prominent. These established actors operate strategically to accommodate hydrogen vehicles in a viable niche furthering the shared vision of vehicle emission reduction (ideally zero-emission), in conjunction with domestically produced renewable fuel.

The TES and the CEP evolved from the perception that a transition to a more sustainable transportation technology was necessary and might be profitable. This indicates that tensions at the landscape level (essentially the challenges road transport posed to sustainability) shaped actor behavior at the level of the automotive regime. Human agency translated the landscape pressures into niche activities. This section characterizes the dynamics within CEP and matches the dynamics with SNM propositions.

CEP I was a demonstration project run by regime players. The shared vision of sustainable and emission-free mobility carried over from TES provided guidance and direction. In tune with the transition perspective, the main emphasis was working toward achieving long-term sustainability objectives. TES and CEP are joint public-private sector innovation efforts aimed at bringing about transition to a more sustainable transport system. The vision was also crucial to attracting governmental support in terms of inclusion of CEP in the Fuel Strategy and in terms of a financial contribution.

When relating TES and CEP to the SNM proposition of Section 11.2, the following dynamics stand out.

1. TES was based on the expectation that alternatives to conventional vehicle propulsion systems and fuels had to be identified. Daimler and BMW had a clear preference for hydrogen and/or fuel cells as sustainable future technologies, while other actors joining TES were more open for alternatives. Out of 10 alternatives, hydrogen was selected as *the* future fuel. Over time, the expectations surrounding hydrogen vehicles became more shared within a growing network. In CEP II, additional regions and companies joined. Expectations also became more specific based on technological development and testing. Both liquid hydrogen and compressed gaseous hydrogen (350 and 700 bar) were tested, and a growing emphasis was placed on 700 bar technology. With the establishment of NIP and NOW, public and private sector funding for hydrogen and fuel cell demonstration and research and development increased exponentially. Both the hydrogen infrastructure and the number of vehicles grew. More actors with specific technological expertise became involved, creating a more versatile network with a technological focus. Within CEP, expectations grew more specific over time both in relation to fuel and infrastructure characteristics. Again expectations became more widely shared. From an SNM perspective, the expectation dynamics have been conducive to niche development. Expectations are still high and widely shared among the actors. This is illustrated by CEP III and the large-scale introduction of FCVs expected from 2015 onwards.

2. Throughout TES and CEP development, actor networks have been growing. Motor companies took the initiative to create TES, and when BMVBW and the mineral oil firms joined, the social network quickly widened. The participation of the powerful ministry added legitimacy and brought in more resources. Further widening occurred with the involvement of two research organizations, especially the IFEU institute with its critical stance toward hydrogen. In CEP I, more companies joined the niche-activities. There was also customer involvement and public relations work. During CEP II, two regions joined and NOW increased governmental representation. Thus, the network grew to incorporate a wider range of corporate and political actors. Some relative outsiders were also involved, such as bus operators and users. Most participants were powerful and could mobilize substantial commitment and resources. CEP brought together actors interested in fuel cell and hydrogen-ICE vehicles and hydrogen production and refueling infrastructure. CEP might be regarded as a corporatist arrangement driven by public and private sector interests in cleaner cars and sustainable fuel provision. Many network actors had very specific technological knowledge, and the network mainly involved deep relationships, rather than a large variety of stakeholders.
3. Learning in TES and CEP initially focused on technological issues, but later increasingly concerned also policy- and market-related matters. TES involved first-order learning in analyzation of data, but resulted in second-order learning in gearing expectations and mindsets toward hydrogen. Learning in CEP I and II was mostly first-order positive, confirming expectations without stimulus for second-order change of basic assumptions. The most notable second-order learning outcome was the increased focus on 700 bar technology.
4. The positive learning outcomes of TES with regard to hydrogen led to the creation of CEP I, involving increased commitment, resources, networks and refined learning along emerging trajectories. The positive experiences of CEP I, in terms of operating experience with vehicles and infrastructure, as well as collaboration benefits, then brought about the larger CEP II. CEP I also markedly influenced the design of NIP and NOW. Thus, positive learning gave rise to further and scaled-up development cycles; contributing to successful niche-building according to SNM.
5. The integration of CEP in the Fuel Strategy ensured that hydrogen propulsion became part of a wider policy approach supported by both the private and public sectors. With the opening of the Total HFS used to fuel buses operated within HyFLEET:CUTE, the first alignments between the niche projects CEP I and HyFLEET:CUTE occurred. Some of the buses later became fully integrated parts of CEP II. CEP II also benefits from collaborative agreements with

international partner organizations and from its inclusion in NIP. NOW ensures that CEP is aligned with related activities at the national and international level. Notable is the collaboration both between competing companies from different industries in TES and CEP and between the private and public sector. This indicates a level of alignment far beyond scattered innovation activities. The exchange with related niche activities facilitates the formulation of generic lessons for the general niche; thus helping niche development according to SNM.

The analysis has shown that the niche demonstration project CEP has grown in scope and alignment with related activities. CEP is a long-term project firmly embedded in NIP. One of the reasons to conceive NIP as a 10-year program was the recognition that efforts had to be aligned and required continuous and longer term support. The hydrogen niche is growing and enjoys support from powerful corporate actors, as apparent from recent initiatives regarding the introduction of FCVs and infrastructure. Political agents continue to support hydrogen propulsion within the Fuel Strategy. Landscape pressures such as climate change and the depletion of mineral oil are increasing. They put pressure on the current automotive regime and stimulate actor responses. For example, the EU has adopted increasingly stringent emissions regulations and requires new cars to emit less than 100g CO_2/km (tank to wheel) by 2020. This is barely achievable with conventional engines, especially with larger cars.

Gauged against the current automotive regime, FCVs can be regarded as radical innovations, as they involve substantial changes. These changes mainly concern vehicle drive-trains, fuel tanks, fuel production and distribution infrastructure. From a user perspective, FCVs bring about non-disruptive change since they require few changes in user practices (in contrast to BEVs involving reduced driving ranges and longer refueling, i.e., charging, times). Limited attention has been given to user preferences in CEP. There was no specific market research regarding, for instance, consumer willingness to pay a surplus for FCVs. This type of research was left to individual companies. The user experiences were generally very positive. Some drivers criticized the limited range of (previous-generation) cars and occasional problems with refueling.

With the formulation of a vision, the building of social networks and the learning processes at the heart of the TES, key dynamics for the creation of a technological niche are present. With CEP, a demonstration project was announced, a typical feature of emerging regimes. Incumbent manufacturers produce conventional cars and gear up toward producing FCVs. Hence, FCVs can be viewed as a regime-preserving innovation. On the other hand, the production of hydrogen from domestic and renewable energy implies departure from mineral oil at the heart of the established fuel provision regime. Thus, hydrogen usage involves regime change on the fuel side.

11.6. CONCLUSION

This chapter has studied the niche of hydrogen and FCVs in Germany, focusing on the CEP and related initiatives. The chapter contributes to answering the central question of the book regarding whether the automotive regime has reached a stage of transition and what role niche developments play therein.

All the main characteristics of SNM are present within CEP and indicate a growing niche. The strong involvement of actors of a dominant regime in niche development is unusual. FCVs are a niche-preserving innovation supported by regime actors and geared toward offering a more sustainable alternative to conventional cars. FCVs are designed to offer the same performance as regular cars and do not challenge several basic regime characteristics such as private car ownership.

FCV development has been exposed to hype-disappointment cycles. In June 2001, TES identified hydrogen as *the* future fuel. CEP was initiated to demonstrate hydrogen vehicle and infrastructure technologies. However, the arrival of lithium ion batteries enhanced the potential of BEVs. Today, public opinion regards FCVs and BEVs as competing technologies (Romm, 2005) with hydrogen vehicles falling short of attention. International policy and media attention focuses on BEVs, and various BEV niche experiments have been initiated. BEVs are heavily marketed by pioneering entrepreneurs such as Shai Agassi of *Better Place*. The perceived competition between FCVs and BEVs is somewhat superficial, however, since most experts regard both options as complementary in terms of technology and markets.

In principle, *all* FCVs are hybrid vehicles, as they incorporate an advanced battery. Also, FCVs as well as BEVs are electric vehicles. Both share a range of components, such as electric engines and devices for recapturing energy produced when braking, and benefit from technological progress regarding electric drives. The latest Mercedes B-Class FCV, for instance, has been designed in parallel to a new B-Class BEV and incorporates various electric components used in both versions (www.daimler.com). A fuel cell can also be added to a BEV to enhance driving range, making the alternatives even more complementary. BEV and FCV thus constitute related niches, making it sensible for car manufacturers to invest in both vehicle types.

Indeed, most manufacturers see a role for both FCVs and BEVs. FCVs are regarded as a viable alternative for longer range and larger vehicles and BEVs as an attractive option for smaller city cars. Most firms have maintained and extended their commitment to FCVs while investing in BEVs in parallel (www.f-cell.de). While it seems likely that the FCV niche will grow further, it is less likely that FCVs will become the one clearly dominating concept. Rather, FCVs have the potential to form one pillar of an electrified and more diversified transport sector, where BEVs might satisfy other transport needs.

Introducing Hydrogen and Fuel Cell Vehicles in German 247

The availability of FCVs and BEVs may well contribute to increasing the sustainability of road transport. Their success will partly depend on oil prices and carbon-constraining policies. Government has been included in the innovation process and can also be expected to facilitate market introduction. In the past, emission and energy efficiency requirements could be met with ICE-vehicles. In the future, regulation requiring electric propulsion might become acceptable to both government and industry. Regulatory influence might strengthen competitive positions and work to the advantage of sustainable development, provided there is good profit in FCVs and BEVs. Incentive schemes supporting commercialization would reduce the extra costs of early-market vehicles. Given customer reluctance to accept significantly higher prices for clean vehicles, it is likely that incentive schemes will be required to facilitate market-entry.

Despite the zero-emissions nature of the FCV itself, the degree of sustainability strongly depends on the method of fuel production. In contrast to regime innovation characterizing technology development on the vehicle side, clean hydrogen production requires *both* radically new production technologies and markets and thus involves transformative change on the fuel side. So the development of the FCV niche comprises regime-preserving, as well as regime-changing, elements. Fuel provision is no longer a taken-for-granted and remote landscape condition of mobility. Rather, the automotive regime needs to change to incorporate sustainable and largely domestic fuel production. Since FCVs and hydrogen involve interdependent innovations, taken together they may account as dynamics of transformation rather than radical change.

Actors are strongly committed within CEP and are keen to establish Germany as a lead market for FCVs and infrastructure starting from 2015. However, changing circumstances, such as breakthroughs in other technologies developed within manufacturers' drive train portfolios may shift the attention of the automobile industry. In that case, Germany would run the risk of stranded assets. On the other hand, if expectations continue to be met, Germany's pioneering activities could result in a leading position in a growing worldwide FCV market.

REFERENCES

Ball, M. & Wietschel, M. (eds) (2009) *The Hydrogen Economy: Opportunities and Challenges*, Cambridge, England: Cambridge University Press.

Bullis, K. (2009) The secretary of energy talks with *Technology Review* about the future of nuclear power post Yucca Mountain and why fuel cell cars have no future. Available at http://www.technologyreview.com (accessed 8 August 2011).

Bundesministerium für Verkehr, Bau und Stadtentwicklung (Federal Ministry of Transport, Building, and Urban Affairs [BMVBS], Bundesministerium für Bildung und Forschung, Bundesministerium für Wirtschaft und Technologie (Federal Ministry of Education and Research, federal Ministry of Economics and

Technology) (2006) Nationales Innovationsprogramm Wasserstoff- und Brennstoffzellentechnologie (National Innovation Programme Hydrogen and Fuel Cell Technology). Available at http://www.bmvbs.de (accessed 8 May 2006).

Bundesministerium für Verkehr, Bau- und Wohnungswesen [BMVBW] (2001) Transport energy strategy (TES): A joint initiative from politics and industry: Second task force status report to the Steering Committee. Available at http://www.bmvbs.de (13 June 2001).

Chapman, R.M. (1998) *The Machine That Could; PNGV, A Government-Industry Partnership*, Critical Technologies Institute, Washington: RAND.

Clean Energy Partnership (n.d. a) *Bericht 2002–2007*, Berlin: Clean Energy Partnership. Available at http://www.cleanenergypartnership.de (accessed 14 October 2011).

Clean Energy Partnership (n.d. b) *Clean Energy Partnership—mobil mit Wasserstoff (On the move with hydrogen)*, Berlin: Clean Energy Partnership.

Clean Energy Partnership (n.d. c) *Report 2002–2007*, Berlin: Clean Energy Partnership. Available at http://www.cleanenergypartnership.de (accessed 17 September 2009).

Clean Energy Partnership (n.d. d) *7 Schritte auf dem Weg zu sauberer Mobilität: CEP Phase II 2010 (7 Steps on the path to clean mobility: CEP Phase II 2010)*, Berlin: Clean Energy Partnership.

Clean Energy Partnership (2010) *Projektbeschreibung CEP: Stand 02.03.2010 (Project outline CEP: Status 02.03.2010)*, Berlin: Clean Energy Partnership.

Collantes, G. & Sperling, D. (2008) 'The origin of California's zero emission vehicle mandate', *Transportation Research Part A*, 42(10): 1302–1313.

Daimler (2009a) Presse-Information: Autohersteller treiben Elektro-Fahrzeuge mit Brennstoffzellenantrieb voran. Available at http://www.daimler.com (accessed 9 September 2009).

Daimler (2009b) Presse-Information: ‚H$_2$ Mobility'—Gemeinsame Initiative führender Industrieunternehmen zum Aufbau einer Wasserstoffinfrastruktur in Deutschland. Available at http://www.daimler.com (accessed 10 September 2009).

Deutsche Energie-Agentur (Dena), Forschungszentrum Karlsruhe (FZK), Fraunhofer-Institut für System-und Innovationsforschung (ISI), Ludwig, Bölkow-Systemtechnik (LBST), Wuppertal Institut für Klima, Umwelt, Energie (Wuppertal Institut) (2009) GermanHy: Studie zur Frage: 'Woher kommt der Wasserstoff in Deutschland bis 2050?' ('GermanHy: Study concerning the Question: 'Where will the hydrogen in Germany come feom by 2050?' Available at http://www.germanhy.de (accessed 14 October 2011).

Ehret, O. (2009, November) 'Wasserstoffproduktion und-Infrastruktur im Nationalen Innovationsprogramm Wassertoff- und Brennstoffzellentechnologie' ('Hydrogen Production and Infrastructure in the National Innovation Programme Hydrogen and Fuel Cell Technology'), paper presented at the Energie-Symposium, Nutzung regenerativer Energiequellen und Wasserstoff der fachhoshschule Stralsund Conference, Straslund, Germany.

Ehret, O. (2004) 'Technological innovation and its social control: An analytical framework and its application to clean vehicle propulsion', unpublished doctoral thesis, Cardiff University, Wales.

Ehret, O. & Bonhoff, K. (2010, December) 'Introducing hydrogen as a future fuel: Strategies and activities in Germany', Proceedings of the World Hydrogen Energy Conference 2010, Essen, Germany. Available at http://julib.fz.juelich.de/vufind/ (accessed 14 October 2011).

Fabri, J., Heinrich, H., Heinen, J., Heuer, W., Huss, C., Krumm, H., Nierhauve, B., Schaller, K.V. & Wolf, J. (2000) 'TES—An initiative for tomorrow's fuel', in C. Winter (ed) *On Energies-of-Change—The Hydrogen Solution: Policy, Business, and Technology Decisions Ahead*, Munich, Germany: Gerling Akademie.

Flotow, P. & Steger, U. (eds) (2000) *Die Brennstoffzelle—Ende des Verbrennungsmotors?Automobilhersteller und Stakeholder im Dialog (The Fuel Cell-the End of the Combustion Engine? Automobile Manufacturers and Stakeholders in Dialogue)*, Bern, Germany: Paul Haupt.

Garche, J., Bonhoff, K., Ehret, O. & Tillmetz, W. (2009) 'The German National Innovation Programme hydrogen and fuel cell technology', *Fuel Cells*, 9(3): 192–196.

Geels, F. & Raven, R. (2006) 'Non-linearity and expectations in niche-development trajectories: Ups and downs in Dutch biogas development (1973–2003)', *Technology Analysis & Strategic Management*, 18(3–4): 375–392.

Geitmann, S. (2002) *Wasserstoff & Brennstoffzellen: Die Technik von morgen! (Hydrogen & Fuel Cells: Tomorrow's Technology!)*, Berlin: Hydrogeit.

Gronich, S. (2010, May) 'Are plug-in/battery electric vehicles more market ready than hydrogen fuel cell vehicles?' NHA Hydrogen Conference, Long Beach, CA.

Heuer, W. & Scheuerer, K. (2002, November) 'VES—Eine Initiative für den Kraftstoff der Zukunft', Paper presented at the Deutscher Wasserstoff-Energietag 2002 Conference, Essen, Germany.

Hoogma, R., Kemp, R., Schot, J. & Truffer, B. (2002) *Experimenting for Sustainable Transport; The Approach of Strategic Niche Management*, London: Spon Press.

Levin, J. (2010) *Interview at AC Transit*, personal communication, Oakland, April, 9.

Lloyd, A.C. (2000), 'Hydrogen and clean-air regulations in California', in C. Winter (ed) *On Energies-of-Change—The Hydrogen Solution: Policy, Business, and Technology Decisions Ahead*, Munich, Germany: Gerling Akademie.

Mytelka, L. & Boyle, G. (eds) (2008) *Making Choices About Hydrogen: Transport Issues for Developing Countries*, Tokyo: United Nations University Press.

Nierhauve, B. (2002, October) 'Verkehrswirtschaftliche Energiestrategie,' Paper presented at the f-cell: Die Brennstoffzelle Conference, Stuttgart, Germany.

Prodi, R. (2003, June) 'The energy vector of the future', Conference on Hydrogen Economy, Brussels, Belgium.

Romm, J.J. (2005) *The Hype About Hydrogen; Fact and Fiction in the Race to Save the Climate*,Washington, DC: Island Press.

Schot, J. & Geels, F. (2008) 'Strategic niche management and sustainable innovation journeys: Theory, findings, research agenda and policy', *Technology Analysis & Strategic Management*, 20(5): 537–554.

Schwarzenegger, A. (2004) 'Governor Schwarzenegger announces California Hydrogen Highways Network', Press Release from the Office of the Governor. Available at http://gov.ca.gov/press-release/3105 (accessed June 18, 2010).

Service, R. (2009) 'Hydrogen cars: Fad or the future?' *Science* 324(5932): 1257–1259.

Sigfússon, A. (2003, December) 'Iceland: Pioneering the hydrogen economy', *Foreign Service Journal*, pp. 62–65.

Sperling, D. (1994) *Future Drive; Electric Vehicles and Sustainable Transportation*. Washington, DC: Island Press.

Strategierat Wasserstoff und Brennstoffzellen (2007) Nationaler Entwicklungsplan: Version 2.1 zum 'Innovationsprogramm Wasserstoff- und Brennstoffzellentechnologie' ('Strategy Council Hydrogen and Fuel Cells (2007) National Development Plan: Version 2.1 Innovation Programme Hydrogen and Fuel cell Technology'), Available at http://www.now-gmbh.de (accessed 30 April 2007).

Van Den Hoed, R. (2004) 'Driving fuel cell vehicles; how established industries react to radical technologies', doctoral thesis, Delft University of Technology, Delft, the Netherlands.

Zetsche, D. (2010, May) 'Fueling the future: Wasserstoff in der Automobilindustrie', Paper presented at the World Hydrogen Energy Conference 2010, Essen, Germany.

12 Transition by Translation
The Dutch Traffic Intelligence Innovation Cascade

Bonno Pel, Geert Teisman and Frank Boons

12.1. INTRODUCTION

The transitions perspective highlights the systemic nature of current mobility problems. This diagnosis suggests a need for radical change, beyond incremental system improvement that runs the risk of only reinforcing system lock-in. The perspective also points out how radical change can result from the complex interplay of a multitude of changes, that is, how system innovations and transitions emerge from co-evolution (Geels (2005), see Chapter 1).

Investigations of historical transitions have brought forward typical patterns and pathways in the emergence of transitions. Geels and Schot (2007) distinguish transformation, reconfiguration, technological substitution and de-alignment and re-alignment. The normal course of events, however, is stable reproduction. Depending on landscape pressures and the readiness of niche innovations, regular regime reproduction may shift to one of the four transition pathways or combinations of those. The pathways develop through different combinations of landscape change, regime adaptation and niche outsider pressure. Taken together, the pathways tell us how transitions are radical changes that emerge from the interplay of a manifold of innovations and changes on many levels. These separate local changes can be either radical or incremental. The transition perspective values radical changes over incremental changes for their potential for lock-out, but radical innovations do require a receptive regime and favorable landscape developments to join the co-evolutionary game of transitions (Schot & Geels, 2008). By contrast, the odds for incremental innovations are more favorable, as posited with the 'default' pattern of reproduction.

Especially interesting in this regard are innovation 'cascades', incremental changes in one subsystem triggering further changes in other subsystems through 'knock-on effects', resulting in system-wide change (see Chapter 2; see also Rotmans, 2006). This technology-triggered accumulation of incremental change exemplifies how local incremental changes can resonate through a socio-technical system to induce further changes. We would like to pursue this line of thought[1] further from a governance perspective, starting from the situated operations of societal actors. Different governmental agencies as well as private actors are responsible for only a part of the system and therefore tend to develop only incremental initiatives to change the system.[2] They may seek to innovate, but at the same time they have to maintain themselves in the face of complexity (see Chapter 4). Reasoning from the polycentric condition of modern societies, no societal actor can be expected to have the force or even the will to make innovation cascades and societal transitions happen (Pel & Teisman, 2009). They are somewhat mysterious phenomena where subsystems co-evolve to produce system-wide change. To increase our understanding we will focus on the empirical content of 'cascading', that is, on the processes and conditions underlying these potentially transformative cumulations of incremental change.

This chapter highlights cascade formation from a governance perspective. Our research question is this: *How, and under what conditions, can incremental innovations 'cascade' into system innovations, and how can this be guided?*

This question implies first of all a change of perspective. We return to the source of innovation: innovation attempts. Whereas the cascade metaphor can easily draw all our attention to the stream as a whole, descending, branching and combining with other streams to create system innovations and a new transition pathway after a considerable amounts of years, our question creates a more uphill perspective. The perspective is guided by our fascination for what the source of the stream is. How was the source initiated and how did it manage to become powerful enough to create a stream of actions and changes to evolve into a new pathway? In that sense we start from a 'micro' research approach but with a close eye on the micro-macro linkage that is essential to transitions research, the focus on (countless amounts of) small resources coming together, sometimes creating a noticeable river of change.

We start by looking through the eyes of the initially lone agent confronted with a huge mobility system (Urry, 2004)but still brave enough to intend to change the course of development in the future—hoping to become the proverbial butterfly in the weather system who can be decisive in the process of remote tornado formation. Assuming the reflexive awareness that he acts under polycentric conditions, he can expect actors in neighboring subsystems to resist changes that interfere with their own actions and habits (Luhmann, 1997). Under the polycentric condition, innovations will not be adopted and diffused in any straightforward way. Innovations are translated by actors (Bijker & Law, 1992; Czarniawska & Joerges, 1996), that is, they are appropriated, interpreted, selected upon, resisted or modified. Because of polycentric translation processes, it is unfortunate to speak of innovation diffusion: Unlike gases, the innovations change fundamentally when being dispersed (Akrich, Callon & Latour, 2002a).

Tracing translations generates insights on the appropriation dynamics that shape an innovation's 'journey' (Van de Ven et al., 1999) through a differentiated society. As mentioned, to conceive of this as diffusion would downplay how innovation attempts are adapted in various ways. The physical metaphor does indicate instructively, however, how system innovation and transitions, as higher order innovations, depend on aggregation processes. Teisman (2005) describes how innovation attempts insufficiently geared to recipient actors' own ambitions repeatedly go up into smoke: After initial acceptance the translators become disenchanted with the external innovation attempt, once it proves not instrumental to the internal organization. When societal actors thus fail to consolidate an innovation into their operations, it evaporates.[3] By contrast, such consolidation, such *crystallization* of the otherwise volatile, is essential for further cascade formation. And considering that crystallization is more the exception than the rule, we identify the conditions for crystallization as essential for cascade formation.

We will apply our research approach in an action field that is widely recognized for its transition potential: Intelligent Transport Systems (ITS). Whether ITS brings merely efficiency enhancing 'incremental' change (Dynamic Traffic Management) to lubricate a car-dependent system or constitutes more radical system change (automated vehicle guidance, the 'digital panopticon', self-organizing traffic) is still debated.[4] Moreover, the acknowledgement of its transformative potential gives rise to further discussion about whether this is desirable or not (Adams, 2005; Urry, 2008). Indeed, the uncertainties and particular socio-technical complexity of ITS create a need for a social science perspective (Chapter 2).

ITS is a miscellaneous group of information and communication technologies (ICT) applications in the transport system. We will describe how in 1996 a subdepartment of the Dutch Transport Ministry launched a policy paper on travel information. The document envisioned a system innovation toward 'informed choice' for travelers, enabled by public and private innovation processes. The case highlights how societal actors translated the vision in different, sometimes interfering ways. Fifteen years later we can see an accelerating cascade formation, but the initiators encountered soon enough their dependence on other actors' translations and especially the crystallization of those. We will give concrete examples of evaporation and crystallization and highlight the relevant changes in system states. If governance is to enhance cascade formation, it has to develop sensitivity to the changing system states that are more or less conducive to crystallization.

Section 1 develops the theoretical framework to study cascade formation. In Sections 12.2 and 12.3 we give empirical accounts of the Dutch 'traffic intelligence cascade'. Analysis of the still evolving and even accelerating 'innovation cascade' follows in Section 12.4. Finally, Section 12.5 presents conclusions and governance recommendations.

12.2. CASCADE FORMATION AS TRANSLATION PROCESS

The transition perspective points out how incremental innovations bear the risk of winding up as pseudo-solutions (Rotmans, 2006) but may also combine and evolve into system innovations. Such cascades exemplify how incremental innovations can trigger further innovations both in technological and social dimensions. From a systemic perspective, the cascade is a sequence where change in one subsystem activates another subsystem to change in a reinforcing way, thus creating a positive feedback loop. Such a process has been described by Geels (2006), using the example of jet engines for airplanes. There cascading is aided by strong linkages among technical subsystems (such as airplanes and airport infrastructure).

Linkages among subsystems are also created through the interaction of actors among different subsystems. We look at these interactions as translations. In modern societies actors tend to entertain different ideas about problems, solutions and measures of success (Koppenjan & Klijn, 2004). As a consequence they differ in the way they perceive and receive an innovation attempt: What is an 'incremental' change for the one, may be a disruptive revolution to the other. After an innovation is initiated, it will be received by many actors who may not only choose to buy it or not, but they may also test, copy, resist or modify it, pass it on or initiate their own innovations in response. As innovation attempts can be expected to be 'translated' (Bijker & Law, 1992; Callon, 1986; Callon & Law, 1982; Czarniawska & Joerges, 1996) in different ways, they are often not merely adopted and 'diffused' (Akrich et al., 2002a, 2002b).

Innovations are therefore unstable entities that change through the adaptations by a variety of actors. These appropriation dynamics create turbulent 'innovation journeys' (Van de Ven, Polley, Garud & Venkataraman, 1999). The course of these journeys emerges out of the interactions between innovators and recipients. In polycentric translation processes the translator is therefore as important as the initiator: They are mutually dependent for stabilization of the innovation into a viable end form and for further cascading.

Analyzing cascading as a series of translations highlights the appropriation dynamics and the heterogeneity of innovation 'journeys'. This does bring along skepticism about technological trajectories, paradigms and path dependencies: Bijker and Law (1992) ask rhetorically, "Do technologies have trajectories?" The voluntaristic focus on heterogeneity and malleability thus risks to obscure under what circumstances innovations stabilize or not and how 'higher order translations' occur (Smith, 2007). On the other hand, the trajectories and path dependencies from evolutionary economics would not clarify sufficiently the empirical content of cascading: a typical problem of micro-macro linkage (Blauwhof, 1994). The translation framework does give foothold for such linkage, however, in three respects: In the first place, it approaches innovation as a process of constant redefinition, with end products provoking new developments. Investigation of translation sequences can follow how innovation journeys meander and co-evolve into wider cascades, system innovations and transitions. Second, it highlights how 'translations' involve changes both in the innovation itself and in the actors translating. Third, it directs attention to the processes of stabilization. We focus on the crystallization points that keep an innovation journey going and prevent a translation sequence from evaporating into an incoherent diversity that fails to achieve any system innovation.

Investigations into governance of complex systems have brought us to conceive of innovation cascades in a way that confirms the relevance of translation processes, yet steers clear from voluntarism. To the contrary, the acknowledgement of deep governance complexity (Teisman, 2005; Teisman, van Buuren & Gerrits, 2009) has led us to the understanding that translation processes take place within highly dynamic governance contexts that are seldom conducive to the crystallization—or anchorage, in administrative terms—required for cascade formation.

To conceptualize the conditions for crystallization and further cascading, we use the notion of system states. Teisman et al. (2009) distinguish four system states: inert, chaotic, dynamic and stable. *Inert* process systems are guided by autopoietic self-organization; they absorb outside impulses by translating them into their own perspective, buffering their change potential. Translations, if made at all, support existing actions and routines. At the other extreme, *chaotic* process systems may be characterized as the archetypical tower of Babel. In this system state, each innovation initiative will be embraced as interesting. Translations pile up at high speed, but initiators and translators have no mechanism through which to connect meanings into a jointly meaningful innovation journey. Fragmentation is the norm, meaningful connections and boundary spanning are scarce and innovation attempts evaporate.[5] This inability to converge is similar to the 'hot', highly controversial processes Callon (1998) distinguishes from less ambiguous and convergence-prone 'cold' situations. *Stable* process systems are best understood as displaying efficiency improving reproduction. Incremental innovations will be the norm. In this system state, translations serve to optimize existing sets of action. The system adapts to changes in its environment, while staying within the boundaries of existing rules and ways of action. This fits with the concept of incremental innovations. *Dynamic* process systems are characterized as the most favorable states for cascading and eventual transition. Here previously unconnected subsystems seek to connect, negotiate and attune their translations, thus creating the opportunities of 'Neue Kombinationen' ('New Combinations' (Schumpeter)) and cascading.

System states are not just a quality of the system targeted for system innovation as a whole but also of its subsystems. The quest for system innovation and transitions implies governance of a nested-systems phenomenon. Cascading appears whenever translations take place among dynamic subsystems to such an extent that the system as a whole becomes dynamic. In addition, cascading requires that along the sequence of translations, the innovation crystallizes in subsystems. Crystallization is unlikely to happen under conditions of inertia or chaos; under these system states the actors or subsystems concerned will not be interested or be able to consolidate. System states change under a manifold of pressures, often accidental to a certain innovation: general government changes like devolutions and appraisal practices for example (Chapter 5), but also market turbulence can easily trigger a shift from inertia to chaos. These shifts make it especially difficult for societal actors to oversee cascade formation. As dynamics relating micro-level translations to their evolutionary context they help us understand the course of translation sequences. Cascades are evolutionary improbabilities that situated actors cannot oversee. Fate can be helped a little by seizing upon these changing process system states. The linkage between system states and translation sequences as conditions for crystallization and cascading will be used as an analytical tool to understand the case presented in Sections 12.2 and 12.3.

12.3. TRAVEL INFORMATION INNOVATION CASCADE (I): THE TRAVEL INFORMATION 'TRAJECTORY'

ITS form a group of innovations that itself is a product of translation: The revolutionary development of ICT was not driven by transportation objectives, but was later adapted into various attempts to improve the transportation system. Sussman (2005) distinguishes Advanced Traffic Management Systems (ATMS), Advanced Traveler Information Systems (ATIS), Advanced Vehicle Control Systems, Commercial Vehicle Operations, Advanced Public Transportation Systems and Advanced Rural Transportation Systems. And whereas ATMS can be considered the typical incremental system improvement on a continuous development toward fine-tuned, responsive traffic management, Advanced Vehicle Control Systems is a more radical attempt. Eventual automated vehicle guidance would amount to a radical shift, challenging the car as the hallmark of individual freedom. Finally, ATIS enhancement of travel information is an ITS translation with as yet uncertain effects on the car-dominated mobility system. ATIS applications stimulate 'informed choice' for the individual traveler. This concept allows for various elaborations, depending on the motives of the innovators. Governments may want to inform citizens about alternatives to car use and offer guidance; commercial actors will be more interested in informed consumer choice. Translated as one-dimensional comparison of travel times, information provision may indeed reinforce car-dependency (Lyons, 2001). On the other hand, information also has an 'emancipatory' potential, when displaying the time lost looking for parking space, environmental performance and options for intermodal travel.

ITS can be translated in various ways. The various translations are not mutually exclusive, however, as will become clear in the following case study. It shows the translation sequence following an attempt to stimulate 'informed choice'. The sequence formed branches that co-evolved. The case study formed part of a multiple-case study on innovation attempts in the traffic management action field, analyzed as initially parallel, yet sometimes intertwined translation sequences (Pel, in press). Empirical research was conducted from an 'uphill' perspective as argued for previously; the empirical accounts were based on interviews with key actors and document analysis. In the following we describe first the trajectory toward 'informed choice' as set out by a subdepartment of the Dutch Ministry for Transport. The next section highlights the perspectives and actions of other involved actors.

In 1996 a subdepartment of the Dutch Ministry of Transport and Water Affairs launched a policy paper entitled 'Travel information' (Ministry of Transport & Water Affairs, 1996). The document sketched how in 2010, the traveler would be enabled to make 'informed choice' on travel mode, route and time of travel. 'Informed choice' was to be supported by improved provision of travel information that was reliable, consistent and integrated. Improvements were aimed for in three domains: traffic management (ATMS), traffic information services to customers (ATIS) and integrated information provision on public transport.

The initiators pursued these goals by establishing a new institutional architecture: The state-led arrangements were supplanted by a model creating room for private sector innovation (Ministry of Transport & Water Affairs, 1996). A centralized repository for travel data was to integrate the 'travel information chain' of data acquisition, processing and eventual dissemination of information. Acquisition thus far depended on the governmental road monitoring systems and on on-site observations (of the police, for instance), but the innovators anticipated the advent of in-car systems to allow for drastic improvement—both in data acquisition and in customer-oriented information provision. This innovation was expected to be in better hands with the private sector. Government would facilitate a market for travel information provision by disclosing and processing data at cheap rates and by retreating from information provision to customers. Data processing remained in governmental hands, thus allowing for consistency control. Integrated information on public transport would be added to the database in a later stage (Adviesdienst Verkeer & Vervoer, Advisory board Transport) AVV, 1996).

In 1998 the Traffic Information Centre became operational, materializing the ambition to arrive at an integrated 'information chain'. In the first few years, the initiators could see little activity on the newly created market, however. The Verkeers Informatie Dienst (VID) (Traffic Information Service), was an exception, exploiting information services to individual drivers and businesses (VID, 2010). The advent of the Internet allowed them to develop customized information services, on top of radio broadcasting. But whereas VID development matched the expectations of the ministerial innovators, the desired development of in-car systems did not take off. In spite of the public-private knowledge exchange meetings and pilots, private sector actors proved insufficiently interested to continue investments beyond the experimenting stage. In 2003 another pilot started to assess the usefulness of Floating Car Data (FCD) for real-time travel information provision. The province of Brabant in the Netherlands stepped into the ministerial 'Intermezzo' innovation program (Rijkswaterstaat, 2004), and a consortium of an ICT company and a telecom provider had developed a system to derive a 'view on the road' from the movements of cell phones. The system was quite successful in filtering out the disturbing signals of non-traffic sources. And even when the penetration rate proved too little to meet traffic management accuracy standards, the evaluations noted plenty of room for improvement (De Wolff, 2005). The advent of in-car navigation systems held the promise of GPS-signals becoming available. Still, the Brabant province launching customers did not find fellow road managers ready to join in, and their FCD trajectory came to a halt a few years later. The pilot continued in another form, though: The technology and its developers were taken over by TomTom, who did manage to round the business-case with their successful navigation platforms. In 2007 they launched their HD (High Definition) Traffic system to help users to find the optimal route to their destination, a reward to the 1996 bet on entrepreneurial innovation and FCD development.

Meanwhile, governmental actors on both central and decentralized levels had joined forces in the establishment of a national database for road data (National Data Warehouse for Traffic Information [NDW], 2010). The initiative was a follow-up on the earlier Traffic Information Center, bringing together data that used to be dispersed across road managing jurisdictions. This NDW was to disclose and synchronize all data available: Road managers throughout the country had made considerable advances in monitoring and processing of raw data. The take-off of in-car intelligence was especially promising to add to the stock of data, as unlike the governmental roadside systems, the in-car systems could cover the entire road network. On the other hand, in 2007 the NDW set quality standards on traffic intensities that were as yet unattainable for floating car data systems (NDW, 2008). The 1996 innovators were also disappointed by TomTom's decision not to disclose their privately developed information to this central repository. Anno 2010 public-private deliberations continue on possible differentiation of quality levels, weighing costs against functional requirements (NM, 2010).

Next to these developments in road traffic information, the 1996 'informed choice' vision also featured a parallel trajectory toward integrated information provision on public transport. It was intended as a ploy in a strategy for intermodal travel, to be integrated at a later stage. The public transport trajectory stagnated for a long time, however. In 2009 the minister reinvigorated the trajectory, launching an 'offensive on multimodal travel information' to establish a central repository as a basis for information services (Ministry of Transport and Water Affairs, 2009). This Nationaal DataWarehouse Openbaar Vervoer (National Database Public Transport) (NDOV) would be the public transport counterpart to NDW, independent from the transport operators. The two repositories would be merged into a multimodal system by 2015. By that time the minister also wanted to have disclosed information on costs, environmental performance and parking: information on entire mobility 'chains'.

By the time the 2010 horizon was closing in, the vision for 'informed choice' had been partially realized. The 1996 initiators could mark improvements both in dynamic traffic management and in traffic information services to users but reinvigorate the development of dynamic public transport information. The next section describes the process from the perspective of various other actors translating the 1996 innovation attempt in process systems shifting between states.

12.4. TRAVEL INFORMATION INNOVATION CASCADE (II): SURROUNDING 'TRAJECTORIES' AND DEVELOPMENTS

The 1996 travel information initiative fit well with the general shift in governance toward liberalization. It met with little resistance in parliament as it entailed very modest investments and was welcomed as an innovative initiative contributing to congestion abatement. The police were happy to retreat from information provision, allowing them to concentrate on core tasks. The ANWB, the national drivers' association, was keen on joining in as a way to serve its members (AVV, 1996).

The Traffic Information Center became operational in 1998, but the intended 'information market' did not take off as smoothly. Stepping into the newly created market, the VID found themselves confronted with market distortion: first, the traditionally close intertwinement of the national driver's association with government. This relation was disentangled later and renewed in market-conform arrangements (Ministry of Transport and Water Affairs, 2001). Second, they noticed how Rijkswaterstaat, the executive department of the transport ministry, continued its information provision to users. This led to a legal procedure in 2003, and Rijkswaterstaat was held to the earlier stipulation that it refrain from information provision to users. The appeal to court signaled tensions in the public-private 'information chain'. In 2003 a special commission was established to restore trust and stimulate functioning of the market (Laan & Prins, 2003). The commission concentrated on specifying 'rules of conduct', fine-tuning the demarcation of public and private responsibilities.

Rijkswaterstaat had difficulties to completely abstain from traffic information provision—this was hard in the face of rapidly increasing congestion on the national main road network. And especially as road pricing remained controversial and air quality regulations tightened the already severe constraints on road construction, they were under considerable pressure to ensure optimal use of existing capacity. The 2005 white paper on transport laid this down explicitly (Ministry of Transport and Water Affairs, 2005). To Rijkswaterstaat, information provision was first and foremost a traffic control instrument.

Meanwhile, the business-case for traffic information provision remained nebulous to entrepreneurs. For a long time, the information had been disseminated for free, after all. And however interesting the idea of using

Global System for Mobile Communication (GSM) signals for new purposes, it would only become viable by large-scale application: high investments, uncertain returns. A breakthrough took place once TomTom started to use their portable navigation devices as platforms for traffic information provision: Unlike built-in navigation systems they were low-price, making it easier to reach the critical mass necessary for FCD systems. TomTom also refined the technology, combined it with other technologies and synthesized a view on the road from both actual and historical data. To TomTom, real-time traffic information was part of a whole portfolio of navigation services. The business-case did not depend on the willingness-to-pay for traffic information. TomTom quickly rose to become a successful company, albeit in a highly dynamic market. By 2007 TomTom and their competitors were seeking strategic alignments with car manufacturers and telecom providers. Market leaders TomTom and Navteq each took over map making firms for a few billions of euros. These were late rewards for considerable investments in map digitization. Similar to the investments in FCD, the returns had remained a distant promise and had several firms decide to abandon the project before the time had come to harvest (NRC, 2007b).

The take-off of FCD-based traffic information presented governmental actors with a new traffic information landscape. The FCD systems could further reinforce the advances in dynamic traffic management, the ambitions for which had only grown. The NDW repository was installed to support network-wide traffic management, taken up jointly by Rijkswaterstaat and the various road managers on the decentralized level. Ideally the in-car systems would help to fill in the blank areas left by the costly roadside systems. Such merge of roadside and in-car data proved difficult in practice however, as FCD development remained insufficiently developed to yield reliable traffic intensities. Furthermore, there was the problem of navigation devices guiding traffic over alternative routes. Several municipal road managers erected signs 'turn navigation device off', and the 'unwanted routes' problem evoked alarming newspaper reports (NRC, 2007a). The 'foundation for research on navigation systems' even asked for a ban on the 'kid killers' that had cars 'blazing over school yards' (Stichting Onderzoek Navigatiesystemen, 2010). Instead, the minister of transport urged to settle the apparent inconsistencies through public-private deliberation and data exchange (Ministry of Transport and Water Affairs, 2008). The importance of the latter had been urged for by the service providers, who stressed that road managers should inform them better about actual road conditions—temporary detours, for example.

Finally, the 2009 action plan on public transport information provision responded to a stagnant development. In line with the policy principle of leaving it to the market to generate consumer value, governmental actors had kept their distance. Various public transport operators had implemented dynamic route information panels for their own services, but integration of the information across different types of public transport had been difficult. The operators had managed to unite and establish an agency to operate a public transport travel planner (OV9292, 2010) but had been hesitant to disclose dynamic data. Consumer organizations and the tendering principals might use the information as management information and sanction deficiencies in punctuality. Furthermore, they had doubts about the benefits of multimodal information on travel times—it might only elicit the shorter travel times of the car. With the 9292 cooperation they could at least control information provision, but in 2009 the minister undermined their monopoly to open up the information market.

12.5. TRANSLATIONS IN TRAVEL INFORMATION PROVISION

The 1996 innovation attempt aimed to set the preconditions for 'informed choice'. The initiators envisioned a future of dynamic traffic management, customer-oriented traffic information and intermodal travel practices—a system innovation driven by entrepreneurship and guided by governmental vision. The 1996 initiators knew from the outset that the realization of their future vision depended on the way other actors would take up the challenge. In the following we will analyze the process as a sequence of translations, made under changing process system states. In line with our 'uphill perspective', analysis will follow the three innovation directions envisioned by the initiators: traffic management, traffic information and integrated information provision on public transport.

The innovation attempt had a good start, meeting with political endorsement for its market-conform, innovative approach to combat congestion and development of consumer services. The police gave their approval, released from an activity that was not their proper task. Also the driver's association embraced the initiative, seeing opportunities for services to its members. The establishment of the Traffic Information Centre materialized the set up of a public-private information chain, supported by converging translations—a first crystallization point in a dynamic political subsystem that was ready for change.

The VID seized upon the business opportunity, and also the driver's association became active as a service provider. The 1996 initiators had hoped to create a dynamic traffic information chain, but market development stagnated under chaotic conditions. VID's development of 'business opportunity' translations ran counter to the remnants of the old order: Both Rijkswaterstaat and the driver's association initially failed to adapt to the intended level playing field. The governmental retreat from information services to users had gotten lost in their instrumental translation. The 2003 law suit was a clear signal of interfering translations, and the initiators from the ministry responded with the public-private commission on 'rules of conduct' to mediate: an attempt to stabilize the chaotic process system, and preparing for a shift toward a more innovative dynamic state in which the divergent translations could be attuned.

Rijkswaterstaat translated information chain optimization into a means to congestion abatement. And while part of this 'instrument' had been devolved to the private sector, congestion levels were only rising. This development pushed Rijkswaterstaat and other road managers back to an instrumental translation, limiting their ability and willingness to connect with divergent entrepreneurial translations. On the other hand, the pressure to improve toward dynamic and network-wide traffic management increased Rijkswaterstaat's willingness to share responsibilities with road managers on the decentralized level. The NDW was a crystallization point of 'congestion abatement' translations. It secured the information base and wedded a network of actors for new innovations in network-oriented traffic management, alignments also useful for future translations toward road pricing.

The 1996 initiative anticipated upon a breakthrough in FCD-based in-car systems. The basic principle of FCD had been known even before 1996. Refinement of the technology took a translation sequence in itself, however. The ministerial innovators organized knowledge exchange meetings and subsidies to stimulate entrepreneurial translation processes, but results kept evaporating into short-lived pilots. Entrepreneurs failed to translate into the required large scale application facing uncertain returns on investment. The entrepreneurial subsystem remained stable for years after the 1996 initiative, without becoming as dynamic as envisioned. Eventually the 2003 Brabant pilot triggered the start of a dynamic phase, stimulating TomTom to wage the crucial translation into a business case. They gathered the technology and know-how to formulate an encompassing strategy to sell portable navigation systems as platforms for miscellaneous services: a crystallization point of several innovative technologies and marketing models that opened up new avenues for innovation. TomTom's successful translation triggered a highly dynamic entrepreneurial process system where new actors joined in, formed alliances and actively investigated new translations. The alliances between map makers, navigation industry and car manufacturers were expressions of an unleashed innovation race for the most commercially successful translation of FCD technology. The dynamics allowed the efforts in map digitization to pay off after all.

When TomTom entered the scene with their customer-oriented information provision, they challenged Rijkswaterstaat's quest for control. The NDW crystallization point was in line with public-private 'information chain' integration envisioned in 1996, but public and private translations still displayed divergences. NDW quality standards were based on traffic management requirements that were difficult to meet through FCD-based systems. Moreover, the very term of 'unwanted routes' signaled tensions between the 'business opportunity' and control-oriented 'congestion abatement' translations. The earlier established public-private commission and various intermediary organizations afforded platforms to discuss the divergence and attune translations, however. In 2008 the minister proved unwilling to resign in control-oriented inertia, pointing out the need for public-private synchronization of translations. He sought to avoid chaos and inertia by getting public and private actors together for a new dynamic wave of 'cooperative systems', that is, combinations of roadside and in-car systems.

Finally, public transport information provision was an element of the innovation attempt that fell behind through inertia. In a process system characterized by increasing market pressure, the public transport operators were reluctant to disclose dynamic information and stuck to cautious translations to retain control over information provision. The fear that disclosure of data would be a liability prevailed. The 2009 initiative to break open the market for information services attempted to enforce a dynamic state after all. 'Informed choice' was to be translated not only by transport operators but also by the service providers. The latter had come up in the more dynamic subsystem of traffic information provision; the dynamics of the neighboring process system could thus spread and break through the inertia in public transport.

The 1996 innovation attempt envisioned a trajectory toward 'informed choice', consisting of improvements in traffic management, consumer-oriented traffic information and public transport information provision. By 2010 they could note that parts of the intended system innovation had been realized and that cascading was taking off at last. Next to more dynamic and 'network-wide' traffic management, a diversity of user-oriented services was being offered in a highly dynamic and expanding information market. The intended rise of in-car systems had taken off eventually.

At the same time, not all aspects of 'informed choice' are as well developed as others, and integration challenges remain. In-car and roadside system integration is still in development, for instance. Moreover, information on public transport is lagging behind, as yet remaining a parallel translation sequence. The emancipatory translation of 'informed choice', the articulation of the relative advantages of public transport use, got reinvigorated only recently. If this renewed attempt to spur dynamics succeeds after all, cascading will come closer to the 1996 vision: informed choice as enhancement of multi-modal travel, away from car-dependency. Finally, it is good to note how the traffic intelligence cascade developed an infrastructure enabling road pricing—the seemingly untranslatable radical innovation.[6]

12.6. CONCLUSIONS

How, and under what conditions, can incremental innovations 'cascade' into system innovations, and how can this be guided? Cascading under the polycentric condition is unlikely to happen in the form of diffusion. It crucially depends on translations by other actors who will modify the innovation. Radical innovation attempts will often be received as disturbance rather than enrichment, however. If appropriated at all, these attempts will be adapted to fit in: The instrumental translation of road managers was a typical example. As also noted by Smith (2007), it is hard to translate what works in the niche into something that also works in the regime. He asserts a paradox of innovation, the successful non-evaporating and thus surviving translations tending to be the relatively shallow ones. Stated more positively, incremental innovations have a higher chance to become translated and keep a cascade going. The transferability of innovations, and the ability of receiving actors to translate an innovation attempt, is crucial for its success. The innovation itself is of secondary importance; as far as it is translated at all, it will change in the process. This is why the beginnings of cascades into system innovation should be sought both in 'niches' and in 'regimes'; in either case the interactions between these actors will be crucial.

The 1996 attempt to stimulate 'informed choice' showed the importance of transferability. The initiative allowed for many different translations into manageable 'incremental' innovations: Road managers translated the innovation as a means for traffic control, entrepreneurs like VID and TomTom seized upon the business opportunity and public transport operators set up, albeit hesitantly, a travel planning service. This apparent ambiguity of the innovation was an asset for its transferability and cascading.

Still, good transferability was in itself not enough for cascading. Cascading also depended on the conditions under which translations took place. FCD development, the development of an information market and integration of public transport information were translations that only took place when the relevant subsystems moved from inert to stable and eventually dynamic process states. These shifts were only partly predictable, let alone controllable: The traffic management subsystem was put under pressure by rising congestion levels that could not be met through capacity expansion and/or road pricing. Similarly, the entrepreneurial subsystem only became dynamic once new business cases were articulated in which travel information was only an element. Similarly the market developments that brought navigation industry, map digitization industry, telecom providers and car manufacturers together were shifts in the process system that the 1996 innovators could not foresee. Also, the inertia in public transport information depended on a range of developments that made operators sensitive to public monitoring of punctuality and reluctant to disclose dynamic data.

Due to inertia and chaos some envisioned translations failed to happen or took longer to develop than hoped for. The ministerial initiators foresaw a bright future for entrepreneurial innovation in in-car systems, but until 2003 their stimulating efforts failed to provoke dynamic cascading. And apart from evaporation they also encountered interferences between translations. The ambitions for dynamic network-wide traffic management spurred innovations in technology and in boundary-crossing management, for instance, with the NDW as a crystallization point; a cooperative platform for future road pricing.[7] The control-oriented translation proved difficult to combine with entrepreneurial translations, though. First there were the market distortions experienced by the latter. The interference even had to be settled in court, and a special commission was established to attune translations and restore trust. Later on, the 'unwanted routes' problem reinvigorated public-private tensions.

In this case, the minister intervened again to synchronize translations, explicitly acknowledging the entrepreneurial translations as integral parts of an expanding and turbulent process system. With the vision of 'cooperative systems' as synergetic combination of roadside and in-car systems, he sought to secure the dynamic state the 1996 innovators had hoped to bring about.

Cascading is an emergent phenomenon. It is an evolutionary improbability, especially when taking the polycentric condition into account (see also Leydesdorff, 2000, and Rip, 2006). Cascading depends strongly on social actors' particular appropriations of an innovation attempt. The translators are therefore as important as the iniators, and high transferability is essential. System innovations do not need to start from radical innovation attempts, as long as incremental innovation attempts lead to sufficient crystallizations to ensure ongoing cascading. It may seem to the initiator that crystallization only happens 'when the stars are in position', seeing cascading to be in the hands of others. The shifting states of subsystems only underline the capricious environment an initiator finds himself confronted with. Still, this does not mean that system innovation cannot be guided. It means that it is of paramount importance to identify the subsystems relevant to a projected cascade, to assess the factors accounting for their state and development and to develop the essential timing for intervention and attunement of translations. This is similar to what has earlier been described as seizing upon 'policy windows' (Kingdon, 1982).

After almost 15 years, the 1996 initiators can look back on a turbulent cascade formation process. The process was marked by the setbacks, breakthroughs and surprises that are to be expected for any innovation 'journey', but also displayed cascade formation toward system innovation and transition. Considering that transitions are long-term processes generally taking three to four decades, we can draw a half-time conclusion: First of all, the translations are converging toward the future vision of 'informed choice'. The advances in dynamic traffic management and traffic information still do not add up in any straightforward way, but the era of cooperative systems (in-car and roadside systems integration) is coming closer. The establishment of a system for integrated and dynamic travel provision for public transport proved more difficult to achieve. The 2009 policy initiative holds the promise to speed up this innovation process however. And as the traffic information service providers broaden their activities, they have started to stir up this process system. Second, the previous crystallizations notwithstanding, cascade formation also remains to be marked by fragmentation. Developments in process systems take place at different speeds, for example. And as improved travel information continues to open up new translations, further cascade formation is bound to be paved with future interferences.

A prominent source of potential future interference is surveillance, that is, travel information translated as information *on* travelers. Several authors have pointed out how a rapid system innovation is taking place (Dodge & Kitchin, 2007; Urry, 2008), with an Orwellian dystopia as an attractor. Our

Transition by Translation 265

polycentric analysis leaves little room for determinism however. These halftime conclusions only underline the difficulty to predict simultaneous yet only partly coordinated change. Polycentric analysis also does not allow for unilateral judgment on the important normative issue raised previously, the dystopian dimension of the 'surveillance' translations. Instead, we highlight the variety in and dynamics of translations. Guidance of cascading is possible through attunement of translations, informed by careful observation of the changing states of the subsystems in which translations take place.

NOTES

1. We take up the idea of cascades as accumulating sequences of incremental change, incremental change as a trigger of escalating further changes that amount to higher order system innovations and transitions.
2. In the Netherlands, for instance, government has been working on a radical regime shift toward dynamic pricing of car use for more than 30 years. Several moments of a breakthrough (the well-known momentums) were indicated. But after that, each of the attempts faded away.
3. This consolidation challenge led strategic niche management scholars to broaden their focus to niche-external processes and windows of opportunity for niches to ascend to regime level (Schot & Geels, 2008). As we approach it, consolidation is important to keep the translation sequence going. Consolidation by marginal 'niche' actors is important, even when less likely to contribute to system innovation 'cascading'.
4. As we have argued elsewhere, these assessments of 'incremental' and 'radical' change are neither obvious nor innocent. In this respect transitions research could benefit from Critical Systems Thinking (Pel & Boons, 2010).
5. Such chaos may be the result of actors operating under conditions of uncertainty without the presence of a focal point (Schelling, 1960).
6. Chapter 7, pertaining to developments in the United Kingdom, discusses road pricing as an eternal niche, constrained not by technical but political factors. On the other hand the author also notes how the in-car systems afforded a latent function for road-pricing: ITS stimulation as a 'soft implementation strategy for pricing'. This strategy fits well with our emphasis on transferability.
7. Following up on the previous note, the NDW removed part of the pivotal political constraints on road pricing.

REFERENCES

Adams, J. (2005) 'Hypermobility, a challenge to governance', in C. Lyall & J. Tait (eds) *New Modes of Governance: Developing an Integrated Policy Approach to Science, Technology, Risk and the Environment*, Aldershot, England: Ashgate.

Akrich, M., Callon, M. & Latour, B. (2002a) 'The key to success in innovation part I: The art of interessement', *International Journal of Innovation Management*, 6(2): 187–206.

Akrich, M., Callon, M. & Latour, B. (2002b) 'The key to success in innovation part II: The art of choosing good spokespersons', *International Journal of Innovation Management*, 6(2): 207–225.

AdviesDienst Verkeer & Vervoer (Advisory board transport), *TIC Definition study*, AVV Rotterdam, 1996.

Bijker, W. & Law, J. (1992) *Shaping Technology/Building Society; Studies in Sociotechnical Change*, Cambridge, MA: MIT.
Blauwhof, G. (1994) 'Non-equilibria dynamics and the Sociology of technology', in L. Leydesdorff & P. van den Besselaar (eds) *Evolutionary Economics and Chaos Theory; New Directions in Technology Studies*, London: Pinter.
Callon, M. (1986) 'Some elements of a sociology of translation: Domestication of the scallops and the fishermen of St Brieuc Bay', in J. Law (ed) *Power, Action, and Belief: A New Sociology of Knowledge?* London: Routledge.
Callon, M. (1998) 'An essay on framing and overflowing', in M. Callon (ed) *The Laws of the Markets*, Oxford, England: Blackwell.
Callon, M. & Law, J. (1982) On interests and their transformation: Enrolment and counter-enrolment, *Social Studies of Science*, 12(4): 615–625.
Czarniawska, B. & Joerges, B. (1996) 'Travels of ideas; organizational change as translation', WZB working papers FS II 95–501, Berlin.
De Wolff, P. (2005) Evaluatie en opschaling van het DVM-project "Brabantse Wegen Beter Zichtbaar." (Evaluation and upscaling of the DTM-project 'Better View on Brabant Roads') Available at http://www.brabant.nl/~/media/Documenten/l/logica_evaluatiebrabantsewegenbeterzichtbaar%20pdf.ashx (accessed 28 March 2010).
Dodge, M. & Kitchin, R. (2007) 'The automated management of drivers and driving spaces', *Geoforum*, 38(2): 264–275.
Geels, F.W. (2005) *Technological Transitions and System Innovations; A Co-evolutionary and Socio-Technical Analysis*, Cheltenham, England: Edward Elgar.
Geels, F.W. (2006) 'Major system change through stepwise reconfiguration: A multi-level analysis of the transformation of American factory production (1850–1930)', *Technology in Society*, 28(4): 445–476.
Geels, F.W. & Schot, J. (2007) 'Typology of sociotechnical transition pathways', *Research Policy*, 36(3): 399–417.
Kingdon, J.W. (1982) *Agendas, Alternatives, and Public Policies*, Boston: HarperCollins.
Koppenjan, J. & Klijn, E.-H. (2004) *Managing Uncertainties in Networks*, London: Routledge.
Laan, J. & Prins, H. (2003) *Advies van de Commissie 'Gedragsregels Verkeersinformatie' (Recommendations from the commission 'Rules of conduct traffic information'*, (Ministry of Transport and Water Affairs), Den Haag, the Netherlands.
Leydesdorff (2000) 'The triple helix; an evolutionary model of innovations', *Research Policy*, 29(2): 243–255.
Luhmann, N. (1997) 'Limits of steering', *Theory, Culture and Society*, 14(1): 41–57.
Lyons, G. (2001) 'Towards integrated traveler information', *Transport Reviews*, 21(2): 217–235.
Ministry of Transport and Water Affairs (1996), Beleidsnota Reisinformatie, Den Haag 1996.
Ministry of Transport and Water Affairs (2001) 'Standpunt Verkeer en Waterstaat over het eindrapport van de Boston Consulting Group (BCG) inzake het onderzoek naar de relaties tussen V&W en de ANWB, HKW/UB/2001/8746' ('Position Ministry regarding the final report by (BCG) on the relations between the Ministry and the national drivers' association'), Den Haag, the Netherlands: Ministry of Transport and Water Affairs.
Ministry of Transport and Water Affairs, (2005). Nota Mobilileit, Den Haag.
Ministry of Transport and Water Affairs (2008) TK 2007–2008, parliamentary report, 31 305, nr. 7, Den Haag, the Netherlands: Ministry of Transport and Water Affairs.
Ministry of Transport and Water Affairs (2009) 'Aanpak multimodale reisinformatie, VENW/DGMo-2009/2606', Den Haag, the Netherlands: Ministry of Transport and Water Affairs.

National Data Warehouse for Traffic Information (2008) Eisen aan de Nationale Databank Wegverkeersgegevens (NDW) systeemketen. Available at www.ndw.nu (accessed 10/10/2011).
National Data Warehouse for Traffic Information (2010) Available at http://www.ndw.nu/en/ (accessed 10/10/2011).
NM (2010) 'Kwaliteitsdifferentiatie', *NM-tijdschrift voor netwerkmanagement*, Volume 2010, issue 1: 13–21.
NRC (2007a) 'Liever sukkelen door dorpen dan stilstaan in de file', NRC 10/01/2007.
NRC (2007b) 'Nederland zet de wereld op de kaart', NRC 06/10/07
OV9292 (Public Transport operators consortium) (2010) Available at http://www.9292ov.nl/9292ov206.asp (accessed 10/10/2011).
Pel, B. (in press) 'System innovation as synchronization; innovation attempts in Dutch traffic management', doctoral thesis, Erasmus University Rotterdam, Rotterdam, the Netherlands.
Pel, B. & Boons, F. (2010), Transition through subsystem innovation? The case of traffic management, technological forecasting & social change, vol. 77 (2010), 1249–1259.
Pel, B. & Teisman, G. (2009, April) 'Governance of transitions as selective connectivity', Paper presented at 13th Annual Conference of the International Research Society for Public Management, Copenhagen, Denmark.
Rijkswaterstaat (2004) *Audit Intermezzo; Markt voor mobiele inwintechnieken?* Delft, the Netherlands: Rijkswaterstaat AGI.
Rip, A. (2006) 'A co-evolutionary approach to reflexive governance—and its ironies', in J. Voss, R. Kemp & D. Bauknecht (eds) *Reflexive Governance*, Cheltenham, England: Edward Elgar.
Rotmans, J. (2006) *Societal Innovation: Between Dream and Reality Lies Complexity*, Rotterdam, England: Erasmus University.
Schelling, T.C. (1960) *The Strategy of Conflict*, Cambridge, MA: Harvard University Press.
Schot, J. & Geels, F.W. (2008) 'Strategic niche management and sustainable innovation journeys; theory, findings, research agenda, and policy', *Technology Analysis and Strategic Management*, 20(5): 537–554.
Smith, A. (2007) 'Translating sustainabilities between green niches and sociotechnical regimes', *Technology Analysis and Strategic Management*, 19(4): 427–450.
SON (2010) Stichting Onderzoek Navigatiesystemen (Foundation Research navigation system) (2010) http://www.stichtingonderzoeknavigatiesystemen.nl/eng/index.php (accessed 01/08/2010).
Sussman, J.M. (2005) *Perspectives on Intelligent Transportation Systems*, New York: Springer.
Teisman, G.R. (2005) *Publiek Management op de grens van Orde en Chaos*, Den Haag, the Netherlands: Academic Service.
Teisman, G., van Buuren, A. & Gerrits, L. (eds) (2009) *Managing Complex Governance Systems; Dynamics, Self-Organization and Coevolution in Public Investments*, New York: Routledge.
Urry, J.(2004) 'The 'system' of automobility', *Theory, Culture and Society*, 21(4–5): 25–39.
Urry, J. (2008) 'Governance, flows, and the end of the car system?' *Global Environmental Change*, 18(3), 343–349.
Van de Ven, A.H., Polley, D.E., Garud, R. & Venkataraman, S. (1999) *The Innovation Journey*, Oxford, England: Oxford University Press.
VID (2010) Available at www.vid.nl (accessed 10/10/2011).

13 The Emergent Role of User Innovation in Reshaping Traveler Information Services

Glenn Lyons, Juliet Jain, Val Mitchell and Andrew May

13.1. INTRODUCTION

The information age is now in full flight. It is providing unprecedented connectivity between institutions and individuals and among individuals. It is also accelerating the capacity to collect, manage, process, present and distribute data and information.

This chapter concerns itself not directly with the way in which the nature of the transport system may be in transition but with the way in which, in particular, information services that support the traveler may be in transition. This may, or may not, in turn affect significantly the ways in which our transport system is used.

The changing spectrum of technological possibilities in the information age in terms of communication and information exchange represents a pervasive technical change at the *landscape* level. Against this landscape the chapter identifies the Intelligent Transport Systems (ITS) sector as the *regime* epitomizing how the transport system has evolved to embrace the information age. This concerns the provision of 'top-down' information services to users that seek, typically, to cater for a (perceived) majority need. In turn the chapter raises the profile of, and considers how, *niche* developments of user innovation, especially associated with user-generated data and information and user co-operation, may or may not be gaining momentum as a groundswell of activity set to impinge upon the regime. Information services in the form of user innovation are arguably more agile and attuned to user needs. Developments in this field are mostly recent and rapidly evolving, and the chapter contemplates the prospects for how and to what extent niche developments in user innovation may see ITS in transition.

13.1.1. The Regime of Intelligent Transport

Technology has always played a major part in the shaping of our transport systems. It has been integral to the development and refinement of new modes of transport in terms of vehicle design and has underpinned development of the infrastructure in terms of construction techniques. As our

The Emergent Role of User Innovation 269

transport system has evolved, increasingly technology has been looked to as a means to manage the flows of vehicles (public and private) and people through the system. The arrival of the information age in its early forms saw the development of a particular field of professional activity in the transport sector, namely ITS.[1] A number of associations now exist worldwide that promote ITS on behalf of member organizations in the public and private sectors.[2] ITS is seen to encompass the use of technology—especially information technology—to support the management of transport systems and vehicles and to support travelers through the collection of data and its processing into information for operators and users. Traveler information services have been one key strand of ITS, with a now vast array of services making information available both before and during journeys and across a range of media. Such services have often been developed through partnerships between public authorities and industry (Lyons & Harman, 2002). Data for information services has tended to exist as a consequence of being collected for traffic and network management purposes.

The ITS field is now long established and is characterized by a top-down approach to its ongoing development as transport and allied information systems have evolved. The phrase 'Intelligent Transport Systems' merits comment. The word 'intelligent' aspires to reflect the way in which the use of information technologies can allow efficient management of the transport system as well as empower travelers to make fully informed choices about whether, when, where and how to travel. Ultimately, especially from a user perspective, ITS aims to facilitate better use of the transport system. An ever present question is whether or to what extent ITS actually achieves 'better use'?

13.1.2. The Prospect of Transition through ICTs

The mid-1990s saw the arrival of the Web and the subsequent mainstreaming of the Internet into society alongside the rapid penetration of mobile phone ownership and use. Accordingly the term information technology has been superseded by the term information and *communications* technology (ICT). This can be seen to be reflective of the increasing extent to which individuals are now engaged, through information technology, with communicating—both with information services and with each other.

The apparent increase in pace with which the information age is unfolding heightens questions surrounding what this might mean in terms of impacts on the operation and use of our transport systems. One particular question forms the focus of this chapter, namely, given the increasingly widespread availability of ICTs and their growing multifunctional nature: *Are we now entering into an era where the transport system user may innovate in using the 'everyday' ICTs around them to address transport challenges and issues they face?* In other words, set against the landscape of technological possibilities, might the top-down regime of ITS now be facing a niche development in the form of ICT-based user innovation? Could this see a transition into a new regime that redefines how ICTs are being employed in the transport sector? Using the transition pathways typology from Chapter 3, the chapter goes on to address the matter of whether user innovation will follow a transformation path (whereby it fills in some gaps but leaves the rules and routines of the ITS regime unchanged) or a reconfiguration path (whereby it has some knock-on effects that bring about changes in the 'architecture' of the regime).

We draw in particular upon insights from an ongoing research and development project in the United Kingdom called Ideas in Transit[3] (funded under the UK government's Future Intelligent Transport Systems initiative). The project is seeking to better identify and understand the (propensity for the) occurrence of bottom-up, grass roots 'user innovation'.

The next part of the chapter develops an understanding of the concept of user innovation and what is likely to distinguish it from the top-down regime. It also considers factors relevant to the context in which user innovation may arise and perhaps flourish. The chapter goes on to illustrate some of the bottom-up innovations that have been emerging of relevance to the transport sector. This is intended to make clear the nature of the niche developments being referred to. The final part of the chapter then discusses the potential for bottom-up innovation to achieve meaningful impacts on passenger transport over and above that being achieved through more conventional ITS approaches.

13.2. EXAMINING THE PROSPECT OF USER INNOVATION

13.2.1. Defining Innovation and User Innovation

Definitions of innovation are associated with the emergence of a product or service into the market place, but the process of innovation may incorporate both creativity and invention. Thus, to pin point what exactly is innovation, we understand firstly, *creativity* is "the production of new ideas or combining old ideas in a new way" (Heye, 2006,253); secondly, *invention* is making an idea real (e.g., a prototype); and finally, *innovation* "is an invention that has a socioeconomic effect; innovation changes the way people live" (Chayutsahakij & Poggenpohl, 2002, 1). It can be seen that the three are strongly related—an idea emerges which can then be converted from concept to reality and in turn applied in society. It is the social impact of innovation that is important in considering how niche development may coalesce in regime change.

The term 'user innovation' was coined by Von Hippel (2005) in relation to specific product innovations by users such as those involved in medical employment or sports participants. What then distinguishes user innovation from other innovation? User innovation implies that users have expertise in the field and are able to respond creatively and innovatively to specific user needs that are not met by traditional locations of innovation (e.g., manufacturers).Thus we developed our own definition of user innovation based on the previous: *the creation and application of an invention initiated by affected individuals that stems from user need or curiosity to address a problem or challenge within social practice.*

Here the distinction between users and producers in relation to the end product is important. Users are firms or individuals that expect to benefit from *using* a product or service *themselves* as distinct from producers who expect to benefit from *selling* a product or service. Users tend to have a more accurate and detailed model of their needs but less knowledge of the solution approach than a specialized producer. Accordingly, user and producer innovations tend to be different—user innovations focus on functional novelty and are associated with a rich understanding of user needs, while producer innovations often focus on *incremental* improvements on well-known needs associated with requiring a rich understanding of technological developments. User innovation does not preclude subsequent commercial exploitation, although this may not be the main motivating factor.

In transport terms and specifically in relation to information services then the distinction between user and producer innovation could be seen as follows: Governments, transport authorities and operators provide centralized information services that aim to cater for all or most users. Journey planners that can interrogate (vast) databases of information on service provision and produce origin-destination travel options constitute the bedrock. Some such journey planners now combine information on more than one mode of travel. Incremental improvements have subsequently included changes to user interface design, added-value information attached to presented travel options, real-time information updates and refinements in how user requirements for travel options can be stipulated. It is well recognized that such systems will struggle to 'please all the people all of the time', which suggests that there will be a prevalence of unmet needs among a diverse user group. Such large systems are encumbered by their legacy approaches and can struggle to be agile in responding to unmet needs and new opportunities for innovation. Meanwhile individuals with such unmet needs, or those believing they can act in the interests of individuals with such unmet needs, may be better placed to focus attention on addressing this in the absence of legacy approach. However, they may potentially struggle to harness the necessary data, data processing and information distribution capability that is associated with institutional approaches.

13.2.2. Factors Influencing User Innovation

Key motivations for a user to innovate are as follows (after Leadbeater, 2006; Lüthje, 2004):

- A need not being met (adequately) by the market
- Having relevant expertise and capabilities (skills, tools, facilities)
- The fun and enjoyment of the development process (and other incentives such as gaining social capital)
- Having an ability to share ideas (and knowledge/expertise) and work co-operatively with others

In several cases such motivations are also enablers of user innovation.

Often, direct financial benefit is not a motivation for the user innovator. This orientation of motivation may lend itself more naturally to the generation of innovations that attune to *true* user needs (and thus have market potential) than commercial motivations in a producer innovation context where *perceptions* of market potential may be less well grounded at an early stage in the innovation process. In other words user innovation should be more likely to concern problems looking for solutions as opposed to solutions looking for problems.

Financial benefits are much more likely to be a *consequence* of successful user innovation rather than a catalyst or motivator. Grabher, Ibert and Flohr (2009) note that "financial incentives are often at odds with the collaborative ethos of communities and undermine credibility, which is built on passion and not profit." This underlines a very different motivation behind *user* innovation (as opposed to innovation per se) and suggests a likely difference in outcomes.[4] Gaining social capital as a motivator relates to the important notion of gift relationships. As Currah (2007, 475) observes, "gifts are driven by the accumulation of alternative forms of capital—for example, social capital . . . or cultural capital . . . Quite simply, the overriding objective in a gift economy is to give away resources to secure and retain status."

The ability to share ideas is seen as a key motivator for user innovation. Traditionally this would have been very much dependent upon spatial proximity. However, the peer-to-peer communication now made possible by the Internet greatly assists the centrality of networks and co-operative behavior in user innovation.

It is instructive to recognize a number of design flaws that the Strategic Niche Management literature (Hoogma, Kemp, Schot & Truffer, 2002) identified on the basis of an analysis from considering various 'experiments' concerning alternative transport approaches. These flaws which may characterize the top-down producer innovation approach (Hoogma et al., 2002) and can be seen to be reflective of the ITS regime are as follows:

- Insufficient user involvement
- Too much focus on technical learning with the starting point being not a local problem but a solution
- Projects too focused to allow 'co-evolutionary' learning to occur (the combinations of policies, technologies and so forth that will impact on travel behavior)

- Experiments dominated by insiders who wanted to maintain the status quo
- Too much of a technology push

A key question is whether user innovation is better able to avoid such design flaws or indeed is able to more effectively expose such design flaws that relate to the existing regime in a way that would encourage support for regime transition.

We recognize that innovation constitutes a process or pathway that is followed. A user innovation begins with a trigger that, through creative thinking, becomes a local/personal invention; this may develop into a 'service' and can lead subsequently to wider adoption and diffusion. At each point in the innovation pathway the innovation can either progress, halt or fail. This innovation pathway relates to context (people, problems, environments, cultures, social practices), process (enablers, barriers) and outcomes (attitudes, behaviors, culture change, costs, benefits). Again in the context of co-operation, while the innovator is key, there will be other people (stakeholders) who along the pathway of innovation development play a part in supporting, funding and marketing.

13.3. EXAMPLES OF USER INNOVATION

The previous section has set out some of the factors and principles that are likely to govern the emergence of the phenomenon of user innovation in transport. This section highlights some examples of innovations—or at least inventions—that have already emerged in relation to supporting and influencing people's travel behavior. One element of the Ideas in Transit project has been to develop an *Innovations Portal*.[5] This is an online resource that highlights hundreds of innovations from many different countries that have been identified by the project and which are, in the main, relevant to transport. The reader is encouraged to visit the Portal to explore the richness of innovations that exist and which are continuing to emerge.

The selection of innovations presented here has been chosen with the intention of giving the reader a clear sense of the types of developments that constitute attempts at user innovation in transport as referred to in this chapter. The chosen examples aim to show some of the diversity of innovation in terms of the problems or challenges being tackled, the centrality of users to the idea behind and creation of the innovation development and how the use of user-generated data and co-operative behavior between users strongly characterizes the innovations concerned.

The examples together also provide a reminder that not all developments necessarily conveniently sit entirely within or outside user innovation in terms of the factors shown previously to characterize such innovation.

274   Glenn Lyons, Juliet Jain, Val Mitchell and Andrew May

Further to introducing this selection of user innovations, the final section of the chapter will discuss how theory associated with user innovation and (early) practice come together to offer insights for the future.

13.3.1. CycleStreets

CycleStreets (see Figure 13.1) bills itself as "a UK-wide cycle journey planner system, which lets you plan routes from A to B by bike. It is designed by cyclists, for cyclists, and caters for the needs of both confident and less confident cyclists." At the time of writing the UK national government is still developing its version of a cycle journey planner[6] with limited coverage. Meanwhile CycleStreets draws upon the success of another user innovation—OpenStreetMap[7] (thus an illustration of an *innovation cascade*). OpenStreetMap describes itself as "a free editable map of the whole world. It is made by people like you." Individuals voluntarily can input geographic information, a large part of which is transport related and in this case identifiable as relevant to cyclists. The mapping data can be used freely by such innovations.

Thus CycleStreets depends upon user-generated content for its existence and development while drawing advantage from the fact that users are tailoring information input to needs they know they have. It has been prompted into existence in part because the available information

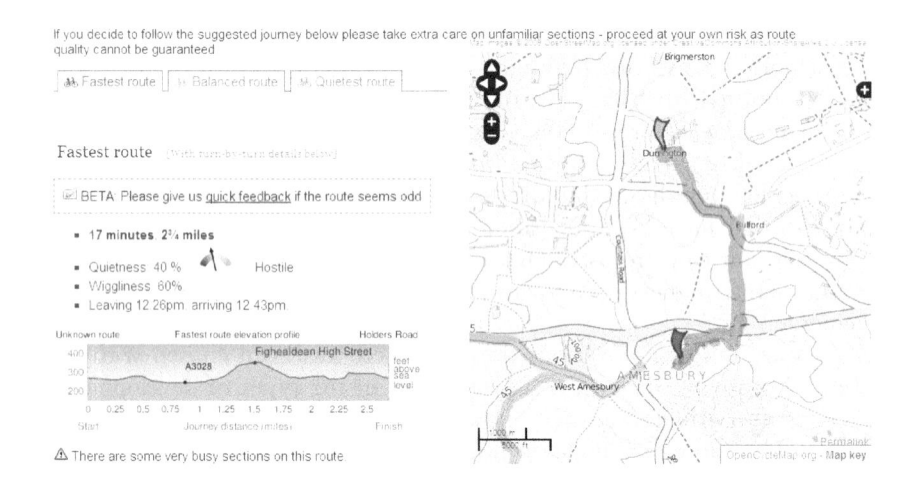

---

[6] See http://www.transportdirect.info/.
[7] See http://www.openstreetmap.org/.

Figure 13.1 CycleStreets (see http://www.cyclestreets.net/). © OpenStreetMap (and) contributors, Cloudmade, OpenCycleMap. CC-BY-SA.

marketplace has not adequately catered for cyclists. Its creators, both regular cyclists, point out that "CycleStreets is being set up as a company (on a not-for-profit basis) or a charity to act as a vehicle for any incoming funds. We also plan to release the code as an Open Source project and enable others to get involved."

The CycleStreets website provides commentary on a recent (April 2010) Freedom of Information request to the UK government that reveals that the government's own 'find a cycle route' service has cost nearly £2.4M to develop to date and has processed around 24,000 requests. In contrast, CycleStreets states that its own project "so far has been achieved on under £12,000, and our two main developers have worked entirely unpaid so far. . . . some 76,107 journeys have been planned in what is still our beta phase."

It can be seen from here that CycleStreets embodies several of the factors relating to user innovation introduced earlier. The previous paragraph notably suggests that user innovation and its ability to harness the 'power of the crowd' can be highly cost effective.

13.3.2. ParkatmyHouse.com

The Ideas in Transit Innovations Portal portrays the history of the next selected innovation—'ParkatmyHouse.com' (see Figure 13.2)—as follows: "[i]n May 2006 a UK tourist in the United States, Anthony Eskinazi, *(nearing the end of his post-university round the world trip)* was failing to find a parking spot near a major sports stadium in San Francisco on match day when he saw an empty driveway close to the stadium. Realising a business opportunity for both homeowners and drivers he set up Parkatmyhouse.com on his return to the UK in June 2006."

Figure 13.2 Parkatmyhouse.com—http://www.parkatmyhouse.co.uk/.

On the innovation's website, Anthony argues that, "with resident permit holder bays springing up all over the country, car park prices making us wince and fines landing on windscreens at a record rate and with no sign of a slowdown, parking is becoming ever more restricted, costly and stressful. At the same time, millions of driveways and car parks—next to major public transport hubs, town centres and theatres for example—stand empty when they could be providing parking for motorists and extra income to homeowners and businesses."

It seems a compelling proposition and is a service now in use. It raises questions as to whether it supports or aggravates government policies for sustainable transport although the website claims "as part of its green initiative" to have "formed an exclusive partnership with Zipcar, the world's largest car sharing company." While the scale of use is not immediately apparent from the website, there is clear evidence that the service is being used, endorsing perhaps the most distinctive aspect of this example—behind what is a particularly simple and yet highly creative idea is a demonstrable user need (now being met). It does appear that this innovation may have been motivated in part by commercial opportunity, and indeed a fee is charged to parking space owners for any successful transactions. However, the origins and perhaps early motivations of the development are grounded in a directly experienced user need. The extent of this need across the car driving population remains to be seen (and may be changing over time).

13.3.3. PickupPal

PickupPal (see Figure 13.3) is based in the United States and on its website highlights a quote about itself from the Financial Post: "This innovative start-up using geopositioning technology could soon be as threatening to the status quo as the automobile was to the horse and buggy." This indicates the aspiration if not prospect of user innovation being a niche development that brings about regime transition.

CEO and Co-founder, John Stewart states that "PickupPal was hatched while I sat stuck in traffic and wondering how all these vehicles could be harnessed to be more efficient instead of just the one-driver, one route function. I began to imagine a scenario wherein each vehicle was a transportation unit and everyone seeking a ride was a unique participant in a virtual 'transportation marketplace'."

PickupPal connects together drivers and passengers to allow ridesharing and includes the means for drivers to make offers to prospective passengers about the charge for the ride.

Ridesharing as a concept is a long-standing phenomenon, but efforts to date to achieve widespread take-up have only met with limited success, reflecting in part people's reluctance to inconvenience themselves as a driver by subscribing to a service and then going 'out of their way' to share their vehicle with someone else. Arguably success has also been constrained by the rather mechanistic ways in which centralized systems try and act as

Figure 13.3 PickupPal (see http://www.pickuppal.com).

brokers. A new generation of ridesharing brokerage services such as PickupPal capitalize on the connectivity of the Internet and mobile ICTs and also upon the existing social networks and networking that takes place: PickupPal "integrates with the most popular social networking tools, making it easier for people to share rides with others who have similar interests or with whom they are already connected via colleagues or friends." Launched in January 2008, PickupPal has over 140,000 members.

Another ridesharing development is called Carticipate[8] produced in San Francisco. Carticipate is a social network for ridesharing and carpooling available on the iPhone and Facebook which takes advantage of the location based services on the iPhone and matches riders and drives in the user's local area.

While such developments may operate on a commercial footing there is a distinct entrepreneurial spirit behind the innovations and, on the part of the creators, an empathy with, and direct experience of being, users.

13.3.4. TrainDelays

Train operators in the United Kingdom have policies about compensation entitlement in relation to delays caused to passengers by late running trains. Passengers have up to 28 days to submit a compensation claim after a delay has occurred. However, it can be cumbersome to identify delays and go through the process of making a claim. A disgruntled rail commuter, Chris Davy, set up a website[9] to make it easy for people to identify delays and submit claims. The service relies on contributors to report delays which can then be shared with others through the service. The service aims to make it easy to then submit claims. "Train Delays is a free service with the sole aim of bringing commuters together to make sure that we all make the claims that we are entitled to."

This example returns us to some of the key factors characteristic of user innovation: a problem or frustration being faced by users that has prompted the idea, an individual who is a user experiencing this problem and who has the expertise and capabilities to do something about it, a motivation that is not derived from financial gain (in terms of the innovation helping others) and an emphasis upon co-operative working among users to help the innovation flourish.

13.3.5. MyBikeLane

An individual cyclist living in New York built and launched MyBikeLane (see Figure 13.4) in 2006 and continues to maintain it single-handedly. He writes on the website "MyBikeLane was conceived after repeated frustration at having to dodge cars illegally parked in the bike lanes. Several near crash experiences as a result of people too lazy to find valid parking motivated me to build this site." The site's aim is to use the power of the community to make the problem of illegal parking more prominent through exposure to the media and public officials in the hope of contributing to pressure to have something done about it. Contributors to the site can upload photos of illegally parked vehicles and indicate when and where the incident occurred and the license plate information.

Both this example and the last are reflective of how user innovation can arise from, and be a means to react to, frustrations about 'injustices' seen to exist within the operation of our transport systems.

Figure 13.4   MyBikeLane (see http://www.mybikelane.com/).

13.3.6. Slugging

This last example is included as a reminder that user innovation is not new in transport but perhaps is now being accelerated or at least being made more visible through the Web. Slugging is an informal carpooling system that started in Washington, DC in the 1970s when high occupancy vehicle (HOV) lanes were introduced. A car needing additional passengers to meet the required three-person HOV minimum will pull up to one of the known slug lines. The driver either displays a sign with the destination or calls out the destination. The people first in line for that particular destination then get into the car and are given a lift to the destination. No money is exchanged because of the mutual benefit; the car driver needs riders to access the HOV lane and the passengers need a ride. Slug-Lines.com comments that "not only is it free, but it gets people to and from work faster than the typical bus, metro, or train. It's unique because it is not a government sponsored commuter program, but one created out of ingenuity from local citizens to solve commuter problems."

User innovations are not necessarily dependent upon the Internet, but the set of examples presented here highlights how significant the connectivity it has brought about is for many innovations that rely upon a 'self-help' collaborative culture among users. The examples reflect targeted achievements, often with modest resource, which can emerge from as little as one person's imagination, motivation and capabilities.

13.4. THE PROSPECTS OF TRANSITION

This final section of the chapter reflects upon developments in the user innovation space associated with transport in the context of the existing ITS regime and looks to the prospect for significant transition from this regime.

13.4.1. Uptake of User Innovations

The creation of the Innovations Portal and its growing list of over 200 identified innovations could suggest a substantial groundswell of niche activity set to impinge upon the transport sector. However, there is a need to return to the essence of what defines an innovation: An innovation is an invention whose take-up influences social practice, that is, changes the way people live or, in this case, travel. One can certainly find, in examining some of the previous example innovations, that there are highlighted instances of individuals who offer praise for the development having had direct experience of it. However, there is precious little indication as yet that these 'innovations' have moved much further than the innovator and possibly early adopter stage in terms of the diffusion of the innovation (Rogers, 1962).

Even in the case of such examples as PickupPal with its more than 140,000 members, there is a distinction to be made between curiosity in the ideas which attract membership (and media coverage—which several of the examples described in the last section have enjoyed) and application of the idea in social practice. Ridesharing schemes have been long known for their diminishing numbers from the stage of stated interest in the prospect, through sign-up to membership to actual ridesharing.

It can be suggested that the Web has made it easier to both (a) make visible creative ideas and inventions and (b) to seek them out which may create an inflated impression of innovation at least in terms of the prospects for take-up and influence on social practice.

Several of the previous selected examples are only a few years old. It could be the case that they have yet to be brought properly to the attention of a large audience of prospective users (albeit that some, as noted, have enjoyed at least limited national media coverage). Conversely, it could be argued that some are long-enough established that were they to be truly holding the prospect of (appreciable) social change then electronic word-of-mouth made possible through ICTs would have seen more dramatic levels of membership/use. The question that follows is this: What are the *future* prospects for large(r) scale take-up of (some of) the user innovations in the transport sector?

In contemplating this question there are insights relating to human behavior to suggest that the answer could be 'rather limited'—inertial forces against change appear to exist. In the field of travel information services, it is now being recognized that most people, most of the time, do not need information in order to undertake their travel (Lyons, 2006). This stems from the fact that much travel undertaken by the public overall is local and routine; it is travel that people are familiar with to the point that their behaviors have become habitualized. Habit becomes prevalent because, it can be argued, many (or most) people are content with satisfactory options and outcomes to their travel. In other words they are adopting 'good enough' travel options rather than optimum (utility-maximized) solutions. People have other things on their minds aside from travel, and this encourages them to look for short-cut decision making in relation to travel.

While a major short-cut is habit itself (i.e., an automated behavioral response such as always drive alone to work in the morning), another short-cut relates to the notion of anticipated regret (Loomes & Sugden, 1982). In this case an individual will only review their travel choice (and in so doing perhaps seeking further information) if they anticipate regretting not doing so with respect to the outcome of their travel choice (e.g., being late for an interview because of traffic could force consideration of travel options).

Added to this, there is a sense that the challenge faced by society at large, and reflected in policy objectives, is not necessarily the challenge faced by each individual that makes up that society. Goodwin and Lyons (2010) have examined public attitudes toward congestion. United Kingdom evidence revealed that in 2005, 87% of people thought congestion was a serious problem for the country as a whole (Department for Transport, 2007). Yet other evidence from 2005 (DfT, 2005) found 63% of adults saying that road congestion was either not a very serious problem or not a problem at all for them personally (in spite of only 8% indicating that they had not experienced congestion in the last month). For many, congestion is now seen as just a fact of life.

Not only could there be limited appetite for ICT-based innovations to help address travel behavior, but there is evidence to suggest that ICT products are capitalizing upon people being on the move and upon being 'stuck in traffic' through their ability to make people's passing of travel time more bearable, enjoyable or productive and to help time *seem* to pass more quickly (see Lyons, Jain & Holley, 2007, for evidence on such matters in relation to rail travel).

Such limitations to the take-up of developments intended to support or influence travel behavior are not unique to user innovations but apply to all such developments. For example, Farag and Lyons (2008) highlight what they argue to be a straightforward yet key observation: The effect of using public transport on using public transport *information* is stronger than the other way around. In other words, the propensity of people to consider using public transport is what affects people's need for information—the availability of information does not in itself prompt people to consider changing their behaviors.

Hence the question of uptake of user innovations may reside in what will motivate people to review their behavior. For example, it may need the cost of motoring to increase sharply before ridesharing services such as PickupPal truly flourish; cycling facilities may need to improve substantially, coupled with heightened awareness of public health, before more people will choose to consider cycling—at which point the need for CycleStreets will become more pronounced; or parking may need to become truly intolerable in urban areas before people will seek out innovations such as ParkatmyHouse.com.

User innovations have distinctive characteristics to set them aside from, and seem more expressly attuned to user needs than, the top-down developments of the ITS regime. However, it can be suggested that this seemingly more socio-technical approach remains rather deficient in terms of its grasp of human behavior, in which case doubt remains concerning the scale of uptake and impact on travel behavior of user innovations.

13.4.2. Transformation or Reconfiguration?

At present, the ITS sector (with its partnerships between government and industry) continues to rely upon its strength in terms of the infrastructure and systems put in place that provide for the gathering and management of data which in turn support (through data agreements) the provision of information services for operators and users of the transport system. An important question is this: Will the user innovations now emerging reconfigure the ITS regime, or will they be transformational (filling in the gaps)?

The picture is not altogether clear. In relation to data there are signs that user innovation is both seeking to substitute for ITS provision and in some cases do things that traditional ITS is unable to. The example of OpenStreetMap is a case in point. National mapping agencies have traditionally been the custodians of geographic data able to support information services. Such data comes with assurances about accuracy and completeness. It has also tended to come at either a cost or with terms of use restrictions. OpenStreetMap is a new model for the generation of geographic data. As the level of contributor activity has grown, so too has the coverage, accuracy and completeness of the mapping. The data is not subject to costs and restrictions for use in the same way as for the national mapping agencies (though see later), and this is seeing a spawning of subsequent innovations such as CycleStreets. OpenStreetMap brings into question the notion of 'completeness'. National agencies can ensure their mapping is complete according to the defined elements that they presume to make up the set depicting mapping detail. Meanwhile, OpenStreetMap allows users to define and extend this set to include new elements specific to what are seen to be relevant to their or the community's needs.

OpenStreetMap relates to geographic information. The opening up of data extends beyond this. There is scope for more user innovation to occur for example if public transport service information is made openly available by transit agencies and public transport operators. There are already early indications that when this happens, innovations can emerge quite rapidly and at minimal or no costs to the agencies or operators. This is in contrast to the slower and more costly traditional approach in which datasets are closely guarded and made available for use in carefully established and agreed partnerships between industry and public authorities to develop information services.

The rapidly developing culture of user generated content that the connectivity of the Web has made possible could have the potential to shake the foundations of the traditional ITS sector unless it adapts. Users have the capability through crowd behavior to generate data quickly and of a sort that may be difficult or impossible for top-down systems to address. Adaptation may take the form of established information service providers making use of user-generated data, but it may also be the case that the market is opened up so widely by the availability of user-generated data that established providers are overwhelmed by more agile and numerous entrepreneurs who are now able to rapidly bring services to market.

So, there are real prospects for user innovation, or at least innovations derived from peer-to-peer and user generated data, to challenge the existing ITS regime. However, at the same time, some of the user innovations are more evidently filling in the gaps in terms of what may remain niche user needs—for example the services aimed at expressing user frustration with service provision such as MyBikeLane and TrainDelays.co.uk. To some extent the challenge to the existing regime may be dampened by the earlier points associated with human behavior concerning what can affect up-take of developments.

Another issue associated with how user innovation may interact with the ITS sector is that of whether or not the two can work together. In other words, is it conceivable for example that CycleStreets could in some way be 'taken over' or supported and managed by government and its suppliers? Could Ordnance Survey, the United Kingdom's national mapping agency, collaborate with volunteer mappers, and integrate professionally sourced and user generated data? In this way one might argue that the establishment can take advantage of users as a source of information that enables better services to be delivered back to users. However, it could be suggested that one of the things that motivates individuals to contribute to user-generated data services is that such services are 'outside of the establishment'—institutionalizing user generated data could fundamentally undermine its vitality.

13.4.3. Looking Ahead

The picture concerning data and its role within the regime and amidst niche developments is not clear, and the interactions with and between actors are not fully understood. However, added to this the picture continues to change. Indeed recent change suggests greater prospects for reconfiguration of the ITS regime. In the United Kingdom in March 2010, Prime Minister Brown made a speech entitled 'Building Britain's Digital Future'.[10] He claimed that government would support "unleashing the entrepreneurial, innovative and dynamic talents we have in Britain." He referred to the next generation "semantic web"[11] and "new enterprises spun off from the new data, information and knowledge that flows more freely." The vision was of "the radical opening up of information and data" with an intention to bring about "making public data public." In stark contrast to how the United Kingdom's national mapping agency—the Ordnance Survey—has operated its business model as addressed previously, the former Prime Minister announced that "from 1st April, we will be making a substantial package of information held by Ordnance Survey freely available to the public, without restrictions on re-use . . . Any business or individual will be free to embed this public data in their own websites and to use it in creative ways within their own applications."

Of course the wider automobility regime discussed elsewhere in this book is much more than only about data and information services. However, in terms of the ITS regime centered upon in this chapter, the way in which information services are being developed could be fundamentally changing: Data is being opened up for third parties (including users) to access freely, other data is being generated by users, there are then prospects for the different types of data to be made to work together and user innovators and the power of the crowd are joining forces to deliver new information services that were not formerly possible. It would seem that at the very time of writing we are indeed experiencing the beginning of a transition in the way we conceive of, develop, engage with and experience using information services.

However, the significance of this regime transition for the higher level regime of automobility is bound into whether it affects, differently or to a greater extent, travel behavior. This itself may well be fundamentally dependent upon externalities such as oil prices, climate change pressures on policy and public health concerns. These may stimulate the public appetite for reappraising its travel behaviors and for entertaining adjustments to social norms (such as the norm of driving alone). In turn, the sorts of services sought by the traveling public may very much be those that user innovation is trying to deliver, based strongly upon peer-to-peer information sharing and needs matching.

ACKNOWLEDGMENTS

We would like to acknowledge the UK sponsors of the Ideas in Transit project who have made possible the development of this chapter, namely the Engineering and Physical Sciences Research Council, the Department for Transport and the Technology Strategy Board.[12] The views expressed in this chapter are those of the authors.

NOTES

1. See also Chapter 2.
2. For example ITS (UK), see http://www.its-uk.org.uk/.
3. See http://www.ideasintransit.org/.
4. Some individuals are motivated to innovate by the technical/intellectual challenge more than by community sharing/belonging. Such individuals can be enlisted (outside of our notion of user innovation) as 'creative customers' by producers seeking to innovate—people who either are allowed to develop or customize a product belonging to an organization or do it anyway. Their innovation is likely to be driven less by need than curiosity.
5. See http://www.ideasintransit.org/wiki/Ideas_in_Transit.
6. See http://www.transportdirect.info/.
7. See http://www.openstreetmap.org/.
8. See http://www.carticipate.com/.
9. See http://www.traindelays.co.uk.
10. See http://www.labour.org.uk/gordon-browns-speech-on-building-britains-digital-future (accessed 2010–03–26).
11. The "semantic web" is an evolving development of the World Wide Web in which the meaning (semantics) of information and services on the Web is defined, making it possible for the Web to "understand" and satisfy the requests of people and machines to use the Web content (see http://en.wikipedia.org/wiki/Semantic_Web).
12. The Technology Strategy Board's role is to promote and support research into, and development and exploitation of, technology and innovation for the

benefit of UK business in order to increase economic growth and improve the quality of life. www.innovateuk.org.

REFERENCES

Chayutsahakij, P. & Poggenpohl, S. (2002) 'User-centered innovation: The interplay between user-research and design inovation', The European Academy of Management 2nd Annual Conference on Innovative Research in Management, May, Stockholm, Sweden.

Currah, A. (2007) 'Managing creativity: The tensions between commodities and gifts in a digital networked environment', *Economy and Society*, 36(3): 467–494.

Department for Transport (2005, August) 'Attitudes to congestion on motorways and other roads', report, London: Transport Statistics, Department for Transport.

Department for Transport (2007, October) 'Public attitudes to congestion and road pricing', report, London: Transport Statistics, Department for Transport.

Farag, S. & Lyons, G. (2008) 'What affects use of pretrip public transport information? Empirical results of a qualitative study', *Transportation Research Record: Journal of the Transportation Research Board*, 2069: 85–92.

Goodwin, P. & Lyons, G. (2010) 'Public attitudes to transport: Interpreting the evidence', *Journal of Transportation Planning and Technology: UTSG Special Issue*, 33(1): 3–17.

Grabher, G., Ibert, O. & Flohr, S. (2009) 'The neglected king: The customer in the new knowledge ecology of innovation', *Economic Geography*, 84(3): 253–280.

Heye, D. (2006) 'Creativity and innovation: Two key characteristics of the successful 21st century information professional', *Business Information Review*, 23(4): 252–257.

Hoogma, R., Kemp, R., Schot, J. & Truffer, B. (2002) *Experimenting for Sustainable Transport: The Approach of Strategic Niche Management*. London: Spon Press.

Leadbeater, C. (2006) *The User Innovation Revolution: How Business Can Unlock the Value of Customers' Ideas*, London: National Consumer Council.

Loomes, G. & Sugden, R. (1982) 'Regret theory: An alternative theory of rational choice under uncertainty', *Economic Journal*, 92(368): 805–824.

Lüthje, C. (2004) 'Characteristics of innovating users in a consumer goods field: An empirical study of sport-related product consumers', *Technovation*, 24(9): 683–695.

Lyons, G. (2006) 'The role of information in decision-making with regard to travel', *Intelligent Transport Systems*, 153(3): 199–212.

Lyons, G. & Harman, R. (2002) 'The UK public transport industry and provision of multi-modal traveler information', *International Journal of Transport Management*, 1(1): 1–13.

Lyons, G., Jain, J. & Holley, D. (2007) 'The use of travel time by rail passengers in Great Britain', *Transportation Research*, 41(A): 107–120.

Rogers, E. (1962) *Diffusion of Innovations*. New York: Free Press.

Von Hippel, E. (2005) *Democratizing Innovation*. Cambridge, MA: MIT Press.

14 Innovation in Public Transport

Reg Harman, Wijnand Veeneman
and Peter Harman

14.1. PUBLIC TRANSPORT—POTENTIAL OR PAST?

By the end of the 20th century, there was a growing awareness that high levels of automobility posed severe problems for developed societies (see, e.g., Harman, 2004; Veeneman, 2002). The immediate attention was on its environmental impact: The concern at first was over localized pollution through emissions, but this gradually turned to a focus on the role of carbon output in global warming. But understanding also grew of how automobility was a factor in social inequity (Social Exclusion Unit, 2003) and might actually limit economic development. Simply reversing the demand for mobility was impossible because it was so built into peoples' expectations. Walking and cycling already cater for many shorter trips and might cater for more. But for longer ones a motorized alternative was needed.

Public transport had been there before car ownership and use started to grow, and the negative side of automobility led to an increasing focus on strengthening the role of public transport to provide a less damaging alternative. This could only work with services attracting more travelers and providing a stronger alternative for the car. This need to improve services took a number of forms. Formal policy documents of government and other public authorities incorporated policies giving priority to public transport and aiming at a modal shift from car use. These were backed up by guidance and regulations. Companies providing public transport used the theme in their marketing for the general public and highlighted opportunities for public authorities to take up. For example, *New Transit* magazine of December 2009 (issue 6) was entitled *The Great Green Opportunity?* (note the question mark). But how far has this led to significant change in public transport use?

The growing interest has to be placed within two broad contexts. One is to define what is meant by public transport. The other is to consider briefly the historical context within which public transport has developed.

Public transport is essentially a publicly available service of multiple occupancy vehicles, operated almost entirely on set routes at pre-defined schedules. Charges are levied on users, according to distance, time, type of service and other factors. Services are provided by operators, who may be private or public companies: These plan and run the routes and networks, provide the vehicles/trains and staff, manage fares systems and promote services. The infrastructure on which services run is sometimes provided by commercial bodies, more usually by public bodies (especially as much of it involves public highways). Other bodies have a strong interest in public transport provision, especially manufacturers of infrastructure, vehicles and control equipment.

Public transport covers a range of modes and services, operated at different speeds and over distances varying from very long distance to very local. Typically long distance services tend to operate at higher speeds, regional and local services at lower speeds (see, e.g., Faulks, 1990; Papacostas & Prevedouros, 1993). The modes can be grouped into three broad categories:

- Interurban (national and regional)—inter-city and high-speed rail, regional and cross-country rail, inter-urban coach services
- Local (sub-regional/urban/local)—suburban and local rail, metro, light rail (primarily on own right-of-way), tramways (primarily on public highways), bus transit systems (primarily on own right-of-way), main (trunk) bus services (on public highways), local (connecting) bus services
- Complementary (local)—community transport (flexible minibus services), taxis and hire cars

The range of modes is sometimes described as a spectrum. There are some areas where more than one public transport mode might be appropriate: For example, some urban corridors might in principle be served by a tram route or a high capacity bus route. But there are quite wide areas for which only one main mode is appropriate, for example, the bus for more complex local networks or light or local rail for heavily trafficked medium distance corridors.

Most public transport provision and travel involves one or more of the three principal modes: train, urban transit and bus.[1]

In most Western countries, including the Netherlands and Great Britain, the development was similar. In the early 19th century, development of steam engines underpinned the development of railway systems, followed by a rapid, partly privately funded growth of the network. At an early stage, national and local governments stepped in to help fund national and local railways and build a broader network from the first trunk routes developed by eager entrepreneurs. The train was a growing success for at least 100 years (see Wolmar, 2007, 2009).

After those years the development of internal combustion engines brought rapid growth in the use of buses and cars. Both proved to be fierce competitors of the train, and since the 1920s many rail lines have been closed and replaced by bus services. From the 1960s private car ownership took off and put pressure on public transport as a whole. Cars directly replaced some bus and rail journeys. But car use also offered the freedom to go anywhere, leading to a far greater dispersal of travel patterns. Over time people's activity patterns and the development of settlements changed to accommodate these much more complex travel patterns and thus they reinforced them. Rail and conventional bus services simply could not compete for such dispersed journeys. This has been a recurrent theme over the last century, in the developed world and now in developing countries (Gakenheimer, 1999).

However, as cities and mobility grew in both Great Britain and the Netherlands, public transport held its own in certain sections of the travel market. This happened especially in those places where high volume led to congestion on roads, and rail systems with their own right of way and high capacities got a second chance (see also Hibbs, 1989).

While many rail and bus operators had started as private enterprises, with growing car ownership and use, the operators struggled. Government stepped in and provided subsidies and growing control, often bringing the operators to the public sector. From the 1980s, several European countries experimented with bringing the private sector back, albeit in various ways. In both Great Britain and the Netherlands private commercial companies now play an important role in public transport development.

14.2. A TALE OF TWO COUNTRIES—GREAT BRITAIN AND THE NETHERLANDS

Public transport in Europe today reflects development during the 20th century from this historical basis. Some key trends emerge, but there are also differences because different countries have taken different approaches. This section looks in turn at railways and at urban transport, covering Great Britain and then the Netherlands (see also Veeneman, 2002).

The development of public transport over the last 50 years has been dominated—some might say overwhelmed—by the growth in acquisition of cars and their use in travel, and the resulting focus in supporting these. This has led to two main strands in evolution of public transport systems. First, public transport operators have incurred growing deficits and run down their services as rapid growth in car ownership and use from the 1960s has caused public transport use to decline. Bus, transit and train operation require a significant level of personnel, and it is not possible to reduce these beyond a certain point without reducing service levels and quality. Thus there are limits to the economies of scale that can be achieved on public transport, and so operations become more inefficient. This has led to the second theme, the increasing public intervention to provide operational support and investment to prevent more severe losses of services.

To an increasing measure these have been justified by concerns over the environmental and social impacts of ever growing car use.

In Great Britain and the Netherlands, the 1980s and 1990s have fostered a perspective where this strong role of government is seen more and more as hampering efficiency and the quality of service. In both countries this has led to more competitive regulatory regimes being developed, aimed at allowing for private initiatives and better incentives for efficiency. This has most immediately affected bus services. In Great Britain, the Transport Act of 1985 opened up the bus services market outside London for any qualified operator to run where they wished; in association with this the publicly owned bus companies were sold into private hands. Within London all bus service development remain the responsibility of the capital's transport authority, now Transport for London, but services are operated by private companies under concessions. In the Netherlands, a similar development followed 15 years later, with the Wet Personenvervoer of 2000 (Passenger Transport Law of 2000), which required local authorities to tender out concessions for service provision (Veeneman, 2002).

The same approach has been taken with railways. Following European Union legislation that started with Directive 91/440, the former single national railway companies were broken up into separate components, the freight businesses sold off as separate commercial bodies, and the passenger operations leased as separate concessions. Here too there are differences between the two countries (see Van de Velde & Röntgen, 2008). In Great Britain the whole passenger network was broken into two dozen units and each has been leased, by the national Department for Transport, to a private company; most franchises are for 7 years. The rail infrastructure was sold to a private company (Railtrack), but when this went into administration, the infrastructure was taken over by a partly national company, Network Rail. In the Netherlands the national rail system is directly awarded to Nederlandse Spoorwegen (NS –Dutch Rail), as are intra-urban services around Amsterdam, The Hague and Rotterdam to their respective local operators. Other local rail lines and networks are tendered out by regional authorities in concessions for around 15 years. These authorities (in total 19) also tender out the bus services in a total of 70 concessions.

In terms of funding, key differences exist. In Great Britain ticket structures and prices are set by the operators for bus services, controlling their income. In London and the Netherlands, a single and publicly controlled ticketing system exists. Operators are dependent on the authorities for their fare box revenues. In the Netherlands, that income can be still dependent on the number of passengers carried, but a complex clearing house is needed to make sure every operator (or authority) gets its share of the sales of the national zonal ticketing system, the strippenkaart (strip ticket). That system is currently being replaced by a chipcard similar to London's Oyster card to allow for easier redistribution of the farebox revenue to all operators authorities.

With the changes in regulation, the actor landscape in public transport changed dramatically within a decade in both Great Britain and the Netherlands (Van de Velde, Beck, Van Elburg & Terschüren, 2008). Local bus operators were amalgamated in larger (inter)national entities and competed for the market, either by competition on the road (in the United Kingdom outside London) by competition for concessions (London and the Netherlands). In addition, the English public transport executives (PTEs) in the main cities outside London lost a great deal of their influence, as did the local transport authorities across the rest of Great Britain. In contrast, in the Netherlands the PTEs had been non-existent but gained a lot of power; the national government decentralized all its control over public transport to 19 regional authorities.

What has been the result, in terms of trends in public transport travel over the last 20 years? This is shown in Table 14.1, which shows percentages of total national travel by the main modes. The key point demonstrated is that public transport remains a small part of total travel in terms of distance: around 13% in Great Britain and 11% in the Netherlands. In Great Britain local modes are more significant than rail, but both have increases sharply over the last decade, following a previous decade of decline; cycling plays a very minor part. Local public transport's share is very low and declining in the Netherlands, where the bicycle plays a significant role in travel, but rail's share is slowly increasing. Car use remains dominant, more so in Great Britain. In both countries total travel (passenger kilometers) grew by about one third during the 20-year period.

Table 14.1 Overall Proportions of Passenger Kilometers Traveled by Mode in Percentages

	1987	1997	2007
Great Britain			
Bus/Tram/Metro	8.8	7.0	7.4
Train	5.5	4.8	6.0
Bicycle/Motorcycle	2.2	1.1	1.2
Car	83.1	86.3	83.9
Other	0.4	0.8	1.5
Total	100.0	100.0	100.0
The Netherlands			
Bus/Tram/Metro	4.7	4.4	2.9
Train	5.9	7.6	8.0
Bicycle/Motorcycle	8.5	8.1	7.7
Car	75.3	74.5	75.8
Other	5.6	5.4	5.6
Total	100.0	100.0	100.0

Note: Dutch data from Centraal Bureau voor de Statistiek (2010). British data from Department for Transport (2008).

Great Britain's public transport administration also differs widely from that in other European Union countries, as for example, in cities and regions in France (see Harman, l'Hostis & Menerault, 2008).

14.3. BARRIERS AND OPPORTUNITIES FOR ADVANCING PUBLIC TRANSPORT

What then are the key policy influences on public transport in the two countries? Where are they similar and in what respects do they differ? Have any niche activities emerged to bring changes to the public transport regime?

In Great Britain the government in office from 1997 to 2010 aimed to develop an integrated transport system and to coordinate this with land use planning. The current government has in principle continued these policies. However, while this approach has been encouraged through key national policy documents, effective mechanisms have never been established at regional and local levels (and indeed the current UK government has abolished the regional development agencies). Instead, there has been a concentration on controlling public spending and avoiding any impression of being 'anti-car'. Some support for local public transport is possible for the Integrated Transport Authorities in major cities and the local transport authorities elsewhere, through their local transport plans; but these are closely supervised by the Department for Transport, who control all funding.

National transport policies and national land use planning policies have been revised more than once; and there are now some differences between England, Scotland and Wales. Throughout Great Britain the authorities responsible for transport are mostly different from those responsible for land use planning, which makes it difficult to achieve sound integration. There are wide variations in geography, from the densely occupied South East region to the remote areas of the Scottish Highlands (Bolden & Harman, 2008).

In contrast, an integrated approach to land-use planning and the transport system has been the basis for the Dutch transport and planning policies since the early 1990s. In the White Papers (SVV II and VINEX) of that era the development of locations for both employment working and housing focused on sound integration with a strong role for public transport. The main idea was to locate economic activities attracting many passengers, like offices, near major stations, while major housing developments should have relatively high densities and be close to existing cities. In addition, they should have good public transport links between the center and the new development. Although not all developments met the standards, this set the scene for economically booming station areas and light rail development to the new outskirts of most larger cities.

Ten years on, decentralization of the responsibility for public transport to regional authorities allowed them to more intricately link urban and regional planning and public transport provision in both cases, by setting service conditions to the operators. In Great Britain, however, the public transport authorities lost power due to the deregulation of 1985 (though they have regained some possibilities with Quality Partnerships). But the regional governments of the Netherlands took full control over public transport. Decentralization meant the obligation to tender out concessions, giving the authorities a great deal of control over service types and levels, similar to metropolitan London in England. Some of the regional authorities took the challenge; others struggled more with the new responsibilities and the possibilities it offered.

The situations in Great Britain (outside London) and the Netherlands led to different forms of innovations. In Great Britain the freedom of the operator to develop its services has led to a continuous process of service changes, new routes and times. This in principle should allow supply to better follow demand: It has certainly allowed operating companies to achieve continual efficiencies in the face of generally declining demand and in some localities to increase service attractiveness. But it has also created uncertainty on whether services will still be available in the coming years, so that travelers and communities have become reluctant to rely on buses.

In the Netherlands, regional authorities demanded other innovations: low-floor buses and low-emission engines. Many authorities even chose to limit changes in routes and times, as they expected the traveler to value the consistency. The innovations in the Netherlands were more intermittent. The regular tendering of concessions allowed the authorities to develop new demands for the next operator to bring in innovations. Critics suggest the fact that many of these innovations constituted hobby horses of the authorities and that the authorities were less interested in tuning supply to demand. The same criticism has been made of London, where the bus regulatory system differs strongly from the rest of Great Britain.

Authorities are struggling with the direction they wish to develop public transport. On the one hand, public transport is expected by policymakers to aid in reducing the negative effects of automobility. To do so, services should be aimed at peak hour commuter traffic, where speed is the key performance indicator for the traveler. This means a less dense network with high frequencies and limited stops. On the other hand, a key public value that policymakers seek from public transport is social inclusion: the ability of those not owning a car to participate in society by offering them mobility. For those access to the network is a key performance indicator. This means a dense network of services with many stops, only cost effective at much lower frequencies and capacities. Not only the demand pattern but also the policy incentive for the two demographics is very different: It is the policymakers wish for car owners to use public transport; it is the pensioners' wish for policymakers to provide basic services.

And then there is the third actor in the triangle of developing services: the operator. In the Netherlands, many authorities are now finding new ways to co-develop the services together with the operator. The authorities are seeking the best of both worlds, with strong incentives for the operator to develop the supply along the lines of demand and possibilities for the authority to innovate toward better sustainability and accessibility. The same approach is being taken in Great Britain, through Quality Partnerships as instruments to better integrate the operator's market orientation and the public authority's powers. Both should help a more balanced form of innovation, with room for initiatives from operators and authorities.

The national government in the Netherlands now funds the majority of transport projects, including public transport. In consequence, the regional authorities develop innovations that are in line with general national policies and solutions, as this provides the highest chance of funding. As a result, the Netherlands has a national ticketing system, based on a national system of zones and a standard ticket (the strippenkaart); this is now being developed to use a contactless chip card. Such an innovation is hard to realize because even though funding might be national, many local and regional authorities play a crucial role in public transport and its ticketing. This interplay between a single payer (national government) and multiple demanders (regional authorities and operators) can influence the possibilities and direction for innovations in Dutch public transport.

Government dominance of local authority actions is also visible in Great Britain but not always on a positive note. In 1997, the Government envisaged major development of light rail, with the target figure of achieving 25 lines within a decade. Despite this interest hardly any lines have been built. At present there are six: one or two lines each in Blackpool (a historic system), Birmingham, Croydon, Manchester, Nottingham and Sheffield. In addition, a new one-line system is being built in Edinburgh, and substantial expansion of the Manchester system is in hand. Several others have completed the required processes of appraisal and public inquiry, with strong backing from the promoting cities and interest by operating companies. But most of them have been rejected by government, who judged that they did not offer "value for money"—a vague term unrelated to specific appraisal criteria.

Expansion of the Greater Manchester network was approved in part only, after a referendum in the conurbation rejected a scheme to introduce road pricing to pay for much of the investment, which government had required. However, the necessary funds have since been found from other sources. A few other schemes, mostly small, are still being considered for various cities. Some cities whose tramway plans were rejected, such as Leeds, are now developing advanced bus systems, running on their own right of way. These are seen as being much cheaper but offering many of the benefits.

The situation reflects a combination of factors in Great Britain:

- The UK government exercises tight control over local councils' actions and funding.
- There is fragmentation in the governance of city and district councils. The major conurbations have an Integrated Transport Authority

responsible for all transport in their area. But there are no single authorities at conurbation or regional level with overall responsibility for all functions and investment.
- Transport projects are appraised mostly on econometric principles, drawing together the identifiable impacts on transport operation and travel through a Cost Benefit Appraisal. Wider benefits of light rail are treated as secondary.

The governance of public transport allows for different forms of innovations. The same though holds for the structure of the market itself. For example, in many European cities, such as Montpellier, Stockholm, Saarbrucken, Lisbon, Sheffield and The Hague, light rail systems have been (re)introduced or extended in recent decades. Many European policymakers saw this as a valuable means of enabling cities to function effectively while reducing the dependence of citizens and of city businesses to rely on access solely by car. From the evaluations carried out, light rail systems have generally been successful (see Sautter, 2009).

The same view was taken in the Netherlands. However, the spatial structure of the country's economic center and the importance of the bicycle both challenged some of the key strengths of light rail. The core of the Netherlands consists of several mid-size cities (100,000 to 1,000,000) close together with a dense network of train infrastructure and motorways. This concentration of a great deal of economic centers with many links creates transport patterns with less focus, a situation which is helpful in making light rail a success. In addition, on the shorter distances, the bicycle is the key form of transport in the Netherlands. Of the trips below 10 km, one in three is done on a bicycle. This means that the relatively small Dutch urban areas are suited well for the bicycle; for longer trips the dense networks for train and car provide excellent connections. The structure of demand and existing supply of transport meant that an innovation like light rail, widely regarded to be an excellent linkage between a city and its wider hinterland, was less successful in the Netherlands.

With policymakers still relying on good public transport services to turn car drivers into public transport passengers, the reality in the last decades has been harsh. Research (Wee & Rietveld, 2003) has shown for the Netherlands that simply improving public transport services will not have a large impact on car users to make the switch. Our society has discovered the attractiveness of the car and has developed itself around the dispersed travel patterns possible with that mode. Furthermore, urban planning has usually focused new developments and regeneration around car travel, thus reinforcing people's dependence on the use of cars (Gakenheimer, 1999). Societal change away from car oriented lifestyles means a great deal of individual changes, and this is not going swiftly. The key reason is that individuals make long-term mobility choices that severely limit the attractiveness of other options on the short run. Living at the fringes of city at low densities makes offering good public transport very expensive and never very good. The substantial fixed costs of owning a car make leaving it by the curb and taking the train expensive (see also Paulley et al., 2006). And habitual behaviour is hard to break when an unknown alternative exists. Many people have made these choices, reducing the option value of public transport. Public transport only comes back in the picture when changes have to be made for other reasons. This has led policymakers to develop modal shift policies around road reconstruction (with free public transport tickets) and moving house (with new developments positioned as public transport oriented). Obviously, these policies still depend on excellent public transport and intelligent linking town planning with the strengths of public transport.

To illustrate how difficult it is to create a modal shift, the municipality of Hasselt introduced free public transport for all in 1997 (see Goeverden, Rietveld, Koelemeijer & Peeters, 2006). Patronage exploded to 11 times more passengers. Only 16% of those new passengers took a car before they used public transport. More than half of the passengers were new travelers; they would have made the trip without free bus transport. The free service can as such be seen as a great result for social inclusion. However its contribution to sustainability and congestion reduction can be doubted, leaving the region in 2000 with still one of the lowest modal shares for public transport in the country (Vlaams Gewest, 2002).

14.4. INNOVATION IN TECHNIQUES AND TECHNOLOGY

Public transport providers in both countries continue to improve their services, through improving existing techniques and technologies and through innovation. The innovations tried, and the results are varied. This section examines two areas in particular to illustrate the issues: the development of light and local rail and enhancing bus services. It then outlines the potential scope for transferring technological developments from automotive engineering to public transport.

14.4.1. Rail: From High Speed to Local Tram

Rail passenger services vary widely in their characteristics. The recent decades have shown two key innovations. At one end of the range are high speed trains, operating over long distances between cities at speeds of 200 km/hr or above. The network has been gradually developing within the larger European countries as the main driver for faster national links. European support followed to further strengthen transnational networks. This was supported by developing a European standard for train control systems (though this is being applied only very slowly). Here the innovations have been significant, with France and Germany in the forefront with the train a grande vitesse (TGV) and the Inter City Express (ICE). This transnational network of infrastructure and services is developing slowly but steadily. Eurostar in Great Britain and Thalys in the Netherlands are bringing fast TGV services but only as parts of international routes.

Arguably, the fact that Great Britain and the Netherlands were relatively late in adapting high speed rail has different reasons. For the Netherlands the national network is far more important than international links. National rail traffic is relatively short distance, given the size of the country. The relative advantages of higher speeds are low. Even for international links these advantages are limited as important international destinations like Antwerp (Belgium) and Aachen (Germany) are just over the border. In Great Britain, British Rail had developed the diesel powered Intercity 125 for operation on existing main lines as their high speed system. It was early and less advanced as the later developments of the TGV and the ICE. Once these continental alternatives were fully developed, the privatized British rail sector did not support the massive investments necessary to roll out high speed trains on such a massive scale.

At the other end are local tramways, running through cities at relatively low speeds and serving stops at intervals of around 800 m. In between there exists a number of categories. In most European countries the precise boundaries for these are not always clear, especially for shorter distance services in the main urban areas: within cities, between cities and satellite towns or between several towns in a district. Sometimes the main types of train may serve all types of operation, perhaps stopping at simple stops in city and suburban areas but at distinct stations in small towns and villages.

A second innovation for many European cities in recent decades has been the reintroduction of rail into the city as light rail (tramway) systems. France and Germany have again been in the forefront. Earlier, rail infrastructure had often been removed following the development of buses. With higher capacities, cleaner operation and separate lanes trams proved to be attractive alternatives (CONNAISSANCE DU RAIL numero speciale 344–345, 2010). This development also is seen on a smaller scale in Great Britain (Manchester, Sheffield). In the Netherlands reintroducing the tram into the city proved hard. In Utrecht attempts failed and current projects in Leiden and Groningen are of a different kind, discussed next. Other urban rail projects consisted of extensions of existing systems in Rotterdam, Amsterdam and The Hague. A key element here seems to be the funding and political control over public transport. In the Netherlands, projects like new trams are mainly funded and decided upon by the national government. This has a tendency to favor national rather than local projects (see Bruijn & Veeneman, 2009). Similarly, in Great Britain decisions and funding for city systems are strongly controlled by central government, who have taken a very cautious approach and appraised projects largely on financial grounds; this has meant that very few of the lines proposed by city PTEs have been approved.

A third innovation is the coupling of existing heavy rail networks with new networks in the city. Under the label of tram-train, European vehicle and infrastructure technology providers have been filling the gap between heavy and light rail in the last decades. Karlsruhe is a leading example of what is termed tram-train: Vehicles have been designed to operate on conventional railway lines as well as city tramways, and regional lines around some cities have been upgraded and linked to the city tramways, so that a frequent through service can operate from catchment towns into the city center. This makes the service much more attractive because residents in areas served by the line can make a through journey to the city center rather than changing at the main railway station; they also benefit from new and accessible vehicles, often run more frequently. Usually this involves extra stations on the line, perhaps coupled with new settlements. The railway lines continue to carry other railway operations too, including some regional trains and freight services. Thus the approach offers scope for combining advanced light rail technologies with careful city planning to create a new product from existing and possibly underused local railway lines.

Also in the Netherlands this development is seen. In 2008 RandstadRail started operation on old heavy rail tracks, the metro network of Rotterdam and the tram infrastructure of The Hague. The Rijn-Gouwelijn near Leiden is under development and is taking the same approach. In Groningen the tram is currently introduced first, with the intention to link the network later to existing heavy rail network around the town. These larger projects with the potential of improved usage of the nationally controlled heavy rail network get more attention and swifter funding from the national government and as such found funding earlier than the tram.

A growing interest in tram-train has been shown in Great Britain. The Manchester tram system has been developed mostly on former suburban railway lines, but these are not shared with railway services, the infrastructure is now used purely for the trams. The only case of track sharing has been the extension of the Tyne and Wear (Nexus) light metro over a conventional line to Sunderland. Interest in tram-train techniques in Great Britain has focused on the scope for reducing costs of providing infrastructure rather than widening markets for public transport travel. Initial proposals are mostly aimed at trial operations to identify the benefits in this respect. But so far no tram-train has emerged that recognizably matches the concept being applied in the Netherlands, Germany and France.

It is interesting to see how policymakers seem to favor these rail solutions over other improvements of service levels of bus services. Richmond (2005) studied this phenomenon. Travelers like the certainty of service they receive with tracks; they see where the service will bring them and are more certain about service provision in the future than with buses. Well implemented bus systems could have the same characteristics at lower investment costs. Still, policymakers generally prefer the relatively higher investment of rail systems.

14.4.2. Buses

The innovation of the bus sector has a different character than the rail sector. In the rail sector innovation is very large in new project developments. The interaction between vehicle and infrastructure means that innovations come in twos: new vehicles with new infrastructure. In addition, rail systems still very much have a local implementation; development of local trains in Great Britain is different from those in the Netherlands. Innovation in the bus sector has a more gradual character. Buses can be innovated autonomously from infrastructure for the entire European or even the world market. As a result buses became cleaner and lighter and developed low floors.

However, buses mostly use the same roads as the car, and this raises the serious problem that most such roads carry high levels of other traffic. In consequence bus services suffer from delay and especially unreliability. Here innovations are sought that link the vehicles closer to the infrastructure. Three approaches have been adopted to address this.

- Information links (usually radio) between buses and control centers enable the control center to be continually aware of how buses are operating. This also allows information on bus running to be transmitted electronically to displays at bus stops and through Internet links, including mobile phone ones, so that intending passengers can be advised when their bus will arrive.
- Short distance radio systems allow for buses to get priority at traffic light influencing to improve circulation and raise operating and service speeds of the bus.
- Modification of highways gives priority to buses. Techniques include reallocating space on the highway (e.g., through bus only lanes) and controlling other traffic (e.g., through traffic lights which always give precedence to lanes with buses in them, the buses being equipped with a transponder allowing them to signal their approach).

These techniques are both feasible and rewarding where bus operators and local authorities work closely together. They have been widely applied in most Dutch urban areas and in London, where they have provided considerable benefits. For tendering authorities, vehicle location and communication technologies provide the added benefit of monitoring the output of operators.

In Great Britain outside London local authorities do not control bus services. Routes and networks are developed by bus operators for the most part: usually large companies which are owned by the main privately owned public transport corporations. These companies, and some of the reputed smaller ones, have considerable experience in utilizing new techniques, especially in order to improve efficiency and attractiveness. However, they are required to operate in competition with one another, and this constrains investment except where it will provide an identifiable return in such systems as electronic control and information. Furthermore they do not control their infrastructure, the public highways. In consequence the main benefits are where they can achieve agreements with local authorities for highway priorities which enable reliable running and for other shared facilities, for example, joint organization of electronic communications and information.

Legislation over recent years in Great Britain has sought to offer more scope for Integrated Transport Authority and local transport authorities to do this. However, proposals often encounter opposition from local businesses and residents, concerned that reducing highway space for other traffic will worsen congestion and undermine local businesses' income. Public authorities are often reluctant to proceed, especially as bus companies are seen to be pursuing private profit rather than providing a service. So improvement of bus services through available techniques is very slow.

One useful trend in Great Britain has been the rapid emergence over the last decade of low floor buses: In 1998 they formed about 1 in 10 of the total fleet, in 2008 over two thirds. This reflected pressure especially from disabled interest groups and concerned politicians, leading operators and manufacturers to adopt low floor designs as standard. Local authorities have supported this, often through including suitable boarding facilities at stops, generally as part of Bus Quality Partnership agreements. However, many bus stopping points and bus stations have not been adapted, and sometimes ineffective parking control means that buses cannot pull into a stop, thus negating the benefits of the new design for passengers. In the Netherlands this innovation was laid down in law, with all buses to be accessible from 2010 though low floor and raised platforms, providing level entry.

The development of communications technologies also allowed for other innovations. As the car grew in popularity, especially in more peripheral areas, bus services were providing too much capacity, while at the same time the provision of a public transport service was still desirable. Demand dependent services have developed with intelligent reservation and vehicle routing systems.

Where circumstances are particularly suitable the development of bus technology has also made great strides in recent decades, as for light rail. Following the developments of a bus rapid transit system in Curitiba, Brazil, many other South American cities discovered the potential for buses to play a major role in densely used urban corridors, including Santiago de Chile, Bogota and others. These systems have high frequencies, wider networks of separate lanes, extensive travel information systems, longer articulated and double articulated vehicles and even off-board ticketing and validation. South American city authorities generally have considerable powers over spatial development and transport.

In Europe buses saw a number of innovations along those lines, although not as extensive as the metropolitan roll out in South America. In the Amsterdam area the ZuidTangent bus rapid transit route was developed largely on a disused tram infrastructure in an effort to improve the accessibility of the Schiphol Airport from Amsterdam's suburbs. In Eindhoven the authority went a step further, supporting development of the Phileas system, a vehicle that combined several different technological innovations. This included a hybrid drive train, all-wheel drive and steering, full low floor and magnet guided steering largely on a separate lane. This has however suffered many implementation problems: Was this perhaps a step too far in innovation?

A final innovation in the bus sector is the development of low capacity vehicles and on-demand services. Mention was made previously of the dilemma facing decision makers between rapid services that would best serve commuters and more slow services that would best serve several vulnerable groups that depend on public transport for their mobility. They can use small bus services operating on flexible routes, often with individual customer booking (dial-a-ride). These low capacity and demand-dependent services have become a growing part of the total public transport offer; they have a higher unit cost but provide an effective quality service. This separates the two different types of demand in different segments of the supply. A specific form is personalized rapid transit, very small vehicles running under automated control on networks of fixed tracks; localized systems run on very advanced technology (e.g., Parkshuttle near Rotterdam).

14.4.3. The Scope for Automotive Engineering Technology in Public Transport

Much investment has been made into environmentally focused improvements in the car industry. This has engaged both manufacturers and governments, often together—the former as researchers and manufacturers, the latter as regulators and through public funding. The innovations cover such fields as materials technologies, drive systems and power sources. On the latter, the key issues are now well understood and the greatest hurdle is the step-change in refueling, or charging, infrastructure required in order to support electric vehicles, plug-in hybrids or alternative fuels.

Public transport rarely benefits significantly from the national programs, however. The dominance of the automotive industry means that it is able to command very substantial funding itself and development is often supported by substantial government funding programs (e.g., the UK government funding of £500 million for low carbon vehicles, including electric vehicle infrastructure). Yet public transport systems may offer greater scope for, and possibly easier implementation of, efficiency improvements, for two reasons. Firstly, there are very few problems with refueling infrastructure, as refueling of vehicles is performed by the operator. Secondly, the usage of

Innovation in Public Transport 301

the vehicle can be predicted to a far greater degree of accuracy than a passenger car allowing greater optimization in the design.

However, public transport systems have seen some valuable steps in engineering. For example, some developments in bus power trains have preceded those in the automotive industry; in particular the use of alternative fuels such as compressed natural gas and biofuels are commonplace due to the independence from a public refueling infrastructure. Trials of purely electric buses are also becoming more common; there may be a comeback for the once popular trolleybus. Some developments however follow the automotive industry; large diesel power trains used are often shared with the truck industry, where vehicles follow a very different usage pattern of continuous speed rather than repetitive acceleration and deceleration.

A key opportunity with power train developments is to ensure that they are part of an optimization of the vehicle as a whole, minimizing weight and maximizing impact protection, rather than as a retrofitting exercise. Technologies developed in the automotive industry are readily available, such as composite crash structures designed to absorb impact energy allowing the rest of the vehicle structure to be reduced in thickness and therefore weight.

Like a bus network, a light rail network does not depend on a pre-existing refueling infrastructure; in fact it has greater freedom. Diesel or other fueled vehicles have the same situation as a bus network, but electric rail vehicles draw their power along the line, so there is no refueling or charging requirement. An alternative charging situation is shown by the Parry People Mover, where flywheel energy storage is used on the train which can be charged at each station stop, thereby removing the need for either refueling at the depot or power transmission along the route.

Storage of energy need not be as a primary source of power but can be used to partially reuse braking energy to accelerate the vehicle, called regenerative braking or "kinetic energy recovery system" (KERS). The scope for regenerative braking on light rail systems is greater than for road vehicles, as the usage or duty-cycle of the vehicle is simply one of acceleration to an operating speed, which could be assisted by the KERS, then decelerating to rest, charging the KERS. The scope is also greater than for mainline rail, as the vehicle is lighter and the stops closer together, giving a greater proportion of the energy used in the acceleration phase. Implementing regenerative braking for light rail is not without challenges: the amount of energy recovered depends on the number of powered wheels; if the energy is fed back into the electrical system it must be able to handle the additional power, and if the energy is stored either on the vehicle or trackside it must be in a safe manner. Some current developments with light rail vehicles, primarily aimed at enabling operation without overhead power lines for part of routes, may make this more readily available.

The greatest opportunity lies not just in applying these innovative technologies but in combining selection of the required capabilities of the vehicle with the planning of the route. This offers the potential to give the most suitable combination of vehicle and energy source to provide the safest and most efficient solution. It is primarily applicable to rail, but the idea of closer development between the vehicle and the route also applies to buses. But this is not really happening in practice: Despite its potentially green credentials, public transport may actually be falling behind the automotive industry in benefiting from technical opportunities. Operators still work primarily within relatively narrow markets for the most part, and the relatively limited market for equipment constrains manufacturers' ability to engage in major research and development programs; they also lack the levels of support funding that car manufacturers benefit from. The city and local authorities who plan and procure a lot of local services are rarely interested in the technological opportunities. In consequence the public transport community is in effect a fragmented one.

14.5. CAN PUBLIC TRANSPORT SECURE ITS POTENTIAL FUTURE?

What role can public transport play in providing alternative mobility in a car oriented world, and as such help mitigate the negative effects? Since the 1970s, public transport has been positioned as a key alternative, often with great expectations of a modal shift from the car to public transport. Policymakers have become key players in public transport in defining the alternative. Public entities on three levels have become key players, on a European level they have pushed a more open market, on a national level they set a regulatory framework and regional authorities have worked within the framework to link policies in the field of mobility to public transport services.

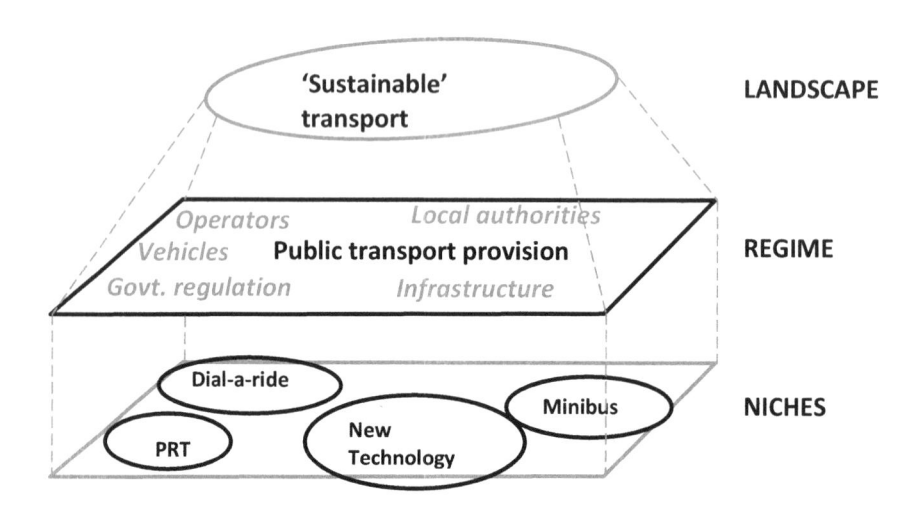

Figure 14.1 The public transport regime.

Innovation in Public Transport   303

Next to the public authorities providing infrastructure, franchising and regulating services, a second group of actors plays an important role: the operators. They use production factors to develop and provide the services that should attract passengers. The interaction between authorities, operators and passengers occurs within limited and regulated markets but is now seen as having considerable potential for meeting the challenges of the future (climate change, energy depletion). Thus its real place is as a key part of sustainable transport (a term which is very widely used but not always specified). The structure of the regime can then be illustrated as in Figure 14.1

As history shows, public transport used to dominate all travel but now plays a relatively minor role, except in certain markets. These include some disadvantaged groups everywhere, such as the old and the poor, disadvantaged in that they do not share in the widely accepted norm of having a car available. The markets also include some geographically distinct movements for better off groups: especially business and leisure travel into and between main cities.

But to play its potential role public transport must provide a real alternative to car use over a wide range of markets (a point often made in policy papers). At present however neither bus, transit nor rail play any significant role in wider general movement, especially over shorter distances and outside main urban centers, where car use dominates and travel patterns are very dispersed. Indeed, simply providing more bus services over a dispersed network is most likely to result in more empty buses running on the streets, and this is no more sustainable than widespread car use. Therefore operators concentrate largely on developing their existing customer bases. If they are to play a much greater role in local and regional travel (and it cannot be taken as a foregone conclusion), what steps might be taken and who would lead on these steps?

Public transport is essentially a mass form of movement, involving vehicles carrying between 20 and 1,000 people, needing to operate in a thoroughly reliable fashion and requiring dedicated or priority infrastructure for this. The potential benefits for sustainable movement are high, but the costs of provision are commensurately high too. This is equally true in principle for a local bus service or for a high capacity suburban railway. Infrastructure and routing must be designed to relate to the structures of urban settlements to maximize accessibility and attractiveness of the public transport systems. The infrastructure is primarily in the hands of public authorities, as spatial planning and highway authorities; thus they have the opportunity to act in support of public transport provision through supporting the development of bus lanes and bus priorities, transit routes and railways, including the location of stations and stops. Here there are clear differences between the two countries: Development of good public transport infrastructure forms an important part of the work of city and local authorities in the Netherlands, but in Great Britain it is left almost entirely to market forces (even in London).

Public authorities can also play an important role in encouraging innovations by operators, including both service development and new technology. This has happened in the Netherlands and London through tendering of concessions. However, British authorities outside London struggle with this role more, with far more limited powers. This could harm the relative performance of public transport in the long run in these areas.

But even where the provision of infrastructure and services is supported by public authorities, the principal responsibility for driving change lies with the operators. Use of appropriate technological systems is most immediately in their hands, and the long-term performance of these systems is dependent on the way in which the operator is using them to provide an attractive service. The operator is the one that has to make the most of the technology by providing the right service in the right place at the right time.

Some activities address areas of low and scattered travel demand. These include personalized rapid transit systems and dial-a-ride and other forms of small bus service. These will almost certainly continue to play a valuable role within certain specific and generally local contexts. They are thus valuable forms of innovation. But, on the experience of recent decades, they appear unlikely to play any significant role in the overall development of public transport to meet the potential opportunities of a changing world.

The real opportunities lie in development of the mainstream public transport systems. These are mostly provided by the main operating companies, albeit in some form of partnership with public authorities. Strong innovations are needed to keep improving the performance of the sector. The automobile is also rapidly innovating toward greater sustainability. As for the motor car, new technology has a major role to play, in conjunction with conditions that allow or even encourage its use.

This requires a holistic approach, addressing facets of planning, operating and engineering together, identifying the roles of various actors and highlighting the importance of shared responsibilities. It is particularly important to consider the relationship between technical developments on the one hand and institutional behavior patterns on the other.

The role of the technology provider is to provide the technological basis, in infrastructure and vehicle systems, for a further improvement of public transport. That technological basis should be developed in two directions. First, public transport has to improve its attractiveness for travelers to deliver on the promise that public transport holds a more sustainable alternative for the automobile. The idea that more public transport per se is more sustainable is naive and outdated: Public transport can only provide a sustainable alternative when it attracts passengers in its own right. Second, public transport technology providers have to innovate to improve public transport's environmental performance.

There is, as discussed previously, plenty of scope for public transport design to benefit from technologies developed by the automotive industry. However, public transport technologies cannot easily follow the pace because in many respects they are regulated more substantially than car technology. This is mostly true for (heavy) rail, where that regulation is partly needed to coordinate between infrastructure and vehicle development, which are very much interdependent on the tracks. An example is the fact that Dutch Railways is finding it hard to reduce the weight of trains for better efficiency, due to regulation by the network manager asking for heavy trains to ensure the proper working of specific traffic control systems.

To optimize the use of new technology and provide attractive services that enable public transport to play a significantly increased role in travel, it appears that three possible steps can be identified through which providers might achieve their potential.

By enhancing the present structures but retaining the same objectives, some gains may be made through using improved technology and techniques and more efficient operating systems. Most operators are experienced in doing this, and the majority is already applying this approach. For example, in Great Britain operators started using buses with low floor levels in the mid 1990s, initially because groups representing disabled and elderly people wanted level boarding and alighting; this was supported by government proposals to regulate floor heights and by some local authorities willing to build easy boarding points at stops under partnership arrangements. However, the concept of buses which could be boarded much more easily proved widely attractive. Most buses in Britain now have low floors.

By seeking to gain scope for revised objectives, such as changes to the regulatory framework to reduce constraints, and using this to develop innovative approaches and adoption of new techniques, operators should be able to make significant gains in both existing and new markets. The tram-train concept has revolutionized the role of local rail services round a number of European cities, enabling regional lines to play a much stronger role in travel around the conurbation, especially but not exclusively for access to the main city center. Examples include development of the Randstadrail systems around The Hague and Rotterdam in the Netherlands. This has required development of regulations in parallel with the technology so that lighter trams can run on main line railway routes as well as through urban streets.

Better public transport will not of itself lead to a major shift from car use. The real benefits can only come with major complementary changes to various systems. These might include far greater focus of spatial planning strategies around public transport, such as local and light rail in city areas. The Dutch ABC planning policy already provides a focused example of this. Significant alterations to regulations and taxes can also help to prioritize public transport. With this support public transport providers will have the potential freedom to adopt new technologies and techniques on a far wider basis.

Peak oil is rapidly approaching. Although many car manufacturers are experimenting with electric and hybrid cars, the core of the car sector is still around internal combustion. Innovation here is still an intricate play between oil companies, car manufacturers and the consumer. New players like power companies are stepping into the arena, but it is still unclear in which direction the development will eventually go. Public transport can have an advantage here. The regulation of its technology was often a hampering factor for innovation. On this pivotal moment in history, it might be that the strong role of government, as regulator and authority, might support a swifter change toward more sustainable technology in public transport. The strong market, standardization and modularization can help public transport to innovate quickly. It is that innovation which might provide the basis for mobility when innovation in the automobile industry is peaking long after peak oil has gone. And perhaps it might draw much of its success by taking on and extending technologies which have so far been developed largely within the automotive industry itself.

NOTES

1. The main characteristics (Faulks, 1990) are as follows:
   - Trains are usually taken to mean those operating on the national railway system. Track is laid at 'standard' gauge, 1.435 m. The routes range from main lines with four tracks to local lines with only one track for both directions. The network is controlled primarily from major signal control centers, with local signaling systems on some secondary and minor lines: Most control of routes and points is electronic.
   - Urban transit systems are mostly rail based, usually on 'standard' (1.435 m) gauge track. Almost all provide frequent services. Some major cities have metro lines, entirely on its own right-of-way, some of it in tunnel, often operating longer trains. Most rail transit consists of light rail or tram systems, with modern articulated light rail vehicles. These usually have dedicated track sections (similar to railways) and also sections of operation on public roads. All urban transit systems have stops placed at close distances.
   - Bus services operate over many main and secondary roads and local roads as well. Most local bus services run in cities and larger towns, where networks include multiple routes often operating at very high frequencies. A few services operate on separate busways for part of their journey (sometimes defined as bus rapid transit). Buses are generally powered by diesel engines. Rural areas are served by local buses, sometimes quite infrequently.

REFERENCES

Bolden, T. & Harman, R. (2008) 'Railway and spatial strategies in the south east: Can coordination and delivery be achieved?' *Planning Practice and Research*, 23(3): 303–322.
Bruijn, H. de & Veeneman, W. (2009) 'Decision making on light rail', *Transportation Research A*, 43(4): 349–359.
Centraal Bureau voor de Statistiek (2010) *Statline*, Den Haag, the Netherlands: Centraal Bureau voor de Statistiek. Available at http://www.staline.cbs.nl (accessed 23 April 2010).

Connaissance du Rail (2009) *Les tramways francaises en 2009* Valignat, Editions de l'Ormet.
Department for Transport (2008) *Transport Statistics Great Britain 2008*, London: The Stationary Office.
Faulks, R.W. (1990) *Principles of Transport*, Maidenhead, England: McGraw-Hill.
Gakenheimer, R. (1999) 'Urban mobility in the developing world', *Transportation Research Part A*, 33: 671–689
Goeverden, C. van, Rietveld, P., Koelemeijer, J. & Peeters, P. (2006) 'Subsidies in public transport', *European Transport*, 32: 5–25.
Harman, R. (2004) 'Cars in the landscape: Some undefined questions for transport policy', in F. Terry (ed) *Turning the Corner?* Oxford, England: Blackwell.
Harman, R., l'Hostis, A. & Menerault, P. (2008) 'Public transport in cities and regions: Facing an uncertain future?' in P. Booth, M. Breuillard, C. Fraser & D. Paris (ed) *Spatial Planning Systems of Britain and France: A comparative analysis*, Oxford, England: Routledge.
Hibbs, J. (1989) *The History of British Bus Services*, Dawlish, England: David & Charles.
Papacostas, C.S. & Prevedouros, P.D. (1993) *Transportation Engineering and Planning*, Englewood Cliffs, NJ: Prentice Hall.
Paulley, N., Balcombe, N., Mackett, R., Titheridge, H., Preston, J., Wardman, M., Shires, J. & White, P. (2006) 'The demand for public transport: The effects of fares, quality of service income and car ownership', *Transport Policy*, 13: 295–306.
Richmond, J. (2005) *Transport of Delight*, Akron, OH: University of Akron Press.
Sautter, A. (2009) *The Role of the Operator in European Light Rail Projects; Recent Examples and Keys for Success*, Delft, the Netherlands: Technology Policy and Management.
Social Exclusion Unit (2003) *Making the Connections: Final Report on Transport and Social Exclusion*, London: Social Exclusion Unit.
Van de Velde, D., Beck, A., Van Elburg, J.-C. & Terschüren, K.-H. (2008) *Contracting in Urban Public Transport*, Amsterdam: inno-V.
Van de Velde, D. & Röntgen, E. (2008) *Landenoverzicht Spoorordening (Country overview rail regulation)*, Amsterdam: inno-V.
Veeneman, W.W. (2002) *Mind the Gap: Bridging Theories and Practice for the Organisation of Metropolitan Public Transport*, Delft, the Netherlands: Delft University Press.
Vlaams Gewest (2002) *Vademecum Parkeerbeleid bijlage (Handbook parking policy, supplement)*, Bruxelles: Vlaams Gewest.
Wee, B. van & Rietveld, P. (2003, September) 'Openbaar vervoer: mythen en Feiten' ('Public transport: myths and facts'), *Arena*, 9: 74–77.
Wolmar, C. (2007) *Fire & Steam: A New History of the Railways in Britain*, London: Atlantic Books.
Wolmar, C. (2009) *Blood, Iron & Gold: How the Railway Transformed the World*, London: Atlantic Books.

15 Intermodal Personal Mobility
A Niche Caught Between Two Regimes

Graham Parkhurst, René Kemp, Marc Dijk and Henrietta Sherwin[1]

15.1. EMERGENCE OF A POSITIVE POLICY FOR INTERMODALITY

A key feature of the car regime is the 'door-to-door' mobility offered by the system, which minimizes the need to make any part of a trip using a vehicular mode other than car, and also reduces or eliminates any walking required to access or egress the car-road system. Nonetheless, for a range of practical and policy reasons, intermodal niches have variously survived, flourished and developed since the rise of mass car ownership. And in recent years transport policy has seen a new emphasis on intermodality which seeks to extend beyond those niches. In essence, this new emphasis arises from the view that intermodality in passenger transport has a unique role to play in reducing reliance on the car, particularly for those trips for which a complete substitute to the car is not possible.

This chapter seeks to identify and theorize the development of intermodal travel in Europe, through the analysis of policy and practice developments in two example member states in which the sustainable mobility discourse has become increasingly important: the Netherlands, where 'integration' between modes of transport and between transport and spatial planning has been important in transport policy for some decades, and the United Kingdom where, from the 1980s, the free market principles of deregulation and minimized subsidy resulted in a somewhat different emphasis in the development of intermodal niches, although showing convergence with the Dutch approach more recently.

The following section provides a typology of motivations for intermodality; it considers both traveler choices and system constraints as key factors, while Section 15.3 briefly places intermodality within the theoretical context of the automobility regime and its competitive sub-regimes. Section 15.4 then considers the development of the overarching policies for intermodality in the two states, followed in Section 15.5 by detailed consideration of the emergence of three examples of intermodal practice with relevance for sustainable mobility. Section 15.6 considers the experience of these niche practices in the theoretical context of 'regime change', with Section 15.7 reaching conclusions about the future prospects for intermodality to play an important role in sustainable mobility.

A number of clarifications are necessary about our scope. We recognize that intermodality has interactions with multimodality: a multimodal lifestyle based on some car use, alongside the use of other modes of transport, in the sense that people with multimodal lifestyles may seek the benefits of intermodality more often. However, parallel multimodality is beyond the scope of the chapter, which focuses on the serial combination of modes within the same journey. In addition, we do not consider the important role of walking in mobility: Strictly speaking many simple car and public transport trips are already intermodal, as they involve a walking leg at least at one end of the journey. We do not cover these journeys here as they are generally not regarded as intermodal trips in the technological sense, nor within the consciousness of most travelers, transport planners and politicians. They certainly require special consideration in planning and design, but they do not constitute a complex socio-technical sub-regime in the same way that the higher technologies involve interests such as operators, vehicle manufacturers and maintainers, nor facilities for vehicle storage and refueling. Lastly, many journeys involving different legs are made exclusively using one vehicular mode: We do not cover interchanges between buses or between trains. Although in practice these interchanges exhibit many of the same practical problems, they nonetheless exist within the same sub-regime, and innovation generally meets a smaller scale of barriers, such as those that arise through lock-in and path dependence. We also recognize that, at the boundaries, defining transport modes is becoming more complex, as different kinds of hybrid modes such as tram-trains, taxibuses and shared vehicles emerge. Each innovation brings its own theoretical and empirical challenges to conceptualization and understanding. Not all of these phenomena can be given significant coverage in a single chapter and must wait for future authors.

15.2. MOTIVATIONS FOR INTERMODAL NICHES

Intermodal passenger travel can be categorized in three ways:

- Use of a private vehicle (bicycle, car) to access a transport interchange, with that vehicle then being parked at the interchange, normally until a return trip is made—including widespread examples of 'park and ride' (P&R) with interchange onto bus and rail stations for commuting, shopping or longer range trips and, less usual, one-way car hire, perhaps to access an airport.
- Access to a public transport mode as a passenger in a vehicle which does not remain at the interchange—common examples involve the use of local buses or taxis to reach a railway station, or 'kiss & ride' (K&R)—a practice whereby one individual, perhaps from the same household as the traveler, or a business contact, drives him or her to the interchange as an 'escort journey', as part of a trip chain.

- A rarer kind involves the vehicle of the first mode being physically conveyed by that of a second mode—most routine of these examples is perhaps the carriage of folding or fixed-frame bicycles in railway carriages, on the back of cars or in special compartments or trailers of buses. The conveyance of vehicles across bodies of water by ferries is also common.

The effectiveness of intermodal travel depends on the 'seamlessness' (Stokes & Parkhurst, 1995) of transport systems, where key logistical attributes are as follows:

- The extent to which the vehicles of the different modes can be brought into physical proximity to minimize the effort and time costs of interchange distance.
- Temporal proximity within schedules of the different modes to minimize waiting times.
- Quality of interchange infrastructure, particularly in terms of ease of walking (e.g., step-free access) for the mobility impaired or those encumbered with luggage, and quality of the waiting environment.
- Low transaction costs, so that intermodal travelers perceive no additional financial or practical costs over and above what would apply to a single-leg journey, with the ideal being a 'multimodal ticket' priced at a similar level as a notional mono-modal means of making the same journey.
- The availability of information systems to update travelers—although relevant for mono-modal journeys, dynamic information can be particularly important for travelers using different modes which may be subject to different kinds of disruption.

In practice, however, the traditional approach of transport planners has been to regard transfers between vehicles as tending to incur significant transaction costs due to time delay and physical and psychological effort. For example, Wardman (1998), from a meta-analysis of British studies, concluded that the fixed penalty, expressed in terms of the perceived equivalent amount of in-vehicle time, was worth on average 21 minutes for an interchange between buses and 37 minutes between trains.

From this perspective, travelers will tend only to accept intermodality if they have no alternative; perhaps because they are dependent upon public transport networks which can serve a limited number of origin-destination pairs with direct services or, if they do have alternatives available, because intermodality confers significant compensating advantages. The intrinsic benefits of intermodality include:

- Cost savings, for example, from lower parking fees (the basis for encouraging peripheral P&R facilities), or to take advantage of low-cost, subsidized public transport

- Time savings, due to the journey time advantage of water, rail or air transport over some road journeys due to the presence of high-speed transport systems or because those modes provide a means of overcoming geographical barriers (e.g., bodies of water, mountain ranges)
- Effort or comfort benefits, such as avoiding the fatigue of long-distance driving through the use of 'motorail' services or being able to use the in-vehicle time more productively

In considering the total demand for travel, as the opportunities for these advantages are currently in limited niches, for many journeys the need for them to be intermodal if made by modes other than car has tended to be a strong, sustaining factor of the automobile regime. Nonetheless, in recent years a new instrumental policy emphasis has been placed on the importance of intermodality at different levels of governance. The European Union white paper on transport of 2001 (Commission of the European Communities, 2001) identified intermodality as a key measure, explicitly at that time only in respect of freight policy. The promotion of passenger intermodality can, however, be seen as implicit as part of a policy to promote limited-stop, high-speed rail development as a more sustainable alternative to growth in aviation and long-distance car travel, a policy which had its roots in the socioeconomic cohesion objectives of the Maastricht Treaty, but is increasingly identified as a sustainable mobility measure. From 2007, the LINK Project was supported through the 6th Framework Program of research, specifically examining intermodal passenger travel. The project found that descriptive evidence on the state of passenger intermodality was poor (Link, 2009), but it was suggested that most intermodal travel is 'unstructured', in the sense that it is planned and arranged by the individual rather than promoted by private or public-sector actors as a seamless journey opportunity.

At the member state level, a positive policy for intermodality had already been more strongly articulated both at national and local levels. In the United Kingdom, for example, the 1998 white paper on transport (Department for the Environment, Transport and the Regions, 1998) sought to promote intermodality as part of an integrated transport policy vision, "so that each [mode] contributes its full potential and people can move easily between them." Here, modal 'potential' can be seen as summarizing several of the concepts driving a more positive vision of intermodality.

There has been growing realization that poor intermodal conditions (as well as poor within-mode integration) are important factors in encouraging individual decisions to own and use private cars. They are a factor (but not the only factor) in the 'flight' of travelers from public transport. The importance of door-to-door modes of transport increases when people start to work and establish families. Once people have a car, intermodal choices also become car-based, with people opting to drive as far as they can, which undermines the sustainability of intermodal modes of transport.

Combined with this desire to halt a spiral of decline is a more proactive policy stance which has arisen along with concerns that our mobility patterns need to become more sustainable. This perspective has seen the introduction of policies to reduce the use of carbon-intense, space-inefficient modes such as private cars with low passenger occupancy and airplanes. However, due to diverse factors including the low density of spatial development in many car-dependent societies and the limited acceptance by travelers of significantly increased journey times or physical effort, rather than seeking to switch a journey entirely from one mode to another, compromise policies often seek to replace part of an 'unsustainable' journey with a journey leg completed on a relatively space and energy efficient mode. Hence, as will be considered further in a later section, motorists are encouraged to leave cars at the outskirts of towns they are visiting and make just the final part of the trip by public transport, rather than policy seeking the much harder modal shift of encouraging the traveler to walk from the home to a public transport service.

Where greater intermodality does increase the extent to which the 'active' modes of walking and cycling are used, this is in line with the broad sustainable community objectives within public health policy—to increase physical activity rates in societies suffering from the diseases of affluence, notably high rates of obesity and heart disease.

A further facilitating factor which can be seen to be promoting intermodality is the availability of new technologies which make new modal transfers practically possible, or significantly reduces their transaction costs. In this chapter we will consider how the rise of 'public bike' schemes specifically linked to the public transport system enable a relatively fast mode in urban areas to complement public transport, which also extends the effective catchment of those systems. The greater availability of information technologies at lower cost has facilitated this development as well as 'smartcards' to enable efficient, seamless and silent payment for a range of transport modes as the traveler passes between them.

The following section seeks to understand the emergence of these new and intensified intermodal travel practices through the transition perspective outlined in Chapter 3.

15.3. THE SUB-REGIMES AND NICHES OF INTERMODALITY

In contrast to the focus of many transition studies which identify single regimes, in the case of transport we have the automobility regime and the public transport regime, each of them with sub-regimes. The classic public transport regime includes railways, buses, trams and metros as collective forms of transport which operate on the basis of timed schedules and a set of regulations (safety, economic, environmental). Each of them has regime-like features (alignment, stability, relative autonomy and power). The various systems are operated by specialized companies in either the public or private sector. The various public transport modes serve both people who own a car and those who do not, but in different ways. For the latter category public transport is often a necessity; for car owners it is usually an option. Beyond the dominant car regime and the secondary public transport regime, other regimes and sub-regimes exist with their own regulations, organizations and often quite specific groups of adopters, including walking, bicycles (including the availability of public bikes) and taxis. Between the extremes of necessity, travel choices are influenced by myriad factors relating to the conditions of travel and the psychological efforts travelers are willing to invest in overcoming barriers and constraints, in a context in which some transport modes and types of journey require more investment than others, with the dominant regime of car use often being the simplest, lowest effort option.

Hence, intermodality is a niche caught in a world of the automobility regime and various subaltern regimes. Intermodal initiatives are generally the outcome of actions by mobility providers who operate from a context based in the traditions of one modal regime, and therefore with a particular set of commercial and policy objectives, and a particular evidence and belief base about the intended and actual users of intermodal systems and the potential for the overall evolution of intermodal niches within passenger transport systems. Successful policy entrepreneurs will be those able to reach beyond the personal constraints of being situated within a single regime. Regime theory is very relevant to the case of intermodal travel, because of its dependence on the willingness of regime actors having an interest beyond either public transport or cars. Such interests may struggle to flourish, however, in the context of modal regimes with introverted perspectives focusing on the complexities of the ownership of the assets, maintaining revenues and subsidies to cover operating costs and operating legally within a regulatory framework. This is particularly true for high-technology systems such as railways, the latter being labor intensive, driven by mono-modal objectives, and exhibiting a hierarchical organizational culture. In contrast, the taxi sub-regime typically combines private ownership of vehicles, with an unsubsidized, profitable economic model of operation, but with tight regulation of quantity of supply, service quality, fares and safety. In both cases, risk aversion may simultaneously limit any will to diversify away from what is seen as 'core business'. In achieving intermodality, we hypothesize that the policy or technical entrepreneur must identify points of interface between these regimes and the (sub)regimes. It follows that this implies a certain level of knowledge of each mode and sympathetic professional contacts within each modal community.

Intermodality practices are by no means all innovative, of course. A detailed design and engineering literature (Blow, 2005; Transport for London, 2001) exists about how to deliver effective multimodal transport interchanges at particular nodes on the transport system. Mostly this literature applies to the relatively high-profile, high-traffic spaces and buildings where the public transport modes of rail, road, air and water are consciously brought together, often by public sector-led or influenced planning. Such formalized interchanges particularly include airports, ports, major railway stations and, in some places, local urban transport hubs. At many of these, significant consideration is also given to integration of the (sub)regimes with the dominant car regime, in the form of car parking, set down and pick up areas, taxi and hire car facilities. In many ways, though, these centerpiece interchanges reflect the relatively deliverable aspects of intermodality. Major constraints, such as the availability of land in dense urban areas, may be encountered when seeking to increase the physical proximity of different transport modes. However, provided that sufficient finance is available, engineering expertise can often overcome such barriers. New high-speed railway lines can be tunneled under existing 19th century railway stations, such as at Antwerp Central. The technical challenges proposed by major interchange projects may even be embraced as opportunities to deploy iconic architecture which celebrates the transport function, as was the case with London Heathrow Airport's Terminal 5. Again we are offering a hypothesis here, which is that intermodal initiatives become increasingly subject to lock out (in terms of being disadvantaged) the more they concern lower profile locations, lower technology modes (such as cycling), provide for fewer travelers or for lower status travelers. Moreover, while the construction of interchanges generally requires a one-off investment in a major piece of infrastructure, which then embodies certain features of successful intermodality (particularly proximity) in perpetuity, those aspects of intermodality which require ongoing negotiation, transaction and modification, such as the 'soft engineering' of operating procedures (integrated scheduling, inter-availability of tickets) are much more likely to experience disintegration due to the incompatibility of the modal regimes.

A further factor mitigating against intermodality is that, for any of the modal (sub)regimes, intermodal practices represent a minority share of the travel behavior associated with that mode. Hence, where bus or rail may each be seen as distinctive regimes in relation to the car-dominant regime, intermodal travel is a niche tied into these sub-regimes or, indeed, literally occupies a niche overlapping these sub-regimes. Hence, we can hypothesize that the relative marginality of the sub-regimes combined in delivering an intermodal solution will influence the viability and significance of that new practice.

As what constitutes these niches can only be understood in the context of a transport policy dominated by the national tier of governance, we now turn to consider our case study countries of the United Kingdom and the Netherlands.

15.4. THE NATIONAL CONTEXTS TO INTERMODALITY

As intermodality exists as niche practices within regimes and sub-regimes of personal mobility, it is important to hold present the current relative importance of these modal systems in total travel demand, and some summary statistics are provided next. Further sub-sections consider the limited data available on intermodality at the national level, the regulatory frameworks for the transport systems in the two countries and the emergence of explicit policies for intermodality.

15.4.1. Expressed demand for personal mobility

The United Kingdom and the Netherlands show similarities to each other and are fairly typical Western European states in terms of the modal shares of the principal motorized surface transport modes (see Table 1). The United Kingdom shows slightly above-average dependence on the private car; the relative position of the public transport modes in the Netherlands reflects the importance of urban rail systems in some cities.

A notable modal difference not shown previously is that the Netherlands, along with Denmark, has in European terms an exceptionally high average distance cycled per person per annum, while walking rates are close to average for the European Union of 15 Countries. The United Kingdom instead lies near the extreme end of a long tail to the cycling distribution, and also shows a somewhat lower walking level to the European Union of 15 Countries average (Table 2).

Although on a per trip basis cycling is important, accounting for 28% of all trips in the Netherlands,[2] from the perspective of energy conservation and climate change emissions the total distance traveled is more significant, and walking and cycling even in the exemplar case of the Netherlands accounts for a relatively small total compared to the average 14,950 km

Table 15.1 Percentages of Passenger Kilometers Modal Share of Principal Motorized Land Modes

	Cars	Bus and Coach	Rail Systems
The Netherlands	83.2	6.8	9.9
United Kingdom	86.4	6.4	7.5
European Union of 15 Countries	83.1	8.7	8.3

Note: From Commission of the European Communities (2009), Table 3.3.3.

Table 15.2 Distances in Kilometers Traveled on Foot and by Bicycle per Person per Annum

	Cycled	Walked
The Netherlands	848	377
United Kingdom	75	355
European Union of 15 Countries	188	382

Note: From European Environmental Agency (2008), Figures 11.1 and 11.2.

316   Graham Parkhurst, René Kemp, Marc Dijk and Henrietta Sherwin

traveled by Dutch people per annum. British people travel even further (15,870 km^3) using overall a somewhat less efficient combination of modes. Car ownership levels in both states are close to the European Union average of 460 cars/1,000 inhabitants, while a study of the spatiotemporal severity of highway congestion in the European Union of 15 Countries identified the United Kingdom as having the worst incidence and placed the Netherlands third (Commission for Integrated Transport, 2001).

15.4.2. Extent of Passenger Intermodality

Comparable statistics for intermodal transport are only published at the European Union level for freight. In the United Kingdom, the National Travel Survey publishes information about the 'main mode' used for trips only. Similarly, in the Netherlands, data on multimodal trips are not systematically collected, and the National Mobility Monitor does not provide information on the number of multimodal trips. However, evidence for the Netherlands indicates that trips involving more than one mode other than walking grew from just under 2% of all trips in the mid-1990s to nearly 3% by 2008.[4] The main mode for more than two thirds of these trips was a rail system, for a quarter, bus, and relatively few by car or bicycle.

As 'trunk' transport systems, notably rail and air, rely in particular on feeder modes to access their nodes, multimodality tends to be monitored, and, increasingly, managed, through initiatives such as the United Kingdom's Station Travel Plans. In the Netherlands, 40% of rail travelers use a bicycle to get to the train station (up from 30% in 2003).[5] The UK National

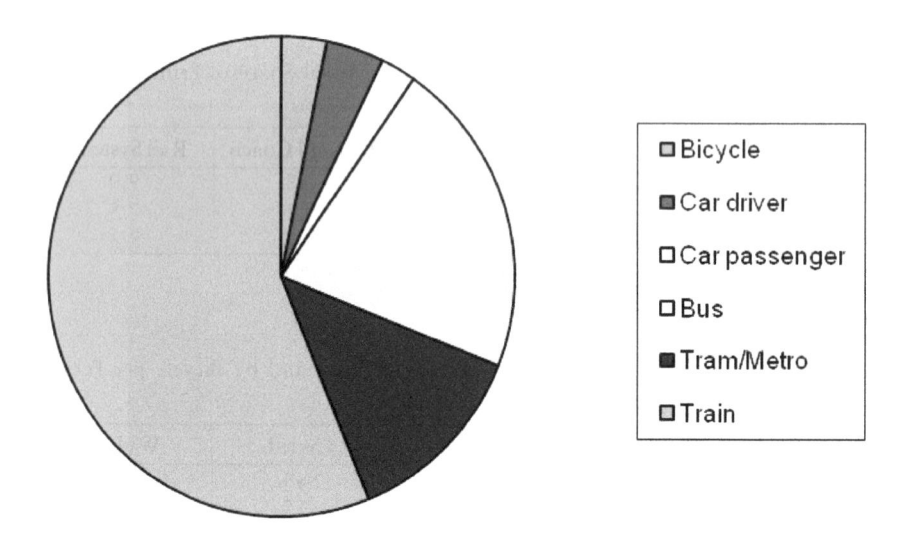

Figure 15.1   Main mode in intermodal trips in the Netherlands.

Table 15.3 Main Access Mode to UK Rail Stations for Certain Journey Purposes

Mode of Access %	Commuting	Business	Leisure	Total %
Walked	58	41	50	54
Bus/coach	10	7	12	10
Car (parked at or near the station)	9	14	8	10
Kiss and ride	6	8	9	7
Bicycle	2	1	1	2
Taxi	1	8	5	3
Light Rail/metros	14	19	12	14
Other	0	0	1	0

Note: From Department for Transport (2007), National Rail Travel Survey.

Rail Travel Survey examines access to rail stations in detail. Although there is considerable variation between stations, overall more than half of trips to stations are walked, a quarter is made by another form of public transport and the rest by private modes, mainly car (see Table 3).

15.4.3. Regulation of the Transport Systems

Although the Netherlands and United Kingdom are both associated with liberalism in the context of European economic regulation (Abbati, 1987, as cited in Glaister, Burnham, Stevens & Travers, 1998), they do nonetheless exhibit important differences in the extent to which these principles have been applied to the transport sector, with some observable consequences in intermodality policy.

In the United Kingdom, the local bus and intercity coach networks outside London were both privatized and deregulated in the 1980s, according to principles of competition not only for the market, but within the market, with other modes of public transport (rail, tram) potentially being competitors. In this context, intermodality will only tend to be pursued if perceived to aid in operator profitability, unless the public sector seeks to use scarce subsidies specifically to promote these objectives. Buses in London were subsequently franchised in 10 spatially defined route bundles, as were the railways in the 1990s, with competition existing for the market and, in the case of the railways, occasionally in the market, where rival companies happened to operate parallel railway routes between pairs of major cities. Where intermodality requires facilities at the main railway terminus stations, this requires agreement with both the quasi-private infrastructure manager *Network Rail*[6] and the relevant Train Operating Companies (TOCs), which take revenue risk and directly control the smaller stations. Taxis are regulated by local authorities in terms of fares and usually in terms of the quantity of licenses issued and quality of vehicle as well.

In the Netherlands, the transport sector was privatized and partly reorganized in the 1990s. Bus services are provided by private companies through concessions. As in the United Kingdom, the integration of different modes of *public* transport has been poor, owing to different companies being involved and lack of incentives for integration, although competition is for the market not in it. The sub-regimes of buses and other modes are more integrated in the four large cities where public transport is offered through local publicly owned transport companies. In other cities the buses are owned and run by big (mostly internationally operating) transport organizations who won concessions for bus services. In 1995 the Dutch railway company Nederlandse Spoorwegen BV was separated into two companies: the infrastructure company Prorail owning and servicing the physical rail infrastructure and the national train company Nederlandse Spoorwegen NS which runs the trains, thereby maintaining the horizontal integration which was removed in the UK rail privatization.

15.4.4. National Policies on Intermodality

In the course of the 1990s attention for intermodality—or *chain-mobility* as it is commonly known—increased among Dutch national transport policymakers. Chain mobility was one of the centerpieces of the white paper *Environment and Economy* of 1997. In 1998 the Ministry of Transport, Public Works and Water Management (Verkeer & Waterstaat [V&W], 1998) produced the white paper *Service Provision and Chain Mobility* which announced measures to promote intermodality, including adaptation of the fiscal regime, provision of infrastructure for physical transfer, stimulus of traffic telematics, co-financing of pilot projects, funding research into market aspects and the promotion of behavioral change through exemplar projects. V&W funded the building of 10 'transferia' (i.e., high-standard P&R facilities) at a cost of €40 million. In 1999 a national 3-year program on chain mobility (MOVE) was launched, providing subsidies for research, development and demonstration of *seamless mobility*. In total €8.6 million was spent on subsidies, divided over 87 projects ranging from car sharing, public bicycles, travel information services and taxi services to a business mobility card which combined public transportation, the use of rental cars and taxi use. This funding program stimulated a range of innovative activities, with the innovators often being small companies.

In 2000 V&W commissioned a major study of the market for intermodal personal mobility, which produced six reports (Adviesdienst Verkeer en Vervoer, 2002). The findings dampened some of the positive expectations for intermodality in terms of prospects for growth and associated environmental gains. The MOVE program was superseded by a 1-year program on chain mobility and a 2-year program on mobility management, within which intermodality was not seen as a key dimension. After 2002, the national government took a laissez-faire approach to intermodality, leaving it to the market and local authorities while road construction became a priority under the three right-wing governments from 2002 to

2006. Chain mobility was supported on an ad hoc basis (e.g., within road construction projects).

Between 2005 and 2007 there was no national innovation support program for chain mobility, but it was given renewed attention in the Mobility Action Plan 2008 which announced a significant extension of P&R capacity at train stations (around 7,500–10,000 additional car parking spaces). Bicycle racks and sheds at train stations were also to be expanded, and connections between the various types of public transport were to be addressed, including the improvement of travel information services. The government assigned a budget of €30 million for this initiative for 2008 to 2012. A new support scheme for intermodality is the €1.5 million innovation support scheme "Innovative mobility from and to the railway station" for small businesses.

The case of the United Kingdom shows several similarities with the Netherlands, although with some contrasting chronology. Laissez-faire toward interchange dominated national policy from 1979 to 1997 under Conservative administrations concerned with allowing consumers to decide what was important about transport services via the fare-box and enabling commercial efficiencies and service rationalization to cut subsidy costs. However, as in the Netherlands, laissez-faire was accompanied by permitting local authorities and other agencies to pursue intermodality as a matter of local policy. Hence, in some cities where cycling was popular (e.g., Oxford, Cambridge) investment in some rail-cycle integration facilities occurred. Similarly some towns in which buses continued to have an important modal share invested in passenger interchanges. Most distinctively, a number of local authorities, initially particularly in historic cities, provided dedicated bus-based P&R services.

As noted in Section 15.2, the 1997 to 2010 Labour administration adopted a pro-intermodal policy. However, the decisions of the administration not to make significant changes to the basis of public transport regulation meant that the ability to deliver intermodality as a matter of public policy was limited. Local authorities were charged with auditing interchange facilities as part of their preparation of local transport plans and could seek to enhance facilities through their funding bids. Similarly, a number of national funding initiatives were not focused on intermodality but could be expected to promote it to some degree, such as the Sustainable Travel Demonstration Town (£10 million 2004–2009) and Cycling Demonstration Town (£8.5 million 2005–2008) programs which sought to concentrate spending on a number of related measures in particular towns to develop evidence and expertise around effects such as benefits of scale which it was hoped other local authorities would seek to emulate.

An exception to the national trend concerns the case of London, where buses had not been deregulated and where political reform after 1997 led to the election of a London Mayor with considerable financial and planning powers, a Greater London Assembly and an extensive executive transport body 'Transport for London'. Transport for London has direct control over all public transport services in the Greater London Area and has consolidated and extended popular practices such as multimodal ticking and intermodal information provision.

From these national overviews, three areas where there is some commonality of innovation experience in terms of intermodality are now identified as case studies for the examination of factors in success, opportunities and barriers: access to trunk rail, short-range P&R and services to facilitate intermodal travel.

15.5. CASE STUDIES OF INNOVATIVE INTERMODALITY

15.5.1. Rail Intermodality for More Sustainable Mobility

Heavy rail systems are a natural focus of intermodality, due to the limited penetration that even relatively dense heavy rail networks can have in space, therefore creating a demand for access facilities. The market potential for railway services depends to a considerable extent on the quality of the total journey from residence to place of activity and vice versa (Rietveld, 2000). However, the rail industry has in the past concentrated on the journey on the rail network itself rather than the access journey to the rail network, despite this leg of the journey having been identified as a barrier to rail travel (Passenger Focus, 2007) and in the context that it is often cheaper and more effective to improve aspects of whole-trip service such as journey time by investing in access facilities rather than on the rail network itself. Additionally, as envisaged by the European Union perspective on increasing the sustainability of the transport system, the space and energy efficiency of the rail system suggest it could play a greater role through substituting car and air trips, provided that a greater number of travelers find it easy to reach railway stations (Sherwin, 2010). In both our case study nations, innovative rail-related intermodality has included rail-taxi integration, rail-flexible bus integration and bike-rail integration through the promotion of bike hire.

Rail-taxi Integration

In the Dutch case, in 1992 a system of shared taxis was created using existing privately owned taxis, with financial support from Nederlandse Spoorwegen (NS) which invested €70 million in the service from 1992 to 2003 with the objective of attracting extra patronage to rail. The system was branded as *Treintaxi* and operated despite heavy opposition from non-participating taxi companies. Treintaxis transported travelers in possession of a train ticket to destinations within a designated area around the station (approximately 8 km) for a fixed price of initially 5 gulden (€2.25). Taxis would wait for up to 12 minutes for other passengers to join (up to a maximum of four), with the driver responsible for determining an optimal routing. Outbound travelers needed only to go to a dedicated rank while for return trips it was necessary to telephone at least 30 minutes prior to collection. The simpler procedure when exiting from the rail system meant that 80% of trips were made away from stations and only 20% to them.

Treintaxi was very popular for a number of years, but demand then declined from 3.8 million trips in 1997 to 2.2 million in 2003, due to competition from (subsidized) regional taxis introduced by public transport companies to replace bus lines which were not economically viable. In 2003, NS withdraw their financial support for Treintaxi, resulting in a sharp reduction in participating stations.

Taxi integration has not occurred in the United Kingdom at the national scale, but there have been local innovations. The Bicester North Taxibus operates from a specific station on the London Marylebone-Birmingham route. The station car park has capacity constraints, and fares on the taxibus (£2.20 single/£3.50 return) are cheaper than parking for the day. Off-peak services operate only in response to telephone bookings. Initially rural and urban taxibuses designed to integrate with the rail timetable were operated with partnership funding from government and local authority grants and support from the TOC, although the rural service was subsequently discontinued due to low levels of patronage. By 2004 the urban route was covering half its costs from fares and contributed to a 9% increase in rail use from Bicester North station, while modal shift from car to taxibus to access the station was evident by around 30 parking spaces having been made available for new rail users (Enoch, Potter, Parkhurst & Smith, 2004). The dual mode of operation means that the vehicles have to be registered both as taxis with the relevant local authority for demand-responsive services and as buses with the traffic commissioner for peak scheduled services.

In the mid-2000s a scarcity of taxi supply at one of the main London terminus stations (Paddington) resulted in a pragmatic initiative for taxi sharing emerging. A separate taxi queue for travelers prepared to share was instigated within which the customers were grouped according to their central London destinations. There was some debate as to whether the practice was legal, as users could be deemed to be paying an individual fare organized by a third party, which would require the taxi to operate legally as a bus service, but the practice ended before this was tested, as the pressures caused by limited supply reduced.

Bus-rail Integration in Rural Areas

In rural areas both bus and rail services are often infrequent, and poor intermodal integration can render bus-rail trips unfeasible. Due to low overall demand, integration can rarely be enhanced by increasing frequency to compensate in the case that one mode experiences schedule reliability problems; therefore flexibility needs to be introduced into the interchange to enhance connection. Given that railways require critical scheduling across extensive networks, the flexibility tends to occur in respect to the bus services. A simple example of how intermodality may be enhanced by local innovative practice was found from a bus-rail link route in Southwest England (Enoch et al., 2004). If the bus service linking to the station was running late, the bus drivers adapted the route to serve the rail station before the town center. While seeming common sense, though, such practices are technically in contravention of the legal requirement for standard public bus services to follow the registered route in sequence. In other locations buses operate with semi-flexible timetables (e.g., linking Charlbury station, Oxfordshire) so that the bus service will wait for a delayed train up until certain published time parameters.

In the Netherlands coordinated bus-rail integration has mostly attracted attention in peripheral areas too. In the 1980s and 1990s a round of budget measures were applied to encourage bus-rail integration, initiated by the national government. In various regions the results were attractive, most notably in Groningen where tariffs and tickets were integrated (i.e., one ticket and price for a combined trip), and schedules and connections were streamlined. Another notable example is Achterhoek, where coordination even resulted in a merger of the regional bus and rail companies. Currently, regional authorities are responsible for public transport, including some rail services. This has furthered integration through integrated concessions in regions such as Limburg.

Bike-Rail Integration

Bike-rail integration (BRI) in the Netherlands is already supported by substantial NS-owned parking facilities at stations. Until recently there was no national policy to promote BRI, although a current Dutch government *Action Plan for the Railways* identifies €20 million for 20,000 extra bicycle racks near stations (V&W, 2007).

Despite this recent official support, innovations have occurred in the Dutch context. Due to the high potential demand and limited space on trains, NS discourages bike carriage on trains through charges and peak-time restrictions (except for folding bikes). To serve train travelers, in 2002 a bike rental system called OV-fiets was introduced at the main railway stations, initially as a joint pilot initiative of Fietsersbond (the national bicyclists organization) and Prorail under the MOVE program. NS subsequently took over the system, which it saw as a means of controlling the cost of providing for cyclists, but only after accepting conditions that it would invest in growth and introduce automatic hire facilities.

The system continues to grow rapidly: In 2009 alone the number of clients grew by 31%, and 40% more trips were made, with the typical user making 10 annual trips, half for business. NS expects a further increase from 0.67 to 1 million trips in 2011. The subscription and rental costs are

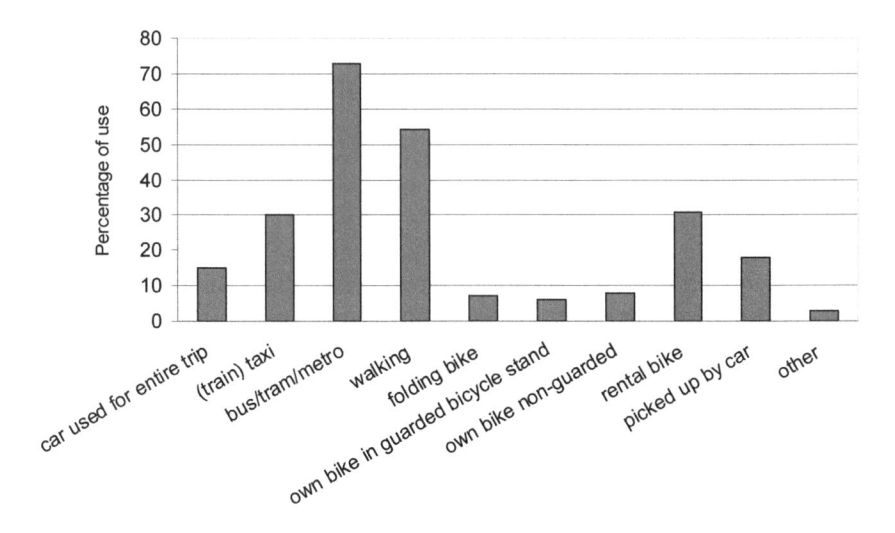

Figure 15.2 Previous transport choice before use of OV-fiets (Fietsersbond, 2003).

both low (in 2010 €9.50 per annum, €2.85 per day). A further innovation is that from a few stations, electric scooters can now be rented (for a cost of €15 per day). Based on a user survey of previous access mode, OV-fiets mainly substitute for public transport (Figure 15.2). Substitution of an entire car driver trip was mentioned by 15% of respondents, while 19% mentioned that a car passenger 'kiss and ride' trip had been replaced. However, it is unclear what the actual substitution percentages are because people could give multiple answers. Nonetheless, the scheme has been associated with an increase in train trips to the value of 4.5 extra trips per member per annum (Fietsersbond, 2003).

The United Kingdom in contrast, with its very small existing level of BRI perceives significant potential to expand this niche, and it is part of official policy on *Delivering a Sustainable Railway* (Department for Transport, 2007) as 60% of the population lives within a 15-minute cycle of a station. One of the barriers to increasing BRI is the fragmented ownership of the rail industry noted previously and the contractual basis of the franchise agreements, which are not consistent across the network. In practice each TOC has its own cycling policy and attitude toward BRI compared to promoting access by other modes. Indeed, greater car access to the railway is also part of national policy, despite the road network environs of stations often being highly congested and the investment required to encourage cycle access being typically 40 times lower than increasing car parking capacity (Green & Hall, 2009). An ongoing mismatch in demand and supply at railway station car parks, though, led to an independent review for

Government in 2009, with the outcome suggesting that a transition may be occurring: An investment of £14 million was announced for packages of rail station cycling facilities, including the funding of 'cycle hubs' at 10 major rail stations and 10,000 extra station cycle parking spaces across the country. The new cycle hubs will follow the Dutch model and include extra cycle storage facilities, repair services, hire schemes and improved cycle access to stations.

As in the case of rail-taxi integration, though, the United Kingdom has not yet seen a national scheme for rail-linked bike hire similar to OV-fiets. Pilots and experiments have occurred at a range of stations including in Cheltenham, Reading and Bristol. In contrast, the first major multipurpose urban 'bike sharing' scheme in the United Kingdom began in London in 2010 with 6,000 bikes at 400 docking stations but with a specific policy *not* to target the rail-related market as the promoters feared the system would be unable to provide for demand if even a small share of the million or so travelers arriving in London each weekday wished to use the system. Nonetheless, in practice the system is designed to provide for 40,000 cycle trips per day, and it is likely that many of those users will have arrived in central London by public transport, even if the system is only passively marketed to this group.

15.5.2. P&R

Low cost peripheral parking facilities linked by public transport have been built on the edge of certain British cities since the early 1970s and Dutch cities since the early 1990s with the objective of encouraging intermodal car-public transport journeys which result in the 'interception' of cars before they enter the urban area. Therefore P&R can encourage trips to the commercial center but without bringing increased road traffic, congestion and pollution into the urban area. However, although broadly similar in function, the British practice almost universally links the car parks with dedicated (and therefore additional) bus services, while the Dutch approach mostly provides the parking capacity on established rail and urban rail lines making use of existing services. The location may be close to the destination or far away from it; a transfer point may serve various destinations or just one, such as a coastal resort or an amusement park (see Figure 15.3).

A laissez-faire approach to spatial planning in the 1980s and 1990s contributed to rapid growth in car use and car dependence. Combined with deregulation and privatization of bus services, which meant that local authorities had little direct control over them, short-range P&R emerged as one of the few urban traffic restraint tools which promised to be effective without deterring trips. Hence, while P&R had been implemented in only a few places in the 1970s and 1980s, significant policy diffusion occurred in the 1990s, particularly after a decision by national government that 'packages' of individually small but inter-related urban transport measures could

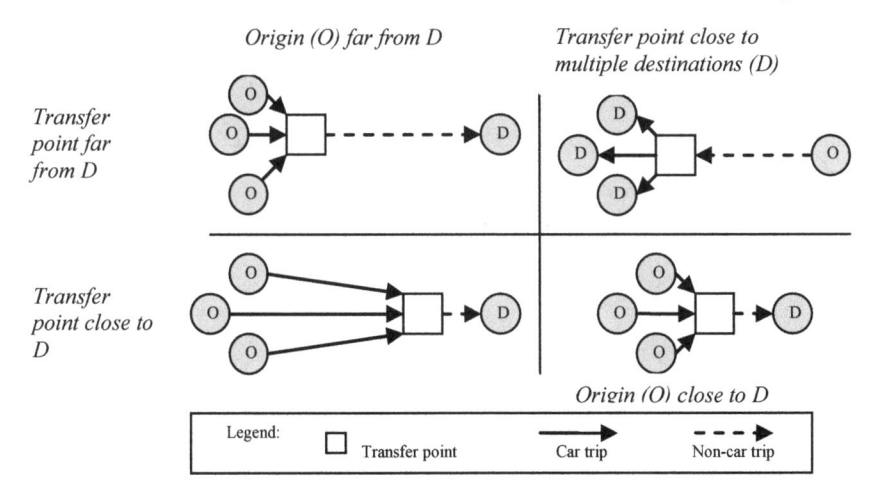

Figure 15.3 Conceptual models of intermodal journey types (Van Binsbergen & Bovy, 1996). O = origin; D = destination.

be funded. Hitherto there had been a presumption in favor of individual, larger investments, such as road schemes. However, an unintended consequence of the UK model of providing a separate, higher quality, lower cost network of buses marketed to travelers with cars available is that a significant minority of P&R users have switched modes from mono-modal bus and rail journeys to car-bus intermodal journeys. As a result, most UK P&R schemes have increased, rather than reduced, car traffic overall, and must be considered as undermining progress toward more genuinely sustainable mobility (Parkhurst, 2000a). Some of the pioneering UK examples are also open to the criticism that P&R provision has in practice increased parking supply, not simply relocated it (Parkhurst, 1993). Despite these controversies, the 2000 Government's *Ten-Year Plan for Transport* foresaw that the then 70 national dedicated schemes might increase to 170 by the end of the decade (Department for the Environment, Transport and the Regions, 2000), although in practice the total had only reached around 100 by 2007 (TAS Partnership, 2007). An alternative model for future development has been proposed, which would place bus services as the main mode of the intermodal niche, rather than the egress mode from a main leg served by car, through placing P&R sites where they can achieve much earlier interception (see Figure 15.4).

In the Netherlands, there are around 400 P&R sites. In the 1990s, V&W funding supported a new type of P&R with easy access by cars: the transferium. Nine transferia were opened between 1995 and 2002. Some of these are destination-based, that is, near city centers (e.g., the *Arena* transferium offers access to the city of Amsterdam, Ajax football stadium and an entertainment complex); others were oriented toward outbound traffic, and

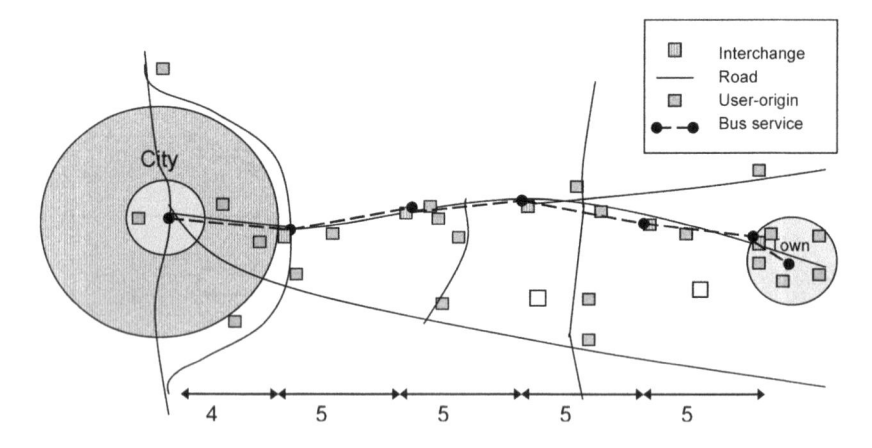

Figure 15.4 Concept for park and ride with bus service as the main mode for serving multiple destinations (Parkhurst, 2000b).

these are typically found in rural areas (e.g., Sittard, Hoorn). Transferia may be used for a short trip (to enter a city) or for making a long-distance trip with a train.

As in the case of the United Kingdom, Dutch P&R policy has been criticized for having limited effect; by 2002 most sites were receiving somewhere between a few dozen and a few hundred users per day (CROW, 2004). After 2002 (when national support stopped) and province and city authorities became responsible for building them, the number of destination-based sites slowly grew, although the number of destination-based sites serving city centers was still not more than a few dozen. The car restraint effect that had been hoped for was mostly not achieved. In Rotterdam, a transport study at one P&R site revealed that 33% of the current users previously traveled by public transport or by bike (Mingardo, 2008). And as in the case of the United Kingdom, in three cities (Amsterdam, Utrecht, Rotterdam), P&R was found to have contributed to an overall increase in city center parking supply (Dijk, 2010).

15.5.3. Services to Facilitate Intermodal Travel

One of the most important factors in the traveler perceiving a seamless interchange between transport modes is a charging structure which does not place a penalty on intermodality and enables quick and easy payment, ideally once for the entire journey. Intermodal ticketing is particularly important between public transport modes, although is increasingly relevant for private modes where parking, charging and hire fees may be due. For example, car-arrivers at many UK stations can now pay for parking via mobile phone once waiting for the train on the station platform. The system reduces cash transaction costs for the car park operator and enables a rail-specific tariff to be charged to enable the car parking capacity to be prioritized for rail users, rather than people parking to access functions which simply happen to be near the station. From the traveler's perspective, the critical car-to-platform leg of the journey can be faster, and once the access code has been obtained from the platform, payment can even be made from on board a train.

The best known and one of the longest running examples of intermodal ticketing in the United Kingdom is the London Travelcard, introduced in 1983 to 1984, which now enables travel for one or more days on railway, underground, light rail, bus and riverbus services in the London area. As might be expected, Travelcard boosted both bus travel overall (by 16%) and underground use overall (by 10%; Gilbert & Jalilian, 1991), however a secondary effect was to encourage modal shift from bus to rail, with mono-modal bus trips being transferred partly to underground due to the cost barrier being reduced (White, 2008). Traditionally the tickets have been in the format of standard UK 'bankcard' size railway tickets and can be purchased from far outside London as part of a day-return rail ticket to the capital. In recent years London has also adopted the 'Oyster' smartcard system which can provide for pay-as-you-go travel on London public transport systems up to a maximum capped price. Both the initiatives in effect give free travel once purchased, or once the cap is reached.

Outside of London the 'Plusbus' initiative is now widespread, enabling rail passengers to purchase a bus 'add-on' valid for single or multiple bus trips in one or both towns connected by the rail trip. As well as reducing the number of ticket transactions, the fares generally provide a modest discount, which from the operators' perspective is an efficient means of attracting additional patronage, as the combined cost of the full separate monomodal fares would be expected to deter some travelers from making the trip using public transport at all.

In terms of intermodal ticketing the Netherlands has been a global frontrunner with a *national* payment system for buses, trams and metros in the form of the strippenkaart (strip ticket) in place for many years. A smartcard-based system was introduced in 2009 mainly to control costs and assist in revenue allocation to operators. The card can also be used in the national railway system, which means that most public transport trips can be paid for via this single card.

A rail-related example of integrated ticketing combined with intermodal journey planning was offered by Transvision, a subsidiary of NS, between 1996 and 2002. The mobility card named Odessey was aimed at business travelers and cost the equivalent of €135 per year, with arrangements being made by phone and the bill for the travel costs plus an arrangement fee sent to the traveler's organization. In addition to facilitating rail travel, Transvision had contracts with Treintaxi, conventional taxi companies, car rental companies and 'Rent-a-Driver'. Despite the annual charge and fees, Odessey was not cost-effective for NS, and it was replaced with 'NS business card', a card with intermodal functionality limited to the train and Traintaxi at a cost of €30 per annum. Benefits to the holder include a 20% off-peak discount on rail fares, prebooking by telephone and internet and use of OV-fiets for free.

Mobility Mixx, founded around 2000, continues to facilitate mobility services in both multimodal and intermodal niches. Initially founded by a public bus company (Connexxion), Mobility Mixx received support from the MOVE program but was subsequently taken over by Lease Plan (a car leasing company). Like Odessey, the services are targeted at business travelers, with intermodal options combining rail, taxis, rental cars, public bikes and P&R site payment. The passenger kilometers facilitated by Mobility Mixx grew from 15 million in 2007 to 45 million in 2009, with 75% of these made on journeys with rail as the main mode and 25% with pool car (provided by Mobility Mixx) as the main mode. Of the train trips, 5.2% were combined with access by public bicycle and 3.7% with access by car (via P&R). Mobility Mixx recently joined forces with Travelcard Nederland, which specializes in fuel payment card accounts and Yellowbrick, which offers payment services for parking via SMS in a growing number of cities.

15.6. UNDERSTANDING THE INTERMODAL INNOVATIONS AS TRANSITIONS

The experiences and developments described in Sections 15.4 and 15.5 in the Netherlands and United Kingdom show that intermodal travel is a niche phenomenon both in terms of use and in terms of there not being a well-developed regime behind it of providers, knowledge, business models, spokespersons and a community of experts. A limited amount of technical and professional cohesion is evident from best practice publications and voluntary, non-statutory organizations such as Linkforum at the European level. Both jurisdictions show that policy interest is unstable and often implicit: In the Netherlands a period of official promotion and significant funding, embodied in particular by the transferia program, was replaced by a focus on mobility management more generally; in the United Kingdom the whole ethos of promoting intermodality did not fit well with the dominant political hegemony of 'allowing the market to decide', and even after 1997 rhetoric was more common than specifically targeted funding measures. As a result, both rail and bus-related initiatives have generally remained piecemeal, tentative (with a number of trials ultimately ended) and over-dependent on local factors, such as the presence of policy entrepreneurs or particular coalitions of actors. More recently, however, both the Dutch and UK administrations have recognized the importance of the BRI niche in particular with specific funding lines.

In both countries, the perceptions and actions of local authorities have been important in promoting intermodality, as is clear from the pattern of adoption of formalized, destination-oriented car-public transport interchange facilities. It is also notable that in those cases where successful schemes of intermodality have been introduced, the impetus sometimes came from regime 'outsiders' joining up with traditional transport companies. An example is OV-fiets: proposed by a cyclists' association working with Prorail but eventually brought 'inside' the regime through direct NS control. Another example from outside the three case studies of this chapter is the startup company for car sharing club *Greenwheels*, created by two students who teamed up with the NS (as one of the car sharing applications is the 'last mile' from the station) and later also with the Rotterdam public transport company RET).

Although there is evidence that some intermodal measures have been successful in terms of broad objectives (notably the Dutch OV-fiets and the intermodal ticketing initiatives more generally) or unsuccessful (transferia in the Netherlands, dedicated bus P&R in United Kingdom), these findings have not resulted in hype-disillusionment cycles for intermodal travel. It seems likely that this is due to the fact that expectations have been low or negative (and in this sense there might be a self-fulfilling prophecy here) but also because intermodal activities are not core to any single regime. This situation can be contrasted with the emergence of citywide public bike systems in Europe (which are mostly not proposed as intermodal facilities but as alternatives for monomodal journeys). Several public bike schemes, led by pioneering schemes in Lyon and Paris, have developed a very high public profile and scale, attracting significant public subsidy and advertising revenues, despite the (limited) available evidence suggesting relatively few trips have switched from private car use (Sherwin & Parkhurst, 2009).

In terms of the transition patterns identified in Chapter 3, intermodality is a hybrid system used as an 'add-on' to address particular problems of the car regime or public transport. Neither regime is threatened or undermined by it; indeed, at least while intermodality is a marginal or minority activity, the main regimes are arguably strengthened, as the systems are not aggressively competing with each other. Treintaxis and public bikes introduced by the Dutch Railways help travelers to make the last leg of a journey but not have a significant impact on the dominance of car use. Destination P&R facilities improve access to towns, but although there may be a significant increase in bus trips as a result, there may also be growth in car travel (and car dominance) overall.

By virtue of enabling more seamless interchange, technological developments such as smartcards and better information provision (internet-based and increasingly on a real-time basis) are contributing to intermodal travel. Other supply-side developments that can be expected to contribute to intermodal travel are improved public transport (such as the high-frequency railway schedules planned by NS and the possibility of the United Kingdom adopting high-speed rail), more public bike and electric bike schemes, and mobility management more generally, as travelers with multimodal lifestyles are more likely to need to combine modes to make particular journeys. On the demand side, information campaigns may lead to travelers becoming more rational about travel choices, but economic factors are likely to be important, such as the possibility of road pricing or peak oil making car use more expensive. Planning and regulation are also relevant, with car-free housing and widespread traffic calmed zones becoming more popular.

However, the greater extent of development of intermodality in the Netherlands, where there has been stronger central government support compared with the United Kingdom, suggests that it will rarely emerge as a significant phenomenon without national government support and coordination from willing actors (which cannot be assumed). No national equivalents to OV-fiets or Treintaxi or Greenwheels are in sight in the United Kingdom. Even in the Netherlands, however, where transport coordination is more possible, major organizational barriers exist, such as the fragmented systems of mobility providers and the fragmented system of providing public transport concessions. Dutch law also forbids the Ministry of Transport from subsidizing service operating costs (except in the case of the railways). To both overcome these barriers and promote sustainable mobility, a long-term coordinated approach is necessary involving the creation and development of convenient car-public transport interchanges which result in public transport being the dominant, not simply the egress mode; public bicycles for short-term rental integrated with public transport; integrated ticketing; dynamic information and booking; and individualized forms of public transport (such as demand-responsive services) interacting with traditional collective, scheduled public transport. However, given that travel behavior is context-dependent, with each context involving different transport actors, intermodality cannot be planned from the top but must be planned and/or grown organically from below, supported by 'carrots' and 'sticks' from the 'top'. The success of the OV-fiets depends on special Dutch circumstances of a good train system, a good cycling infrastructure, cycling being part of the cultural normal and business destinations reachable from a train station with a bike. In other words cycling-public transport integration is a niche between regimes, while in the United Kingdom a significant, national cycling regime ended in the 1950s and has yet to fully re-emerge.

15.7. CONCLUSIONS

This chapter looked at intermodal travel: the involvement of authorities and mobility providers and the willingness of travelers to engage in it. We found that it is a niche which is caught between the regime of car-mobility (where people drive from door-to-door) and the regime of public transport. Although users may benefit from more seamless mobility, the creation of systems of intermodality requires cooperation from various transport actors, none of whom is strongly inclined to invest in it. Recently, passenger intermodality has been receiving more attention by policymakers, particularly in the context of linking to the railways, but it is still not the core of policy, few reliable descriptive data exist and from an academic perspective, it remains an under-studied topic. And although policy attention for intermodality is growing, there is no powerful advocacy coalition speaking on behalf of intermodality and rarely are economic interests behind it. Where the interests and resources are aligned and coordinated, which we see most clearly in respect of destination P&R schemes linked to particular towns and national intermodality schemes linked to the Dutch railways, growth can be rapid and sustained. Critically, though, intermodal niches need to lie between well-established regimes: BRI schemes in the United Kingdom have suffered from the absence of a coherent cycling regime.

Intermodal travel is interesting from a transition point of view as it may constitute a potential mechanism of resilience, enabling society to 'divest' from car dependence once this regime is no longer sustained, with cars increasingly used in combination with other modes of transport, alongside the use of feet and bicycles for short-range trips and public transport for longer trips. This hoped for effect does not happen easily. Car-restraining policies are needed to complement investment in intermodal interchanges.

An important issue for the book is whether intermodal travel will produce sustainability benefits. If it does, it can be expected to win political support from local authorities and political parties, which may put intermodality on the sustainable development agenda. Compared to the use of cars, intermodal travel offers prospects for alleviating seven problems: energy consumption, climate change emissions, road congestion, road accidents, poor air quality, lack of usable public space and noise from motor vehicles. However, a real danger with better technical systems of intermodality—if introduced in the context of the wrong policy frame—is that they may result in *more* unsustainable mobility. While this may be desirable from the perspectives of particular individuals, it is generally undesirable from a societal perspective.

The transition perspective proved useful for understanding the (slow) development of intermodal travel, the importance of entrepreneurs in the field of mobility and the need for transport authorities and mobility providers to share knowledge and experience. The approach also highlights the need for car restraining policies and other control policies, not just to pressure current regimes into change but also for making sure that intermodality produces sustainability benefits. The analysis brings to attention the urgent need for developing a *regime* for intermodality in terms of organization, standards, concepts and routines which can deliver those benefits.

NOTES

1. The authors thank Wijnand Veeneman, Rob van Es, Atty Visch, Bram Munnik, Paul Dam, Paul Pilgram, Nathalie Leclerc and Rick Lindeman for providing us with information. We also thank Glenn Lyons for his comments on an earlier draft of this chapter.

2. Figures are for 1997, reported in Rietveld (2000).
3. Elaboration on population data for 2005 from Commission of the European Communities (2009) and annual personal distance traveled for 2004 from European Environmental Agency (2008).
4. Elaboration on data from Mobiliteitsonderzoek Nederland (2009).
5. http://www.prorail.nl/Publiek/Stationsomgeving/Pages/OV-Fiets.aspx
6. Legally, Network Rail is a private company limited by guarantee. Network Rail has 'members' rather than shareholders and any profits are reinvested in the industry.

REFERENCES

Adviesdienst Verkeer en Vervoer (2002) *De markt voor multimodaal personenvervoer;Onderzoek naar de markt- en beleidspotentie van multimodaal personenvervoer* [The Market for Multimodal Personal Mobility; the Market Potential and Its Attractiveness for Policy], Rotterdam, the Netherlands: Adviesdienst Verkeer en Vervoer.
Blow, C.J. (2005) *Transport Terminals and Modal Interchanges: Planning and Design*, Oxford, England: Architectural Press.
Commission of the European Communities (2001) *European Transport Policy for 2010: Time to Decide*, Brussels, Belgium: Commission of the European Communities.
Commission of the European Communities (2009) *EU Energy & Transport in Figure*, Brussels, Belgium: Commission of the European Communities, Directorate-General for Transport and Energy.
Commission for Integrated Transport (2001) *Study of European Best Practice in the Delivery of Integrated Transport: Key Findings*, London: Commission for Integrated Transport.
CROW (2004) *Overstappunten—Ervaringen met Park and Ride in Nederland* [Transfer Points. Experiences with Park and Ride in the Netherlands], Brochure 10: van parkeerbeheer naar mobiliteitsmanagement, Ede, the Netherlands: CROW & MU-Consult.
Department for the Environment, Transport and the Regions (1998) *A New Deal for Transport: Better for Everyone*, London: Department for the Environment, Transport and the Regions, p. 8.
Department for the Environment, Transport and the Regions (2000) *The Ten-Year Plan for Transport*, London: Department of the Environment, Transport and the Regions.
Department for Transport (2007) *Delivering a Sustainable Railway*, London: Department for Transport.
Dijk, M. (2010) *Innovation in Car Mobility—Coevolution of Demand and Supply Under Sustainability Pressures*, Maastricht, the Netherlands: Universitaire Pers Maastricht.
European Environmental Agency (2008) *Climate for a Transport Change*, Copenhagen, Denmark: European Environmental Agency. http://webarchive.nationalarchives.gov.uk/+/http://www.dft.gov.uk/pgr/regional/policy/coll_intermodeinnovationsindeman/intermodefullreport.pdf (accessed 7/10/11).
Enoch, M., Potter, S., Parkhurst, G. & Smith, M. (2004) 'INTERMODE: Innovations in Demand Responsive Transport', unpublished report for Department for Transport and Greater Manchester Passenger Transport Executive.
Fietsersbond (2003) *Hoe bevalt de OV-fiets? Klantenonderzoek* [A Customer Inquiry of the Satisfaction of the "OV fiets"], Utrecht, the Netherlands: Fietsersbond.

Gilbert, C.L. & Jalilian, H. (1991) 'The demand for travel and for travelcards on London Regional transport', *Journal of Transport Economics and Policy*, 25(1): 3–29.

Glaister, S., Burnham, J., Stevens, H. & Travers, T. (1998) *Transport Policy in Britain*, Basingstoke, England: Macmillan.

Green, C. & Hall, P. (2009) 'Better rail stations: An independent review', unpublished paper presented to Lord Adonis, Secretary of State for Transport, London: Department for Transport.

Link (2009) *Identification of needs for further research*, unpublished project report, Deliverable D23b, European Forum on Intermodal Passenger Travel. Available at: http://www.linkforum.eu/docs/214/LINK_D23b_Identification_of_needs_for_further_research.pdf (accessed 10/10/11),

Mingardo, G. (2008) 'Effects of park & ride facilities in Rotterdam: A research on the effects of the Rotterdam P&R policy on the economy, accessibility and quality of life of the city', Report No. 2008-2, Rotterdam, the Netherlands: EURICUR (The European Institute for Comparative Urban Research).

Mobiliteitsonderzoek Nederland (MON) (2009) *Mobiliteitsonderzoek Nederland 2008*, Den Haag, the Netherlands: Rijkswaterstaat.

Parkhurst, G. (1993) 'A Comparison of Policies Aimed at Controlling Car Use in the Historic Cities of Oxford and York', Proceedings of PTRC Summer Annual Meeting, Seminar A: Environmental issue, 83–95, PTRC education and Research Services Limited, london.

Parkhurst, G. (2000a) 'Influence of bus based park and ride facilities on users' car traffic', *Transport Policy*, 7(2): 159–172.

Parkhurst, G. (2000b) 'A longer-range strategy for car-bus interchange: the Link and Ride concept', *Traffic Engineering and Control*, 41(8): 319–324.

Rietveld, P. (2000) 'Non-motorised modes in transport systems: A multimodal chain perspective for The Netherlands', *Transportation Research Part D*, 5: 31–36.

Passenger Focus (2007) *Getting to the Station: Findings of Research Conducted in the East of England*, London: Passenger Focus.

Sherwin, H. & Parkhurst, G. (2009) 'Bike-sharing schemes: Opportunities and challenges of implementation', paper presented at the 6th Annual Cycling & Society Symposium, University of Bolton, England.

Sherwin, H. (2010) 'Bike-rail integration as one sustainable transport measure to reduce car dependence', doctoral thesis, University of the West of England, Bristol.

Stokes, G. & Parkhurst, G. (1995) 'Change through interchange: Making complex journeys easier', report to Rees Jeffreys Road Fund (Rep. No. 838), Oxford University Transport Studies Unit, England.

TAS Partnership (2007) *Park & Ride Great Britain* (4th ed.), Skipton, England: TAS Partnership.

Transport for London (2001) *Intermodal Transport Interchange for London*, London: Transport for London, Integration Department.

Van Binsbergen, A.J. & Bovy, P.H.L. (1996) 'Intermodaal personenvervoer. de overstap van theorie naar praktijk' [Intermodal personal mobility. From theory to practice], *Verkeerskunde*, 47(10): 22–27.

Verkeer & Waterstaat (1998) 'Dienstverlening and ketenmobiliteit' ['Service provision and chain mobility'], government white paper, Den Haag, the Netherlands: Verkeer & Waterstaat (Ministry of Transport, Public Works and Water Management).

Verkeer & Waterstaat (2007) 'Actieplan spoor' ['Action plan railways'], government white paper, Den Haag, the Netherlands: Verkeer & Waterstaat (Ministry of Transport, Public Works and Water Management).

Wardman, M. (1998) 'A review of British evidence on the valuations of time and service quality', unpublished research paper (No. 525), University of Leeds Institute for Transport Studies, England. Available at http://www.eprints.whiterose.ac.uk/2086/ (accessed 7/10/11).

White, P. (2008) *Public Transport: Its Planning, Management and Operation*, New York: Routledge.

16 Findings, Conclusions and Assessments of Sustainability Transitions in Automobility

Frank W. Geels, Geoff Dudley and René Kemp

The chapters in this book analyzed change and stability for various dimensions of the automobility regime and subaltern regimes of public transport (bus, train, tram, metro). On the one hand, the chapters highlight the stability of the automobility regime, because of lock-in mechanisms related to sunk investments (in infrastructure, people, factory plants), routine mobility patterns, protection of vested interests by incumbent actors and belief systems from policymakers, transport planners, traffic managers and so on. On the other hand, the chapters show a groundswell of change initiatives such as green propulsion technologies, socio-spatial innovations (e.g., deceleration, localism), car-free city centers, intelligent transportation systems, congestion charging, improved public transport and sustainable mobility policies. The analytical challenge is to assess and understand how both tendencies relate to each other and what the possible implications are for future sustainability transitions in automobility. Is a transition underway? Will there be a transition to just greener cars or a transition away from car-mobility?

Section 16.1 addresses these questions by using the multi-level perspective (discussed in Chapter 3) to organize and interpret the empirical findings from the various chapters. We do not use a positivist style of hypothesis testing or making correlations between dependent and independent variables. Instead we make an interpretive analysis that combines theoretical sensitivity with empirical expert assessments. At the end of the section we offer a few scenarios, one about our expectations for the next two decades, one which is more long-term and one which is a dystopia. Section 16.2 discusses policy recommendations with regard to sustainability transitions in automobility. This discussion is not organized in the form of prescriptive 'recipes' but in the form of policy lessons which relate to policy fragmentation, multi-level governance, the embeddedness of policymakers in broader systems of governance, together with the mutual dependencies between policymakers and civil society and between policymakers and industry. In section 16.3 we discuss how empirical findings indicate opportunities for further elaboration of transitions theory. The book closes with concluding comments in sub-section 16.4 about the usefulness of the transition

336 Frank W. Geels, Geoff Dudley and René Kemp

perspective. We do not end the book with a single prediction. Transitions are not deterministic processes, and the transition perspective is not a truth machine. Interpretive disagreements and contestations (among both 'real world' actors and academics) are part and parcel of 'transitions in the making'. The transition perspective helps to give equal attention to the forces of stability and those of change. But the future is not given to us and neither is it simply for us to make.

16.1. CONCLUSIONS ABOUT SUSTAINABLE TRANSPORT BY APPLYING TRANSITIONS THEORY

The empirical chapters focused on different dimensions of the automobility regime and were informed by different theoretical backgrounds. The chapters therefore arrived at different assessments of the drivers, barriers and possibility of sustainability transitions. Table 16.1 provides a summary of the various assessments.

We make our own interpretive analysis of empirical findings in the various chapters by applying the multi-level perspective and conceptual mechanisms, which were discussed in Chapter 3. To assess the chances and shape of possible sustainability transitions, section 16.1.1 first discusses the various change initiatives at the niche level, where we analyze if they provide the seeds for transformative change. Section 16.1.2 then addresses secular trends at the landscape level that either stabilize or create external pressures on the automobility regime. Section 16.1.3 subsequently assesses the degree of stability of the automobility regime and possible 'cracks in the regime' that form 'windows of opportunity' for sustainability transitions. Section 16.1.4 combines the niche, regime and landscape analysis to make an assessment of the chances for a sustainability transition and its possible form (e.g., a 'green' technology path or broader transformation of mobility systems). We develop three scenarios and assess their likelihood and drivers.

16.1.1. PROMISING NICHE DEVELOPMENTS

Building on chapters in the second part of the book, we discuss niche-innovations in the following broad areas: (a) intermodal travel and public transport, (b) cultural and socio-spatial niches, (c) user innovations in information and communication technologies (ICT) related traveler information services, (d) demand management practices, (e) intelligent transport systems (ITS) and traffic management, (f) green propulsion technology for cars. Over the last 5 to 10 years, niche-innovations in the last two areas have received the most attention and resources from powerful actors (car industry, computer and telecom industry, transport planners, policymakers). We

Findings, Conclusions and Assessments of Sustainability 337

Table 16.1 Summary of Main Conclusions From the Book's Empirical Chapters

Chapter, Author(s), Discipline or Theoretical Angle	Dimension of Automobility Regime	Assessment of Sustainability Transition
4. Dudley and Chatterjee; transport studies	Transport policy discourse, transport modes	The automobility regime is still fairly stable, but some cracks have appeared: (a) Parts of the government have become less committed to stimulating automobility, (b) automobility growth has leveled off, (c) urban areas become new loci for transport innovations.
5. Docherty and Shaw; transport studies and political science	Governance	The scope for government transport policy has diminished because of two transitions in governance structures: marketization (privatization, liberalization) and devolution of power to 'lower' jurisdictions
6. Wells, Nieuwenhuis and Orsato; business studies, transport studies	Car industry	The car industry has much inertia. Incumbent firms are contemplating a (slow) greening of cars but not fundamental regime change.
7. Goodwin; transport studies	Infrastructure management and policy	A transition has already occurred in policy discourse, namely a shift from 'predict-and-provide' to demand management (improvement in public transport, traffic calming, smart mobility choices). Quality of life and urban considerations are more important drivers than environmental sustainability.
8. Zijlstra and Avelino; critical theory	Spatial planning	The automobility regime is stabilized by landscape values (such as private property, obsession with time and speed, individualization, neo-liberalization). Fundamental transition requires rethinking of basic principles. Radical socio-spatial niches (modal shift, deceleration, localism, sustainable urban planning) could form seeds for this.

Continued

338 *Frank W. Geels, Geoff Dudley and René Kemp*

Table 16.1 Continued

9. Sheller; cultural studies and mobilities studies	Mobility cultures	Improved bike lanes, congestion charging, light-rail systems, car sharing, high-speed rail and electric vehicles have not yet challenged the dominance of cars. A fundamental transition is more likely to derive from landscape changes such as ICT, pervasive computing and the move toward a Web-based society which may lead to cyber-mobility that "transforms the very way in which drivers sense and make sense of the world."
10. Orsato, Dijk, Kemp and Yarime; business studies and innovation studies	Battery electric vehicles	Electric vehicles are now on the verge of breakthrough, because of stimulating government policies, technical improvements (lithium batteries), serious commitment from established automakers (e.g., Renault-Nissan) and new entrants (BYD, Tesla), new business models (Better Place), support from electric utilities and market demand from fleet operators (e.g., postal services).
11. Ehret and Dignum; innovation studies	Fuel cell vehicles	The development of fuel cell vehicles progresses steadily. Especially in Germany, which is seen as a lead-market. Innovation and demonstration projects are progressing well, with support from industry and policymakers.
12. Pel, Teisman, and Boons; governance studies and complexity theory	Intelligent transportation systems	Traffic management is being transformed through the incorporation of various ICT devices. While these changes may start as incremental improvements, innovation cascades may eventually bring about larger changes.
13. Lyons, Jain, Mitchell and May; transport studies	User innovation and traveler information systems.	Users are creatively tinkering with ICT devices to develop new traveler information services. Most innovations are currently small and oriented toward filling gaps in existing systems. But they could perhaps reconfigure the provision of traveler information.
14. Harman, Veeneman and Harman; transport studies	Public transport (bus, train)	Public transport has grown in the last decade, but remains small compared to automobility. Privatization has made transport operators the core actors. These actors are not very innovative with regard to environmental sustainability innovations.
15. Parkhurst, Kemp, Dijk and Sherwin; transport studies and innovation studies	Intermodal personal mobility	"Intermodality is a niche caught in a world of regimes." This niche depends on actors from subaltern regimes (train, bus, taxi). These actors are not very committed, because they see intermodality as an add-on rather than core business opportunity.

therefore suggest that niche-innovations in these two areas have relatively more momentum than in the other areas.

Intermodal Travel and Public Transport

The niche of intermodal travel has been carried by dozens of (experimental) projects in the last decade. Most of these projects failed or remained very small. Parkhurst, Kemp, Dijk and Sherwin (Chapter 15) characterized intermodality as "a niche caught in a world of regimes." They found that "both rail and bus-related initiatives have generally remained piecemeal, tentative (with a number of trials ultimately ended) and over-dependent on local factors, such as the presence of policy entrepreneurs or particular coalitions of actors." The intermodal niche remained small for various reasons: (a) time losses because of transfers; (b) support from regime players (railways, taxi companies, bus companies) has been lukewarm, because they see intermodal travel as add-on activity, not as core business, (c) "there is no powerful advocacy coalition speaking on behalf of intermodality and rarely are economic interests behind it." Parkhurst et al. conclude that "car-restraining policies are needed to complement investment in intermodal interchanges." Without those, the building of park and rides may lead to an increase in car traffic.

The relative market shares of public transport, which is a regime of its own (with sub-regimes) rather than a niche-innovation, have increased in the last decade in urban areas and for long-range trips from one city to another. Harman, Veeneman and Harman (Chapter 14) found that the decline in the 1980s and 1990s has been halted and turned into moderate growth, although "public transport remains a small part of total travel in terms of distance." Privatization has changed the relationships in public transport: As the government adopted a hands-off approach, private transport operators became the core actors. With regard to efficiency-oriented innovations Harman et al. conclude that "the freedom of the operator to develop its services has led to a continuous process of service changes, new routes and times." But with regard to radical technical change and innovations oriented toward collective goods (such as sustainability), their assessment is more negative because the concession systems and mind sets work against that.[1] Policymakers therefore remain important for innovation in public transport. Through subsidies, direct investments or other measures policymakers were involved in the creation of new high-speed railways that compete relatively successfully with cars and planes on medium-long distances, while light-rail systems are expanding in many urban areas. There are also attempts to modernize bus systems, for example, through special bus lanes, real-time information panels and short-distance radio systems that prioritize buses at traffic lights. Public transport, in particular buses, has also become a testing ground for green propulsion technologies, for example, compressed natural gas, biofuels, battery-electric and fuel cells (often sponsored by national or European authorities). With regard to the mindset of policymakers, Harman et al. comment on the preference for rail-oriented innovations (which are visible and infrastructure-related): "It is interesting to see how policymakers seem to favor these rail solutions over other improvements of service levels of bus services.... Well implemented bus systems could have the same characteristics at lower investment costs. Still, policymakers generally prefer the relatively higher investment of rail systems." With regard to the future sustainability transitions, Harman et al. conclude that "on this pivotal moment in history, it might be that a strong role of government, as regulator and authority, might support a swifter change toward more sustainable technology in public transport." But they also conclude that "simply improving public transport services will not have a large impact on car users to make the switch." A modal shift from cars to public transport would probably require a "far greater focus on spatial planning strategies" and "significant alterations to regulations and taxes." Low-density land-use (common outside cities and at the borders of cities) makes offering good public transport very expensive and generally unattractive. In contrast, public transport has the greatest potential in high-density areas and for interurban travel.

Cultural and Socio-spatial Niches

There are cultural and socio-spatial niches that challenge basic assumptions of the automobility regime. Focusing on the United States, Sheller (Chapter 9) discusses several practices that deviate from 'normality', for example, new political projects such as Transit Oriented Development (TOD) or Complete Streets/Liveable Streets, which both lobby for changes in zoning laws and city planning. Car sharing, public bike-sharing schemes and urban initiatives such as Smart Growth America (which critiques automobile-centered transportation and sprawling land use and calls for smart transportation including TOD, walking and bicycle-friendly communities). These practices are supported by social movement networks and community organizations which draw on counter-discourses that challenge the dominant order (e.g., sustainability, health, anti-consumerism). Although these practices have contributed to the emergence of a new discourse of 'sustainable mobility', Sheller assesses that "it still remains questionable to what extent these cultural shifts will impact on the overwhelmingly automobile-centered pattern of majority mobility." In Europe, transit oriented development is more common place, and in the Netherlands livable streets have a history of 40 years.

Zijlstra and Avelino (Chapter 8) place more hope on radical socio-spatial mobility niches, in particular modal shift, deceleration (e.g., through traffic calming or homezones which include soft edges, a staggered street axis, and visual narrowing of the space), sustainable urban planning (e.g.,

Findings, Conclusions and Assessments of Sustainability 341

compact cities, smart growth, clustering of important destinations around public transport hubs) and localism (with self-reliant local communities reducing the need for transport flows). They advocate these radical niches, because they include "the questioning of paradigms" and the "questioning of the very concept of mobility." So, only comprehensive changes in basic principles constitute transitions for Zijlstra and Avelino. In their view, a 'greening of cars' is a "continuation of the current regime and not a transition toward sustainable mobility." They are not very optimistic about the possibilities of such radical transitions, saying that the counter-movements and radical socio-spatial niches have been around for decades without having much discernable impact. On the other hand, they expect the niches to grow in importance and anticipate the possibility that the financial crisis and ecological crisis become linked with each other, fueling debates about the need for fundamental change. We agree that these radical socio-spatial niches contain seeds for wholesale transitions, but we see limited momentum in this direction, in terms of social networks, investment and wide public support.

User Innovations in ICT-related Traveler Information Services

ICT facilitate a range of user innovations with travel information services. Lyons, Jain, Mitchell and May (Chapter 13) discussed examples such as CycleStreets (a digital UK-wide cycle journey planner system), ParkatmyHouse.com (a Web site that lets people park their cars at someone else's driveway), PickupPal (a brokerage service for ridesharing), TrainDelays (a Web site that makes it easy for people to identify and submit claims for delayed train rides), MyBikeLane (a Web site that exposes people who park cars on cycle lanes). While these innovations represent "a substantial groundswell of niche activity," the authors assess that "there is precious little indication as yet that these 'innovations' have moved much further than the innovator and possibly early adopter stage in terms of the diffusion of the innovation." Furthermore, it seems that most innovations are add-ons rather than transformative: "Some of the user innovations are more evidently filling in the gaps in terms of what may remain niche user needs." As we gradually move toward a 'broadband society' with ubiquitous computing and electronic devices, these innovations may perhaps gather pace and facilitate broader transformations (see also Sheller, Chapter 9). But Lyons et al. suggest that such broader transformation depends on other developments: "It may need the cost of motoring to increase sharply before ridesharing services such as PickupPal truly flourish." "This itself may well be fundamentally dependent upon externalities such as oil prices, climate change pressures on policy and public health concerns. These may stimulate the public appetite for reappraising its travel behaviors and for entertaining adjustments to social norms (such as the norm of driving alone)."

Demand Management Practices

New practices and initiatives have emerged in relation to demand management and better *use* of existing infrastructures. Goodwin (Chapter 7) discusses a range of soft measures (currently relabeled in the United Kingdom as 'Smarter Choices') and initiatives that enable people to choose improved standards of accessibility with less car use. "They include workplace and school travel plans, personalized travel planning, public transport information and marketing, travel awareness campaigns, car clubs and car sharing; teleworking, teleconferencing, and home shopping; residential and leisure travel plans; and initiatives on cycling and walking." Goodwin cites studies that estimate that these initiatives could reduce car use by 10% to 20%. He does not, however, indicate how popular these initiatives are with travelers and policymakers. Our own assessment is that these behavioral-change oriented initiatives are still in the early stages of development with fairly limited momentum. Their voluntary nature and zero costs make them attractive to policymakers but stand in the way of effectiveness, making them dependent on good intentions.

ITS and Traffic Management

The niche of Intelligent Transportation Systems is facilitated by the integration of many 'smart' technologies, such as video cameras, magnetic road detectors, computer networks, communication technologies, electronic signaling devices, navigation technologies (e.g., TomTom) and information devices into transport and highway systems (Geels, 2007b). These technologies are used to gather information, report accidents, control information panels and signs (speed signaling, highway dosage) and manage traffic flows (often from centralized control centers). Pel, Teisman, and Boons (Chapter 12) distinguish a range of configurations such as Advanced Traffic Management Systems, Advanced Traveler Information Systems, Advanced Vehicle Control Systems, Commercial Vehicle Operations, Advanced Public Transportation Systems and Advanced Rural Transportation Systems. Stimulated by technical progress in ICT-devices, this niche has gathered momentum since the 1990s. The niche is also actively supported by policymakers, highway engineers, transport planners and traffic managers, with the hope that it helps reduce congestion problems. The niche is also supported by commercial interests because of its economic potential. Pel et al. (Chapter 12) found that "the market developments brought the navigation industry, map digitization industry, telecom providers and car manufacturers together."

The ITS-niche is accompanied by a range of visions such as (a) *improved efficiency of the road system*, because in-car devices enable efficient route planning and real-time travel information (e.g., on traffic conditions such traffic jams); (b) *automated vehicle guidance*, where interactions between on-board computers and infrastructure would remove the need for humans

Findings, Conclusions and Assessments of Sustainability 343

to drive the cars on highways (cars could be electronically linked into trains that would travel efficiently at high speeds); (c) *bringing fun and entertainment into cars or public transport*, by enabling passengers to watch videos, use the (mobile) Internet or communicate in various ways with the home or office. This possibility, which may be supported by commercial interests in the telecommunications or entertainment business, may lead to a redefinition of 'lost time', for example, in traffic jams. This last development may prove to be a more transformative force than sustainability pressures: "At the niche level, it may not be advocates of sustainability who drive the transition in passenger transport, but more pervasive market forces assembled around personal entertainment, ICT and niche developments within cultures of urban entertainment and infotainment." (Sheller, Chapter 9)

Green Propulsion Technology for Cars

Niches of green propulsion technology such as battery electric vehicles (BEVs) and fuel cell vehicles (FCVs) have acquired substantial momentum in the last 5 or 10 years. According to Orsato et al. (Chapter 10) BEV development has passed a critical threshold. Several developments are responsible for this. Firstly, concerns about climate change and Peak Oil led to stronger regulatory pressure (e.g., European CO_2 regulations for cars) and subsidy programs for the purchase of BEVs in countries such as England, Italy, Germany and Japan, while Denmark and Israel provide tax exemptions for BEVs. A second factor is that the success of the Toyota Prius changed automakers' attitudes: "After 2005, there was a shift in perception, with most car manufacturers investing considerable resources in research and development to catch up" (Orsato et al., Chapter 10); the partnership between Renault-Nissan and Better Place (a new entrant with a new battery-swapping business model for commercializing BEVs) further stimulated positive perceptions. A third factor is that new entrants such as Better Place and Chinese battery company BYD are fully committed to BEV. In contrast to the regular manufacturers, they do not have an interest in internal combustion engine (ICE) vehicles. Fourth, electric utilities have been increasingly involved in BEV partnerships (the list of utilities is long and includes the Swiss Energie Ouest Suisse [EOS], Oregon's Portland General Electric, San Diego's Gas and Electric, Ireland's ESB, Tokyo Electric Power Company). For electric utilities BEVs represent a new market and source of revenues. Utilities can possibly also use the batteries in BEVs as a back-up facility to provide peak load assistance. Fifth, car companies increasingly engage in joint ventures with component suppliers, for example, Toyota and Matsushita (a battery supplier), Nissan and NEC, and Honda and GS Yuasa (a specialized battery maker). Orsato et al. (Chapter 10) conclude that: "Altogether, these developments suggest that a trajectory of electrification of cars is underway."

Hydrogen and FCVs have also been on the agenda since Daimler-Chrysler's pioneering work in the mid-1990s. FCVs have since been supported by corporate research and development and large scale government programs.[2] Although the hydrogen hype seems to have weakened, Ehret and Dignum (Chapter 11) report ongoing development and demonstration projects in Germany, which has been identified as the lead-market for Europe. The German *Clean Energy Partnership*, which involved government and private parties, led to three consecutive projects (2002–2007, 2008–2010, 2011–2016) that aim to move FCVs from research to commercialization. The *Federal Ministry of Transport, Building and Urban Affairs* further supported the *National Innovation Programme Hydrogen and Fuel Cell Technology* with €500 million for demonstration projects, while the *Federal Ministry of Economics and Technology* committed €200 million for research and development projects. The combined public funds worth €700 million are to be matched by roughly the same amount contributed by industry and other bodies running the projects. Based on positive experiences so far, Ehret and Dignum (Chapter 11) make positive assessments: "The hydrogen niche is growing and enjoys support from powerful corporate actors. ... Political agents continue to support hydrogen propulsion within the Fuel Strategy." The authors qualify a shift to FCVs as a technological revolution rather than a socio-technical transition: "FCVs can be regarded as radical innovations, as they involve substantial technical changes in vehicle drive-trains, fuel tanks, fuel production and distribution infrastructure. From a user perspective, FCVs bring about non-disruptive change, since they require few changes in user practices." Regarding the relations between BEVs and FCVs, Ehret and Dignum argue for complementarity rather than competition: "Most manufacturers see a role for both FCVs and BEVs. FCVs are regarded as a viable alternative for longer range and larger vehicles and BEVs an attractive option for smaller city cars. ... While it seems likely that the FCV niche will grow further, it is less likely that FCVs will become the one clearly dominating concept. Rather, FCVs have the potential to form one pillar of an electrified and more diversified transport sector, where BEVs might satisfy other transport needs."

While both BEV and FCV niches have gathered momentum, there are also some reasons for caution. Looking at the ups of green propulsion technologies over a longer period, we find successive hype cycles. There was much public attention for BEVs from 1990 to 1998, triggered by General Motors's innovation efforts with the Impact and the California zero-emission mandate. While the hype weakened in the late 1990s, there has been a revival in the last couple of years with renewed efforts (also from electric utilities, entrepreneurs and governments) to develop and commercialize BEVs. From 1995 to 2005, there was much attention for fuel cells and hydrogen, related to serious innovation efforts from Daimler-Benz, visions about a hydrogen economy (Rifkin, 2002) and large scale government programs. But the hype has declined in recent years, because

Findings, Conclusions and Assessments of Sustainability 345

of controversies and uncertainties about technical barriers (e.g., on-board hydrogen storage), costs and infrastructure (e.g., Romm, 2005). From 2000 to 2005 there was hype around biofuels, with initial ideas of substantial sustainability benefits. In 2003 the European Biofuels Directive formulated ambitious goals of 2% biofuels in 2005 (which was not met[3]), 5.75% in 2010 and 10% in 2020. The biofuel hype diminished after 2005, however, because of concerns about sustainability effects and labor practices of biofuel production in developing countries (e.g., leading to deforestation and additional CO_2 emissions). The food riots in 2007, which stimulated the idea that biofuels compete with food production, caused a backlash which further damaged its public reputation. Nevertheless, increases in biofuel production, including in Brazil and the United States, indicate that public hype-cycles do not always affect techno-economic changes in a direct way. Hybrid-electric vehicles (HEVs), commercialized in Japan in 1997 and in the United States and Europe in 2000, have created a new hype, especially when global car sales boomed after 2003. Although other car manufacturers initially derided Toyota for developing more complex and expensive cars, the unexpected success led to something of an innovation race, with other car manufacturers hurrying to develop their own hybrid models. HEVs have somewhat changed the industry consensus and rules of the game, in showing the prospect of commercial demand for 'green' cars that can substantially reduce CO_2 emissions; this also makes it possible and feasible for policymakers to introduce tougher regulations and weakens the rationale for car makers' resistance against these. HEVs also stimulated the revived interests in BEVs, because of possible crossovers between both technologies (e.g., via plug-in hybrids). Figure 16.1 schematically summarizes these hype cycles.

One of the reasons for this successive hype-cycle pattern is that environmental problems create credibility pressures on policymakers. Because policymakers are eager to show ambition, they are looking for positive stories.

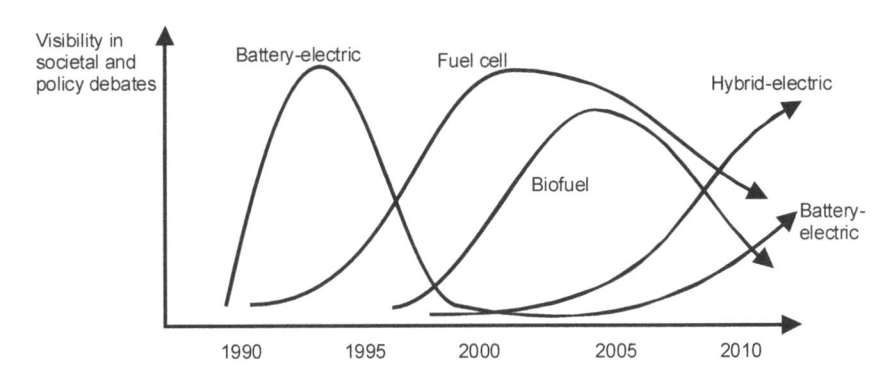

Figure 16.1 Hype-cycles for green car propulsion technologies.

When a technology shows promise (because of a breakthrough or positive soundings from industry), policymakers may jump on the bandwagon and reinforce the expectations. But when setbacks occur a few years later, they often become disappointed and move on to a new promising option. In California, the mandatory sales of BEV under the Zero Emission Vehicle Mandate, has been revised to allow for plug-in HEVs. The Californian policy is exceptional in forcing the commercialization of new vehicles. In Europe, as well as elsewhere, car manufacturers have great influence on the standards to which they will be held. In the past, CO_2 standards have been set so that they could be met by ICE vehicles.

Although niche activities in the green propulsion area have increased substantially in the last 5 years, it remains difficult to predict which technology will win and how long this will take. Expectations about HEV and BEV have increased since 2005, but it is too early to write off hydrogen and fuel cells. Because car companies themselves also do not know which technology will win, most of them use portfolio strategies to prepare themselves for future strategic games, in case 'sustainability' or fuel economy gains further prominence. With regard to speed, it is likely that diffusion of green propulsion technologies will be slow, especially for BEV and FCV, which are both much more expensive than ICEs, and both require new or upgraded recharging infrastructures.[4] Although HEVs do not require a new infrastructure, the fact that they cost a few thousand dollars more than petrol cars of equal size already hampers their diffusion beyond the movie stars or 'green minded' middle classes, which in most countries constitute only a few percent of the population.

Consequently, the diffusion of green cars will greatly depend on taxes or subsidies, tougher CO_2 regulations, learning curves in the technology (which depend on the amount of research and development investments, which in turn depend on the degree of trust and commitment of car manufacturers) and public investments in infrastructure. Enhanced support measures for green cars, in turn, will depend on a cultural sense of urgency about climate change and sustainability or higher fuel prices (perhaps due to Peak Oil), which create more pressure on policymakers. Although there is certainly more talk about climate change than 5 years ago, we are not sure that as yet this is accompanied by a great sense of urgency, because people do not yet experience the (full) effects of climate change or Peak Oil. Diffusion of green cars is also likely to be slow, because it takes time to replace large fleets of cars.

16.1.2. Landscape Developments

The automobility regime is not only facing pressure 'from below' (i.e., from niche-innovations), but also from the broader landscape level in the multi-level perspective. We subsequently discuss stabilizing and destabilizing landscape pressures.

Zijlstra and Avelino (Chapter 8) distinguish the following landscape values that support and promote car dominance: (a) cultural preference for private property rather than collective ownership and use; (b) emphasis on speed and time saving with the car often being the fastest transport mode; (c) individualization, "centered on privatized, unhindered, cocooned movement through public space"; (d) neo-liberalization, with an emphasis on competition, success, freedom to choose: "Here the car is a symbol of modernity and economic progress, it is the embodiment of progress." Because these elements cluster together, the car is stabilized by a broader lifestyle which emerged after the Second World War.

Sheller (Chapter 9) also highlights the stabilizing influence of the landscape level: "Roads and highways dominate the built landscape, and the over-arching culture at the 'landscape' level remains one in which automobility is normalized as freedom and associated with wealth and privilege." The automobility regime is further embedded in social practices related to family, friendship and work relations, and legitimated by "common sense understandings that underlie forms of rationality and valuation, including values such as freedom, security, choice and value itself."

Docherty and Shaw (Chapter 5) note marketization/liberalization and devolution (giving more powers to local and regional jurisdictions) as two macro-trends that changed the broader governance structures of (UK) transport policy. Both trends led to a more hands-off approach by the national government, especially with regard to public transport, but also concerning cars. With regard to marketization, the authors conclude that "the most important legacy is that the complexity of the deregulated transport market makes it even more difficult to implement transport policies aimed at delivering environmental (or social) objectives such as greener mobility." The marketization trend may thus inhibit a major transition toward sustainability. With regard to devolution, the effects are mixed. Transport policy in London has been transformational and multimodal, as indicated by the introduction of the congestion charge, unprecedented investment in bus services and the London Underground and the introduction of the Oyster smart ticketing scheme. But for Scotland and Northern Ireland, Docherty and Shaw found that "substantial programs of inter-urban road building and upgrading" have strengthened the car regime and that politicians did not introduce legislation that was "seen as too politically dangerous and 'anti-motorist'."

Over the past half century, spatial planning has been conducted on the assumption of mass car ownership, with commuting distances generally becoming longer. Thus a mass daily migration in and out of cities became the norm. In addition, commercial and retail developments often take place around major highway intersections on the fringes of conurbations, and in turn this can necessitate circuitous commuting routes for those employed in these places, with public transport alternatives generally being perceived as unattractive. In these circumstances, those making the journeys may consider that they have no alternative to car use, and accept traffic congestion and travel delays as a necessary price to be paid for having a job. The associated decline of local facilities and rural economies also creates a vicious circle, where traveling becomes the necessary price to be paid in order to maintain a desired quality of life.

We suggest some additional landscape developments that generate more mobility and stabilize the existing regime: (a) Globalization increases world-trade and generates more flows of goods and people. (b) The shift toward a network society (Castells, 1996–1998) also generates increasing flows and 'spaces of flows' that facilitate them. (c) Economic growth enables consumers in developing countries (China, India, Brazil) to buy their first car, while households in developed nations move toward multiple cars per household (see Dudley & Chatterjee, Chapter 4).

Destabilizing landscape pressures can come from (a) climate change, (b) Peak Oil and (c) ICT as a pervasive technology and the shift to a digital, pervasive computing society.

Climate Change

Concerns about climate change and CO_2 emissions are already generating some pressure on the automobility regime and its fossil fuel dependence. These concerns form background motivations for government subsidies for innovation programs relating to BEV and FCV (discussed previously) and CO_2 regulations for new cars (discussed next). They also stimulate car makers to explore 'green' propulsion technologies that are not based on fossil fuels. But because climate change is not yet strongly felt and may take another few decades to begin to bite, the pressure is not (yet) very large. Climate change does not (yet) appear to be a major consideration for local or national transport policy (see next). The Kyoto protocol, which had relatively weak CO_2 reduction targets (that many countries did not meet), runs out in 2012, and it is uncertain if it will be replaced with a meaningful new agreement. The European emissions trading scheme for carbon also exerts weak pressures, which in any case do not yet apply to transport. Aviation will be brought under the EU ETS (European Union Emission Trading Scheme) but for road transport this is not foreseen. CO_2 emissions from vehicles are regulated through CO_2 standards, but the standards only apply to new vehicles, and there are no penalties for not meeting the standards. So far the standards can be met with incremental changes. Climate change may also have important cultural impacts. In a world that is increasingly concerned about climate change, gasoline cars could become labeled as climate destroying cars by environmental advocates. They can even become an icon of a climate unfriendly product. A certain part of the population will remain immune to this, but those who are concerned about their own carbon footprint may not want to own a gasoline car or to be seen driving one.

A related complexity is that the translation of climate change into stronger policy pressure assumes that the world community will be able to reach meaningful climate agreements, with CO_2 targets that are subsequently translated into reliable and verifiable sectoral policies. But this assumes a world community that is willing to talk, debate and negotiate on the basis of mutual trust and cooperation. The geo-political landscape of the 21st century could, however, also unfold differently. In a provocative book called *The Return of History and the End of Dreams*, Robert Kagan (2008), for instance, suggests that the 21st century could look like the 19th century in terms of power struggles and coalition politics with new power blocs that involve China, India, Russia or other countries. In that scenario the 21st century could be characterized by distrust, rearmament and struggles for resources (e.g., water, oil, land, biomass), which would make it hard to reach (and implement) global climate change agreements. This is just one example, albeit a somewhat gloomy one, that landscape trends need not all point in the direction of sustainable development (see also Cohen, 2010). It also suggests that future transitions may well be more conflictual than transition research often assumes.

Peak Oil

Peak Oil, which refers to the peaking of maximum rates of *conventional* oil production, will be important for the automobility regime, because more than 95% of all cars run on petrol. There are many uncertainties, however, both with regard to the timing of peak production and the subsequent shape of oil production, which could either follow a bell-shaped curve, with a rapid decline after the peak, or stay on a plateau for a prolonged period of time, because rising prices make difficult-to-get oil more economically feasible, because of enhanced innovation in oil recovery or because of accelerated exploitation of unconventional oil. With regard to the latter, Sperling and Gordon (2009) warn that "the transition to unconventional oil is already under way. Most of the Western oil majors are plowing big money into tar sands, shale, heavy oil, and coal.... The transition will continue and likely accelerate, not just because of economic factors but also ... because oil company culture and business approaches favor unconventional oil over biofuels, hydrogen, and other renewables" (p. 129). Although Peak Oil does not mean that the world is quickly running out of oil, the price of oil and gasoline is likely to rise in coming decades, something that may affect oil consumption and mobility behavior or lead to stronger policy action with regard to renewable alternatives.

With regard to consumer behavior, however, Sperling and Gordon (2009, pp. 158–160) present evidence that indicates that car drivers have become less responsive to moderate fuel price increases. Citing several studies, they report that price elasticity has decreased from –0.30 in the 1970s to –0.04 by the early 2000s. So, if the oil price was to double, consumers would be expected to reduce fuel consumption by only 4%. The reason is that cars are increasingly considered necessities on which consumers depend for commuting, shopping and so forth. Furthermore, people may have limited possibilities to change work, family, leisure and social patterns in the short-run. If prices stay high for longer periods (e.g., several years), people may be able to achieve larger fuel reductions by acquiring more fuel-efficient vehicles, finding a job closer to home or switching to alternative transport modes (if these are available). In the case of high price fluctuations, however, these behavioral changes may be delayed as car drivers may be slow to acknowledge that oil prices will stay high.

ICT and the Shift to a Digital Society

The development and diffusion of ICT is a pervasive landscape trend that facilitates the shift toward an information society with ubiquitous computing. Sheller (Chapter 9) argues that this landscape trend is likely to influence the automobility regime: "The growing integration of ICTs into automobiles and other vehicles is leading to a lacing of technologies of transportation with capacities for conversation, entertainment, information access, navigation, automation, tracking and surveillance. But the transformation goes far beyond vehicles to the planning process itself and to the wider cultural landscape of mobility." ICT may facilitate the rise of cybercars (vehicles enhanced with data processing, information transmission and mobile communications capacities) and digitally enacted environments that enhance the efficiency of the transport system. But, as Sheller argues, ICT may also facilitate broader changes: "Key cultural transitions at each level may come from *outside* the transport arena, instigated by other changes such as cultural shifts in forms of communication, legitimacy, connectivity, sensory perception, embodiment and spatiality." ICT therefore also possesses the potential to bring about significant changes in public perceptions regarding the significance of mobility. First, ICT offers much greater scope for home working. Second, the quality and variety of ICT developments can mean that people spend more of their leisure time at home. Third, the longer term developments of 'virtual reality' may mean that people can have the experience of visiting a particular location, without the inconvenience of actually traveling there. Consequently, while increased mobility became the dominant transport paradigm of the 20th century, it should not be assumed that it will retain this dominance in the 21st century.

While these changes may offer solutions to congestion problems, Sheller (Chapter 9) also raises the possibility of a 'digital panopticon', where centralized planners use travel information for surveillance and control of people. "Research and development of automated transport infrastructures may be promoted by those elements within the state and the scientific community who are also committed to new modes of automated warfare, pervasive sensing and anticipatory surveillance, in short, those for whom security is more important than freedom. This injects into the conversation

Findings, Conclusions and Assessments of Sustainability 351

on transitions a rather more dark vision of the forces that might be driving the wide-scale transition in mobility systems." The broader implication is that transitions in passenger transport may be less driven by sustainability considerations and more driven by either entertainment/information/communication considerations or by "regimes of military technology and state surveillance policy" (Sheller).

16.1.3. Stability and Cracks in Regime

The automobility regime faces pressures from 'below' (from various many niche-innovations) and from 'above' (from landscape pressures). The regime also faces several problems that relate to negative externalities. But the regime is also internally stabilized by lock-in mechanisms that lead to path dependence and incremental changes. This section addresses three issues: (a) regime stability, (b) incremental changes in the regime and (c) cracks in the regime, which may provide windows of opportunity for niche-innovations.

Regime Stability

The automobility regime is not only stabilized by certain landscape trends (discussed previously), but also by lock-in mechanisms on various dimensions of socio-technical regimes: (a) sunk investments in road, urban and spatial infrastructures; (b) sunk investments in plants, skills and people, which create vested interests for industry actors; (c) user patterns and lifestyles, which provide social embeddedness of the car; (d) consumer preferences that benefit cars, for example, convenience, speed, relative cost —many people actually like cars, which on many performance dimensions score better than other transport modes; (e) cultural values and positive discourses around the 'Joy of driving', 'Love affair with the car', 'Can't do without it', 'Expression of freedom and individuality' (Sheller, Chapter 9): (f) vested interests (industry, car lobby, road-building lobby) that resist major change; (g) beliefs from established actors (e.g., transport planners, policymakers, industry actors) that take existing practices for granted and legitimate the status quo.[5] Because of these lock-in mechanisms, the regime is difficult to change in a substantial way. Regime actors instead prefer incremental changes, which we address next.

Incremental Changes in the Regime

The automobility regime is not inert, but 'dynamically stable', which means that regime actors work on incremental changes that stay within the bounds of the existing regime. These incremental changes are guided by the problems and opportunities, *as perceived by regime actors*. Although the automobility regime faces several structural problems (e.g., congestion, parking problems, safety, climate change, health and quality of life issues), our assessment is that *regime actors do not perceive many of these problems as easily resolvable, which undermines the sense of urgency and effective action.* Limiting our analysis to three main groups (industry, policy and consumers), we suggest that regime actors have different priorities of problems and therefore prioritize certain kinds of incremental change.

We suggest that the car industry prioritizes problems as follows (based on Wells, Nieuwenhuis and Orsato, Chapter 6): (a) survival in cut-throat competition, (b) under-utilization of factories and cost pressures, (c) market fragmentation and saturation in developed countries, (d) competition through product innovation and (e) environmental sustainability pressures, mainly from government regulations, and to a lesser degree from consumer demand. The first two problems serve to orient their attention toward survival, profits and cost-savings, for example, through mergers and attempts to improve production efficiency (factory innovation and reorganization). Selling cars in emerging economies is also a top priority. The fourth issue translates into technical innovation oriented toward higher engine performance, more ICT in cars, more safety devices: "Much of the technological effort of the automotive industry is going into areas other than power train or body architecture but in fact into aspects of occupant comfort, safety, entertainment and communications" (Wells et al., Chapter 6). Although environmental issues are only the fifth issue in our assessment of the industry's problem ranking, the car industry does pay attention to the greening agenda. Incumbent car firms are prepared to contemplate "the notion of a partial transition," which is "constructed around green cars, rather than a radical reappraisal of mobility as a whole" (Wells et al., Chapter 6). On the one hand, automakers are working on green niche-innovations such as BEVs and FCVs. On the other hand, the car industry is still committed to the ICE. "The number of patents and new product launches in the period 1990 to 2005 clearly indicates the focus of European automakers on further developments of ICEs, such as the variable valve timing and direct fuel injection systems. On average, around 80% of the patents were awarded to ICE-related technology, against only about 20% for technologies associated with pure battery EVs and hybrid-electric vehicles" (Orsato et al., Chapter 10). Wells et al. (Chapter 6) therefore conclude that: "The pace of change seems to be measured in decades . . . , such that as far as change from within the regime is concerned, there is only the envisaged slow process of greening of cars."

For policy, it is more difficult to come up with a ranking because policy is heterogeneous with lots of internal conflicts. What is being said in public may not reflect what is thought privately. Ambiguity often gets hidden behind firm statements in Parliament and the media. Based on what is being said, and especially what is being done, we suggest the following priorities in the reasons for transport policy (see also Docherty & Shaw, Chapter 5): (a) stimulate the economy and enable demand for mobility by facilitating the smooth flow of goods and people, (b) ensure social equity by facilitating

Findings, Conclusions and Assessments of Sustainability

access to mobility for disadvantaged groups (especially via public transport), (c) addressing negative externalities. In this context , we suggest that policy issues are ranked as follows: (a) congestion, because it has negative social and economic implications; (b) local 'quality of life' problems such as air pollution, parking and space; (c) safety, because transport accidents kill and injure many people; and (d) environmental sustainability such as climate change. This externality ranking explains why most policy programs address congestion, for example, congestion charging, dynamic traffic management and demand management. On this dimension, Goodwin (Chapter 7) argues that a substantial change has occurred in policy orientation, namely a shift from the 'predict-and-provide' paradigm (build roads to facilitated predicted rises in car ownership) to traffic management and demand management, which aim to better *use* existing roads. Because environmental sustainability is less prominent on the national and local policy levels, it receives less attention. Goodwin (Chapter 7) suggests that most policy initiatives "seem not to have been driven by considerations of global environmental imperatives, but by congestion, quality of life, neighborhoods, health, political demonstrations and shortage of money. That is likely to continue: 'Sustainable' transport policies are made politically acceptable and behaviorally effective by aspects other than carbon emissions."

At the European level, however, policymakers have gradually introduced stricter environmental regulations. The recently accepted European regulation 443/2009 stipulates a fleet average target of 130 grams of CO_2 per kilometer for new cars in 2015. This target, which will be reviewed in 2013, could be replaced with a tougher longer term target of 95 grams CO_2 per kilometer in 2020. Through these regulations climate change is beginning to exert pressure on the automobility regime. While the first target can probably still be met with incremental innovations that stay within the current technical regime (Wells et al., Chapter 6), the longer term target will probably require more radical innovations, such as fuel cells, hybrids or BEVs, especially in the case of larger (German) cars (see Ehret & Dignum, Chapter 11). This European policy seems to be more oriented toward the green technology path than the wholesale system transition path.

Consumers evaluate mobility options on various dimensions. Table 16.2 provides a ranking of criteria that consumers apply when they buy a new car. The ranking is a general ranking which will not hold true for all consumers, and there may be country differences. It should not be taken at face value, but nevertheless comes from an authoritative source (*King Review* in the United Kingdom; King, 2007). It shows that costs, reliability, comfort, safety and performance (in terms of speed, journey time, distance) are more important considerations than environmental problems. Consumers also see congestion as an important societal problem. Another ongoing trend is the fragmentation of car markets. For the UK market, Orsato and Wells (2007) report that the total number of car models and body styles increased from 1,303 in 1994 to 3,155 in 2005. The standard sedan family car has been replaced by a range of cars (SUVs, space wagons, mini, sports cars, trucks) that fit different life styles.

Although the automobility regime faces several problems, it still seems fairly stable. The reason for this assessment is not that the problems are small, but that important regime actors are not (yet) fully committed to acknowledging these problems or to placing them on agendas with a high sense of urgency. Among policymakers the need for radical change in the transport sector is not accepted. They are ambiguous about it, as is the transport profession. There is a consensus on the need for an integrated transport policy, but this is something different from an agreement on the need for transformation. Furthermore, the car industry and automobile clubs are powerful advocates in defending the right to drive and fighting off taxes and other measures to limit car mobility, using the media and systems of governance of which they are part. Their power lies in part in expressing a popular view. There is no real debate about the need for transformative change. No powerful societal group is calling for it. Those who do are marginalized (see Chapter 8). Although some *functional problems* that directly affect transport systems (e.g., congestion, parking) are recognized and lead to more attention for ICT, dynamic traffic management and congestion charging, *collective good problems and negative externalities* (such as pedestrian safety, spatial problems and climate change) receive much less attention or are being downplayed (see Chapter 8). In the absence of attractive solutions, they are viewed as less solvable. There is an element of the negative effects (trade-offs) being accepted by society. In particular, environmental sustainability scores relatively low in the problem rankings of the regime actors discussed previously (see also Cohen, 2010, on societal aspirations that rival with sustainability in

Table 16.2 Factors That Are Important to Consumers in Deciding Which Car to Buy

Most Important	Medium Importance	Least Importance
Vehicle price	Performance	Depreciation
Size	Power	Sales package
Reliability	Image	Personal experience
Comfort	Brand name	Dealership
Safety	Insurance costs	Insurance costs
Running costs	Engine size	Engine size
Fuel consumption	Equipment	Equipment
Appearance		Recommendation
		Road tax
		Environment
		Vehicle emissions
		Alternative fuel

Note: From King (2007, p. 60).

Findings, Conclusions and Assessments of Sustainability

the transport domain). Pressure for change on this dimension is therefore not great and currently not oriented toward large-scale systemic change. Furthermore, both consumers and motor interests (car industry, oil companies, motoring organization) exert powerful stabilizing influences (see Chapters 4 and 6).

Cracks in the Regime

Although the automobility regime still has substantial stability, several cracks are appearing. These cracks suggest that the regime may not be as strong as it used to be, although it is still dominant compared to other transport modes.

One important crack is the capacity of physical infrastructure, acting as a constraint on car use, especially in the urban context (Goodwin, Chapter 7). Physical constraints and considerations of urban quality lead city authorities to implement car restraining measures such as one-way streets, parking restrictions and tariffs, traffic calming schemes and the creation of traffic-free pedestrianized centers, which challenge the ubiquity of cars in certain places (see Chapters 4, 7, 8 and 9). Because some cities also play an active role in stimulating alternatives such as bus lanes, bicycles and road pricing, they can be seen as a new actor that challenges some of the established regime elements.

A related second crack relates to perceptions of the car. The introduction of pedestrianized and car-free areas may stimulate "a major re-evaluation of the role of the car in urban areas. . . . Over time these largely local and incremental changes have the potential to shift significantly perceptions in society about the status of ubiquitous car use" (Dudley & Chatterjee, Chapter 4). Additionally, there are some indications of inter-generational shifts in perception with regard to the importance given by the individual to car ownership. Where older generations still consider a car to be the one consumer good they cannot live without, polls in Japan, Germany and other developed markets suggest that younger generation consumers value mobile phones more than they do cars. This would mean that these younger people tend to perceive the car in terms of being 'a means to get from A to B,' rather than having the connotations of status and image familiar in earlier times.

A third crack is that the growth of car mobility (in terms of passenger miles) seems to have come to a halt in developed countries such as the United Kingdom, where it actually declined somewhat (see Chapter 4).[6] While there may be various reasons for this trend (e.g., changes in work patterns, demographic changes, shifts in location of home and work), it is striking that this trend occurred in tandem with some increased use of bicycles, bus and rail transport. This would mean we are perhaps entering a period in which attention may shift from the expansion of the car-based system to a system transportation to deal with various problems and externalities.

A fourth crack is some weakening in the commitment of policymakers to the automobility regime. Thus the commitment to expansion, as indicated by the 'predict-and-provide paradigm', has become weaker. Policymakers have become more critical about the building of more roads and are adopting new guiding principles such as demand management, traffic management and 'sustainable mobility' (Dudley & Chatterjee, Chapter 4; Goodwin, Chapter 7). At the same time, the commitment from other policymakers (with more economic and industrial remits) remains strong, as indicated by continued investments in road expansion and the willingness of many governments to bail out the car industry during the financial crisis (with direct subsidies and demand stimulus policies such as cash for clunkers). Wells et al. (Chapter 6) remark that these rescue efforts "were conspicuous for their lack of connection with any sustainability agenda, or even the amelioration of carbon emissions, apart from some very weak measures. Put simply, in the choice between stimulating (environmentally malign) economic recovery and the longer term option of designing a new industry to meet the pressures for sustainability, it is evident that short-termism triumphed." While these rescue efforts suggest that governments take great interest in healthy car industries, the political capital of the latter may have weakened so that interventions may play out differently in future.

A fifth crack is that regime actors are aware of landscape pressures such as climate change and Peak Oil. So, there is some recognition of potential environmental limits to car growth (although this does not stop them from trying to sell as many cars as possible in developing economies). Car companies are also not impervious to ideas such as sustainable mobility, but they understand this primarily as an issue of greener cars.

16.1.4. Future Sustainable Transport Transitions: Three Scenarios

The previous three sections described various developments and trends at the niche, regime and landscape levels of the multi-level perspective. At times, we added our personal assessments as to the momentum or prominence of these developments. Future transitions to sustainable transport are open in the sense that they depend on interactions between the three levels and the underlying choices, strategies, commitments and coalitions of actors and social groups. In this section, we present three possible scenarios, based on different multi-level interactions and actor strategies. The scenarios also vary in terms of which 'cracks in the regime' become bigger and which niche-innovations take advantage of those windows of opportunity. The three scenarios are as follows: (a) greening of cars, intelligent transportation systems and smart grids; (b) intermodal urban transport and redefining the role of the car; (c) continuation of petrol cars, unconventional oil and digital panopticon. In terms of the theoretical transition pathways, discussed in Chapter 3, these scenarios correlate with (a) the reconfiguration path, in which regime actors survive by adopting radical innovations that reconfigure the regime's basic architecture; (b) the de-alignment and re-alignment path, in which the existing regime falls apart because of internal tensions and external pressures, followed by the build-up of a radically new regime; and (c) the incremental transformation path, based on incremental reorientations by regime actors. Although all three scenarios are in principle possible, our analysis suggests that most attention and resources are currently oriented toward the first scenario, which therefore at present seems more likely. The second scenario could be realized if present undercurrents become more dominant. This scenario will be slow to unfold (i.e., after 2030) because it first requires the breakdown of the existing regime and then entails changes in mobility routines and a reconfiguration of urban physical urban structures, which will take a long time to be realized. The third scenario is the most gloomy and is characterized by resource scarcities, power struggles and increasing prominence of security and surveillance. Because the first scenario seems at present most likely, its description is longer than the other two.

Scenario 1: Greening of Cars, Intelligent Transportation Systems, and Smart Grids

This scenario follows from our assessment of the contemporary situation that (a) the automobility regime is still dominant and stable, although maybe less so than 15 years ago; (b) that there are only (moderate) cracks in the regime; and (c) that most promising niches have limited internal momentum; this momentum is larger, however, for green propulsion technology and ICT/ITS.

For the next 20 years this scenario assumes continuing dominance of the car-based transport system, although maybe somewhat less in relative terms; some further growth in public transport modes (train, bus, light rail). Intermodal transport, cultural and socio-spatial changes, and demand management practices will remain small niches (but bigger than they are now).

In this scenario, two cracks in the regime will gradually become bigger: (a) Pressures from Peak Oil and climate change lead to increasing public concerns; (b) governments detracting from the automobility regime, in the sense that they are less inclined to protect the status quo and become more willing to introduce green policies (both CO_2 regulations and innovation programs) aimed at changing the regime. Car restraining policies are not introduced, because governments do not want to antagonize consumers and motor interests. The niche of green propulsion technologies exploits opportunities afforded by these cracks. It is difficult to predict which green propulsion technology will become dominant. Although current expectations of BEVs are high, setbacks may occur and FCVs (or biofuels) may regain the lead in expectation dynamics. The rise of BEVs or FCVs will have knock-on effects on the electricity system and lead to innovation cascades that create new linkages between transport and electricity systems.

Firstly, the large-scale use of BEVs provides new markets for the electricity system and will require a major upgrade of electricity systems and substantial investments from utilities. If hydrogen were to be produced from (renewable) electricity, the same argument holds for the widespread diffusion of FCVs, although there are specific challenges on where hydrogen should be generated (e.g., centrally, at home or via on-board reformers). Secondly, BEV and FCV could also change the functioning of the electricity grid. Consumers could use FCVs to generate power at home and deliver this to the grid, while BEVs can act as storage devices and thus provide partial back-up from which grids can draw power in the case of emergencies (Orsato et al., Chapter 10). This reconfiguration of transport and electricity systems would certainly constitute a major system innovation that requires a shift to smart grids that can facilitate two-way streams in the electricity system. Thirdly, a transition toward BEVs could stimulate further reconfiguration of the car, for example, more electronics inside the car or changes in materials (such as composites or fibers) to reduce weight. A shift away from steel-based cars could also trigger changes in the production of cars, where large and capital intensive metal presses form an important anchor of mass-production (Wells et al., Chapter 6). If these new materials facilitate easier creation of car body shapes, they may enable a shift toward smaller factories or even micro-factory retailing (Nieuwenhuis & Wells, 2003). Fourthly, a transition toward BEVs may be accompanied by a change in business models. The start-up Better Place, for instance, is currently pioneering a model for electric mobility in which car owners lease batteries instead of owning them (Orsato et al., Chapter 10). In addition to recharging batteries, they can also be replaced at changing stations.

The ICT/ITS niche will expand. Momentum is already strong. It is pushed by powerful companies and embraced by the transport community. ICT-based innovation such as dynamic traffic management, information provision and dynamic route planning are being rolled out and ICT-based forms of road charging are coming. Such innovations are primarily aimed at improving the efficiency of the existing system by enhancing road capacity, smoothing and redirecting traffic flows and improving punctuality. But a shift toward cyber-mobility with continuous interactions between on-board devices and smart infrastructures would also "transform the very way in which drivers sense and make sense of the world" (Sheller, Chapter 9). An ITS-based promise that extends further into the future is automated vehicle guidance, where electronic coupling and speed regulation enable individual cars to travel as trains at short distances from each other. Although this kind of promise has been around for many decades, and thus invites some skepticism, it would further transform car-driving and free up time for other activities.

While these ICT based niche-innovations promise to reduce congestion problems, it remains to be seen if they are structural solutions or temporary measures that delay a total gridlock for another 10 years. Nevertheless, even

Findings, Conclusions and Assessments of Sustainability 359

if the current measures provide temporary solutions, the extra time may be important to develop alternatives and prepare the ground for wider change. One wider change, which is also facilitated by ICT, is congestion charging and road pricing, which would introduce market mechanisms to deal with the road scarcity problem. Road pricing would transform roads from public goods into payable services. Public acceptance and political will are the greatest uncertainties for this development. Several British cities have rejected congestion charging in referendums, and Dutch plans to introduce road pricing on highways keep being postponed, delayed and debated.

If congestion problems appear to be unsolvable, ICT may change the perception of lost time. The increasing penetration of ICT in the car will enhance the opportunities for on-board entertainment, communication, fun and information, which may lead to a reframing of congestion: Instead of trying to solve congestion, attention may shift toward better use of the time spent in congestion. ICT may thus transform the car into an extension of the office or the home, where people can work or relax.

Scenario 2: Intermodal Urban Transport and Redefining the Role of the Car

This scenario is not a short-term scenario. It is more likely to unfold after 2030, when four cracks in the regime become so large that the regime begins to de-align: (a) Peak Oil and climate change begin to bite, leading to high public concerns; (b) congestion, space and quality of life problems lead to a much stronger role for urban actors; (c) discourses about both problems lead to changes in public perceptions of the role of cars, and people begin to question the normality of cars; (d) governments further defect from automobility, because of increasing problems and civil society pressures—this is accompanied by tougher CO_2 regulations and car-restraining policies (higher parking fees, banning cars from certain urban spaces, higher taxes on fuel and car purchase).

A redefinition of the perception of the car will be strongest in the urban context, where car traffic is most problematic from a *user* point of view, and relatively good alternatives are available. A crucial role is played by urban authorities, supported by national authorities and a new generation of transport engineers educated in mobility management. Multimodal transport is facilitated by large-scale investments in urban rail systems, cycling infrastructures and transfer facilities on the edge of cities (where people can park cars and transfer into public transport modes). Some cities even begin to work toward compact cities to reverse the trend of urban sprawl. Companies stimulate employees to share rides to work and locate new offices near public transport hubs and interchanges. Mobility companies spring up that offer personalized travel services and leisure travel plans. Car sharing clubs become more popular because people begin to question the need for privately owned cars and because these cars are exempted from car-restraining policies.

Cars still exist but in a larger variety of shapes and forms, for example, small light-weight electric vehicles for urban use and larger FCVs for longer distances. Other types of vehicles and services emerge as public transport companies become much more innovative (spurred on by policymakers): on-demand bus systems (which are dispatched to bus stops specified by group users via information service terminals or telephone), train-tram systems (which fill the gap between heavy and light rail systems), superbuses (long, low, sleek and very fast (250 km/hr) traveling buses) and people movers (automated vehicles that transport people along single stop fixed routes). Some new vehicles and services blur the boundaries between existing 'transport modes', for example, amphibus (a bus-like vehicle that can also navigate water to connect islands or peninsulas) and taxi sharing (when people with more or less similar destinations share taxis). These innovations increase the variety in terms of green technologies, transport modes, user behavior, business models and conceptual categories. Some cities will move strongly in this direction, others will not, depending on local transport circumstances, local politics and culture. The speed and nature of alternative mobility developments will differ between cities, regions and nations.

Scenario 3: Continuation of Petrol Cars, Unconventional Oil and Digital Panopticon

Although tensions and problems in the automobility regime increase in this scenario, there is limited willingness to make substantial changes: (a) Consumers are reluctant to abandon cars, partly because their use has become routine, partly because people prefer cars over other transport modes; (b) oil companies and car companies strongly resist car restraining and tough green policies; and (c) policymakers stick to weak policies because they don't want to go against the public will and powerful economic interests.

Peak Oil is the most prominent problem (crack in the regime). The depletion of conventional oil drives up fuel prices and leads to geo-political struggles. Climate change gets worse, but the international community is unable to reach agreements because of geo-political tensions. This scenario assumes that by 2030 the world will be a harder place, characterized by geo-political tensions between regional blocks (e.g., China/India, United States/Europe, Russia/Iran/Venezuela), struggles between industry and governments (with car industries threatening to move factories if national governments introduce strong policies) and tensions between social classes if rising oil prices threaten social equity in transport.

Policymakers will not push strongly for alternative vehicles but rely on car industry efforts in improving fuel economy. This also reduces CO_2 emissions, but that is not the main driver. There will also be technical change in alternative cars (BEV, HEV and FCV), but these do not take off because consumers are unwilling to pay a few thousand dollars more for green cars. Scale economies therefore do not set in, and the price of these vehicles

Findings, Conclusions and Assessments of Sustainability

remains high. Car companies are not committed to these cars, because they would cannibalize their sunk investments in petrol cars. As the oil price goes up, oil companies diversify into unconventional oil (tar sands, shale, heavy oil), which pushes up CO_2 emissions. Some countries even stimulate coal-to-liquid technologies to escape the global struggle for the remaining oil reserves. Others will opt for biodiesel.

Intelligent transportation systems are not only used to reduce congestion and improve the efficiency of roads and transport systems but also for purposes of surveillance. GPS data and information from on-board devices are collected centrally by national security agencies to trace and analyze the movement of people and goods. The expanded use of surveillance technology is also a possibility in the other scenarios. The different scenarios are not mutually exclusive.

These scenarios are all possible, depending on the kinds of multi-level interactions and actor strategies. But most actors presently seem to work in the direction of the first scenario (as indicated by resources, beliefs, visions, commitments). Because of 'self-fulfilling prophecy' elements, the first scenario presently looks like the most likely scenario, although it can be blown off course by setbacks, landscape developments or unexpected events. The second scenario is also possible but would require several changes in actor attitudes and strategies: a willingness of national governments to introduce car-restraining policies, a much stronger role of local and city governments, much stronger innovation strategies by public transport actors and a willingness of consumers to change mobility routines and use cars less. At present, there are few indications of these kinds of changes. We added the third scenario to indicate that sustainability transitions are not inevitable and that multi-level interactions can also lead into less sustainable directions of greater inequality, social unrest and international conflict.

16.2. POLICY LESSONS

As the previous scenarios indicate, policy is an important dimension in transitions. This is especially so for transitions toward sustainability, where the direction of change is related to societal goals and addressing negative externalities. Policy can facilitate the internalization of these externalities, shape economic frame conditions and modulate ongoing processes at niche and regime levels. We first assess how strong contemporary transport policies are oriented toward sustainability transitions and then make suggestions for improvement.

16.2.1. General Assessment

In Chapter 3 we suggested that transition policy should follow a two-pronged strategy: (a) Stimulate the emergence and diffusion of niche-innovations and (b) enhance pressure on regimes through economic instruments (e.g., taxes, carbon emission trading, road pricing) and regulation (e.g., environmental legislation). If we confront this prescriptive policy strategy with the empirical findings from various book chapters, we conclude that transport policies give moderate attention to the first strategy (supporting innovations) and little attention to the second strategy (restraining and reducing motor car traffic). At the local level, some cities engage in policies that restrain cars and encourage alternatives to cars, often for reasons of urban regeneration and quality of life (see Chapters 4, 7, 8 and 9). Examples are one-way streets, parking restrictions and tariffs, traffic calming measures, bus lanes, special infrastructure for bicycles and the use of road pricing. Several cities such as Ghent in Belgium have created large pedestrian zones in city centers, where cars are banned. Although this niche phenomenon is spreading among European (and some American) cities, it remains a somewhat isolated instance of creating pressure on the car regime.

At the national level, transport planners and policymakers seem to pin their hopes and efforts more on technical fixes than on facilitating broader transformative change. In Germany, for instance, the government is involved in programs with hydrogen and FCVs (Chapter 11). In the Netherlands, Denmark, Israel, California and the United Kingdom, governments have engaged in promoting electric vehicles (Chapter 10). But even though governments stimulate 'green' innovation, they do not want to drive this agenda forcefully. Wells et al. (Chapter 6) comment that the car industry bailout programs "were conspicuous for their lack of connection with any sustainability agenda," despite talk about a New Green Deal. In their analysis of UK transport policy Docherty and Shaw (Chapter 5) conclude that "the extent to which governments are genuinely prepared to tackle the (social and environmental) externalities of the car regime is highly uncertain." They also assess that "although marketization and devolution have generated plenty of innovation in transport policy, this has not always made the strategic transition to a more sustainable mobility system in the future more likely." Instead of being strongly oriented toward green innovation, it thus seems that transport innovation policy is rather piecemeal, both in financial terms and coordination. Although this is more true for certain countries (the United States and United Kingdom in particular), we agree with Sperling and Gordon (2009) who conclude the following: "If alternatives are to take root, we need . . . vastly expanded science and technology research, development, demonstration, and investment, along with consistent, powerful government policies that encourage these investments" (p. 110). On the other hand, cleaner vehicles and faster trains are not the ultimate solution. Part of the solution lies in people using non-motorized forms of transport. For this, no special innovation is needed but things that already exist: bicycles and bicycle lanes for example. The integration of public transport, shared taxis, public bikes and organized car sharing again is not about technology, it is about organization.

Findings, Conclusions and Assessments of Sustainability 363

Based on this analysis, we conclude that current transport policies are only weakly oriented toward a sustainability transition. One explanation is that transport policies are often driven by other considerations (see Docherty & Shaw, Chapter 5), such as economic reasons (stimulate industry and trade by creating efficient transport hubs and infrastructures), social reasons (facilitate people's mobility demands in relation to leisure, tourism, family visits, etc.) and financial/fiscal reasons (raise taxes). Sustainability, and addressing negative externalities more generally, forms only one motivation for transport policy and probably not the most important one. In so far as sustainability is recognized, it is probably fair to say that the need for large-scale transitions has not yet gained much traction in transport departments.

Another explanation is that policymakers are constrained by at least two double binds and dependencies. First, policymakers are constrained by the wider public, who form the electorate they are supposed to represent. Policymakers seem to follow rather than lead public opinion. If they introduce tough policies for which the public is not ready, they risk electoral defeat, lack of legitimacy or public protest and rejection. Chapter 4, for instance, showed how the UK fuel protests in 2000 led to an institutionalized fear not to go against what the public wants and to postponement of ambitious transport policies formulated in the 1990s. In general, governments tend to accept and facilitate car mobility because the car is culturally accepted and embedded in people's lives. In the perception of most car drivers there is no valid alternative to the car, and, in the absence of good alternatives, it is very difficult to pursue car restraining policies.

The second double bind is that policymakers are dependent on industry actors for jobs, taxes, economic growth and new technologies. This dependence and willingness of government to maintain good relations with industry makes them receptive to industry wishes, even when the public good is at issue. With regard to emissions regulations, Wells et al. (Chapter 6) show that policymakers seem to follow what is technologically feasible, relying on industry assessments about this. In other words, the car industry has substantial power to influence the strictness of regulations. This was also visible in Chapter 10 (Orsato et al.), which showed that the state of California implemented technology-forcing policies in the 1990s (zero emission mandate) but had to withdraw and compromise after fierce political contestation from the car industry.

These double binds and dependencies explain why it is difficult for policymakers to introduce tough regulations or car-restraining policies. Policymakers have no privileged position outside the system (a 'cockpit') from which they can pull levers and change the transport system and people's mobility behavior at will. Instead, policymakers are part of the system and are constrained by their dependence on other actors. Even if ministers are committed to principles of sustainability, they are often reluctant to alienate powerful regime actors such as motor interests. In the 1990s, for example, the British government tried to encourage a modal shift and promulgated predict and prevent policies but withdrew after fuel protests and resistance, as Chapters 4, 5 and 7 show. Additionally, the influence of national governments has diminished because macro-trends such as marketization/liberalization and devolution led to a more hands-off approach (Docherty & Shaw, Chapter 5). Devolution of power has also created coordination problems in British multi-level governance systems. Dudley and Chatterjee (Chapter 4) found that national policymakers tended to set ambitious targets (e.g., for roads and congestion reduction) but left the implementation of concrete policies to local authorities, without always providing them with sufficient resources and responsibilities. This design flaw created tensions between governance levels and hindered the achievement of targets.

Although car-restraining policies remain difficult, the last decade did see transport planners beginning to pay more attention to demand side and behavioral changes (see Chapters 2 and 7). But the focus on behavior tends to be framed in terms of psychology and behavioral economics, which adheres to the ABC-program (attitude, behavior, choice; Shove, 2010). Policies aimed at behavior change therefore either emphasize information provision and advertising campaigns (to change attitudes) or taxes, subsidies and other market-based incentives (to change choices based on cost-benefit calculation). While these instruments are important, policymakers pay less attention to important dynamics such as social learning processes, the co-evolution of technology and behavior, political struggles between civil society groups or cultural dynamics that change preferences

16.2.2. Suggestions for Better Sustainable Transitions Policy

Applying our prescriptive policy strategy from Chapter 3, we suggest that transport policy to accelerate a transition to sustainable mobility should be improved in two ways. Firstly, policies need to substantially increase the pressure on the automobility regime, for example, through higher taxes for fossil fuels, stricter CO_2 regulations, higher parking rates or differentiated road pricing, depending upon forms of nuisance, to stimulate cleaner cars and a more efficient use of the road infrastructure but also to create space for alternative mobility. Secondly, innovation policies for the nurturing and diffusion of niche-innovations should be strengthened. There is a strategic choice with regard to timing of both strategies. In the absence of feasible alternatives, regime pressure policies are likely to lead to resistance from industry and the wider public. We therefore suggest that the focus should initially be on the second strategy, that is, developing alternatives and stimulating (broad) learning processes. When alternatives (either new technologies or other transport modes) are more developed and expanded, tougher regime pressure policies can be introduced. The presence of feasible alternatives can then be used to weaken resistance from the industry, by delegitimizing the claim that 'radical change is impossible' and by creating first-movers that may break a closed industry front. Social learning

Findings, Conclusions and Assessments of Sustainability 365

processes with alternatives and public debates may also facilitate changes in mobility preferences and create more social acceptance of alternatives. An example on a small scale is the London Congestion Charge, the introduction of which was preceded by an expansion and upgrade of public transport modes in London (see Chapter 4). Alternatives were thus introduced (and upgraded) before the implementation of strict policy measures.

The bidirectional causality may create somewhat of a chicken and egg problem. With regard to intermodal traffic, for instance, Parkhurst et al. (Chapter 15) conclude that this will remain a small niche unless car-restraining policies are introduced. While this may be true, it is probably too difficult to introduce car-restraining policies when alternative niches have not yet sufficiently developed. This dilemma can be overcome through a gradual co-evolution process, where gradual growth of alternative mobility niches makes it easier to introduce stricter policies. Transition policy should thus be seen as a process of modulating ongoing dynamics through niche support policies and automobile regime control (Kemp, Rotmans & Loorbach, 2007). Via a process of guided evolution, regimes of alternative mobility can be created. A special aspect of mobility is that it is place-bounded and that transport decisions of authorities and travelers defy control from above. Local actors will make the decisions they feel suit them best. The transition to alternative mobility is thus a transition which has to be managed at the local level. Similarly, the transition to modern sanitation (Geels & Kemp, 2007) had to be managed at the local level, but here there was a positive attractor for users (in the form of toilets). Sustainable mobility lacks a big positive attractor as it is largely about constraints and creating conditions for options (bicycle use, walking, public transport) which are not culturally aspiring for the large majority of the people.

With regard to niche-innovation policy, governments tend toward stimulating new technologies, for example, via research and development and demonstration projects. While this strategy is important and should be scaled up, there is a danger of a narrow technical focus. In an evaluation of various sustainable transport demonstration projects, Hoogma, Kemp, Schot and Truffer (2002, p. 192) found recurring problems, such as too much focus on technical learning and their self-contained nature. The projects did not learn about alternative systems of mobility and were mostly discontinued in the absence of immediate success. We therefore strongly argue for broader *socio*-technical policies where niche innovation policies also facilitate social learning (e.g., about behavior and mobility patterns), second-order learning (which enables the questioning of established assumptions about mobility), inclusion of civil society groups, user organizations, different user groups and so on. This is important, first, because it enables exploration of the possibility and desirability of broader systemic transformation in transport systems, which cannot be forced by policymakers but requires social learning and broader experimentation. Secondly, the inclusion of civil society and citizens is important to build support for tougher policies later on. Giddens (2009) in this respect not only suggests that "responding to climate change will prompt and require innovation in government itself and in the relation between the state, markets and civil society" (p. 94) but also that "whatever can be done through the state will in turn depend upon generating widespread political support from citizens within the context of democratic rights and freedoms" (p. 91). The recognition of dependencies of governments on other actors should thus be incorporated in niche-innovation policies.

Beyond this general transition strategy, we want to highlight eight more specific lessons that follow from various book chapters. We do not propose specific policy instruments but aim to show how the transition analysis in this book can generate suggestions for policy strategies and tactics.

The first lesson is the importance of framing. Politics is not only about policy instruments but also about defining problems and framing solutions, that is, giving them meaning in ways that make them more (or less) acceptable. Goodwin (Chapter 7), for example, showed how changes in urban traffic planning started out as a congestion argument but gradually evolved in a health and quality of life argument. Another example is that the mayor of London (Livingstone) increased the public acceptance of the London Congestion Charge by reframing it from a moneymaking tool to a congestion reduction tool (Dudley & Chatterjee, Chapter 4). A third example of the importance of framing is that the fuel protesters in 2000 were more influential in shaping the discourse on high fuel prices than politicians, with the effect that public opinion sided with the protesters (Dudley & Chatterjee, Chapter 4).

The second lesson is that smart implementation strategies should be used to circumvent barriers. Goodwin (Chapter 7), for example, suggests that road pricing may be introduced via a fit-stretch pattern. The first phase ('fit') would consist of incorporation of ICT equipment in cars, paid for by drivers for the purpose of improved traffic information. When many cars are equipped with such equipment, the government could move to a second phase ('stretch') by using this equipment for road pricing purposes. This may require some (possibly subsidized) upgrading or uploading of new software programs, which would allow the equipment to communicate with infrastructure pricing devices. But such a fit-stretch pattern would cost the government much less than a complete pricing system funded from scratch.

The third lesson is that sustainability policies should look for lateral alignments that allow them to piggyback on other motivations such as congestion, quality of life, neighborhoods and health (Goodwin, Chapter 7; Sheller, Chapter 9).

The fourth lesson is that spatial planning has an important role to play in sustainable mobility. In a world designed for car mobility, in which authorities try to accommodate a growing volume of car traffic, car mobility will continue to remain important, even when good alternatives are available. Perhaps the radical niche innovations of shared space, slow cities and transition towns described in Chapter 8 by Zijlstra and Avelino will challenge assumptions

Findings, Conclusions and Assessments of Sustainability 367

on the part of town planners, policymakers and transport experts that car mobility is what people want. On the other hand, some of the radical spatial niche innovations described in Chapter 8 have been around for many years, without much demonstrable influence on the established regime.

The fifth lesson is that policy should be mindful about unintended consequences. Offering people better alternatives to car use may encourage a few drivers to make less car trips, but it may stimulate more travel, a large part of which can be considered low-value travel (of too low value for people to have undertaken before). The creation of Park and Ride facilities at city borders without reducing parking places in towns has had the effect of stimulating car travel (Chapter 15). Free buses in Hasselt in Belgium resulted in an increase in passenger travel (cyclists started taking the bus, and bus users started to make far more trips). Well-intended policies may thus have undesirable system-wide effects.

A sixth lesson is that the notion of sustainable mobility seems to be much weaker than the notion of sustainable energy, which is a clear policy orientation point for energy producers and consumers alike. One reason for the relatively toothless character of the notion of sustainable mobility is that the term is rather diffuse, with no agreement as to its meaning. A second reason, related to the first, is that it is seldom or only weakly used as a yardstick for policy evaluation. Sustainable mobility policy seems to be mainly about 'positives' (e.g., greener cars, better public transport, reduced congestions, bike lanes), while 'negatives' (undermining of urban quality, landscape degradation, etc.) are backgrounded as something we cannot do much about or something we have to accept as negative externalities. The reduction of car mobility is usually not seen as part of sustainable mobility policy. This relates not only to social resistance but also to economic considerations. In contrast to energy, where energy efficiency represents a multi-billion dollar market, there are relatively few jobs in reduced personal mobility. There are also much weaker economic interests behind reduced mobility. This contrasts with the production of green cars, which may generate new jobs and create new markets.

The seventh lesson is that transition policy should be seen as a process which requires (some) leadership, persistence and ability to deal with unexpected events. With regard to the ongoing transition in transport planning, Goodwin (Chapter 7) concluded that "it is likely that the transition itself would require 5 to 10 (and perhaps 20) years of persistent effort, creative imagination, political courage and consistency. During that period there are U-turns, shifts of power, faltering will and external events which blow any plan off course."

A final lesson concerns the need for enhanced multi-level and horizontal co-ordination. In horizontal terms, a common weakness is the failure of a public transport regime to 'punch its weight' through working together for a common aim. Instead, separate public transport modes often consider themselves to be chiefly in competition with each other, rather than with the car.

The overall effect, politically and economically, is to allow the more united car regime to maintain its dominance. In governance terms, transport policy is also often framed in terms of its economic effects, with environmental criteria placed in a subordinate role. This position is reinforced by the sensitivity of governments to what they perceive to be the political importance of the motor regime and the need to keep the voting motorist happy. Even more crucial, however, is the need for greater co-ordination between different levels of governance systems. For example, in the United Kingdom ambitious targets were set for reducing congestion, but implementation was left in the hands of the local authorities, without them being given the necessary economic and administrative resources to ensure success. This may have the advantage for central government of offloading responsibility for difficult and politically sensitive problems but requires on their part considerable practical assistance, together with skillful steering of multi-level governance systems.

16.3. CONCLUSIONS ABOUT SUSTAINABLE TRANSPORT THAT CHALLENGE AND EXTEND TRANSITIONS THEORY

Sustainability transitions in transport have some specificities that indicate opportunities for theoretical extensions of the multi-level perspective. We discuss four of these, drawing on empirical findings from the book chapters.

16.3.1. Interactions Between Multiple Regimes

Transitions theory often focuses on one niche (sometimes a few niches) struggling against a single dominant regime. This is clearly visible, for instance, in the standardized multi-level figure, discussed in Chapter 3. Sustainability transitions in transport challenge that view because they may involve interactions between *multiple* regimes. As indicated previously, the widespread diffusion of BEVs and FCVs will link the transport system to the electricity system, which may require upgrades in the latter and could possibly lead to further changes in the functioning of the grid. A shift toward biofuels would similarly entail new linkages between the agriculture regime, the oil regime (assuming that oil companies will play a role in biofuels) and the transport regime. Also a shift toward multimodal transport, which at present is a relatively small-scale phenomenon, would by definition require interactions between the dominant car regime and smaller subaltern regimes (train, bus, taxi). In addition, as Sheller (Chapter 9) suggested, the shift toward ITS may be accompanied by new interactions between transport, computing, telecommunications, surveillance and military regimes. For transitions theory these findings suggest that scholars should broaden their attention to include interactions between multiple regimes (see Geels, 2007a; Raven, 2007; Raven & Verbong, 2007; Konrad, Truffer & Voss, 2008, for initial explorations of this topic).

16.3.2. Continuing Importance of Regime Actors for Sustainability Transitions

Some transition scholars focus primarily on the S-shaped diffusion curve of niche-innovations and the role of pioneers, frontrunners, new entrants or start-ups. The book chapters indicate, however, that regime actors will be crucial in a future sustainability transition. The reason is that public transport companies and car manufacturers have many 'complementary assets' (Rothaermel, 2001; Teece, 1986), such as specialized manufacturing capabilities, distribution channels and service networks. These regime-specific assets give incumbent regime actors powerful positions vis-à-vis niche-actors, who either face high entry barriers or need to collaborate with regime actors in order to access complementary assets relevant for developing, scaling up and commercializing 'green' niche-innovations. Regime actors hold many important cards in this collaboration game. If 'green' innovations cannibalize existing technologies and markets, regime actors may be reluctant to collaborate or they collaborate primarily for defensive reasons (either as a hedging strategy or to control the pace of technological change).

Parkhurst et al. (Chapter 15) clearly showed how the niche of intermodal travel depends on collaborations with actors from the train, bus and taxi regimes. The chapter also showed that these actors from subaltern regimes saw intermodal travel as an additional business to which they were limitedly committed. These incumbents therefore did not invest large resources in innovative projects and quickly abandoned them when results were disappointing.

Orsato et al. (Chapter 10) and Ehret and Dignum (Chapter 11) also found the involvement of incumbent car manufacturers in the niche of green propulsion technologies (BEV and FCV). Although these car companies are investing increasing amounts of money in green propulsion alternatives (up to hundreds of millions per firm per year), most of their investments, innovation efforts and patents are still focused on improving the ICE (which has considerable scope for environmental improvement). Although car companies have probably moved from purely defensive orientations toward early hedging strategies, they still face doubts about future consumer demand for more expensive green cars (beyond the 'deep green' consumers), government regulations and technical possibilities. So, although car manufacturers show some interest and activity with regard to green propulsion, they may not yet be fully committed.

In order to understand sustainability transitions in transport, it is thus insufficient (and misleading) to focus only on niche-innovations. While pioneers, entrepreneurs and start-ups were important for the early development of FCVs (e.g., Ballard) and BEVs (e.g., Aerovironment, Solectria, CALSTART, Th!nk), the upswing and acceleration of niche-innovations probably depends on the involvement of regime actors. With regard to the strategic reorientation of corporate regime actors, transition theory could benefit from including insights from the business and management literatures. If niche actors work on their own, their innovations are likely to remain small, because of limited resources and entry barriers. Strategic collaborations between new entrants and incumbents are therefore an important topic for transitions theory.

16.3.3. Transition Toward More Diverse Regimes with Multiple Dominant Designs

Transition theory assumes that transitions entail a shift from Regime A to Regime B, usually organized around a dominant technical design. Future transitions in transport may, however, also consist of a shift toward a more diverse regime with multiple dominant designs, for example, FCV for long distance travel and BEV for city travel (see Ehret & Dignum, Chapter 11). This implicitly assumes, however, that the selection environment becomes more differentiated into different markets or operating environments. This could happen if future regulations would, for instance, favor the driving of BEVs inside cities. Alternatively, differentiation could result from the differential impact of tough CO_2 regulations on small and large cars; whereas batteries might be feasible to power small cars, this may not be the case for large cars, which is a reason why German car manufacturers (who are strong in the large car segment) have become more interested in FCVs (Ehret & Dignum, Chapter 11). So, perhaps single dominant designs were characteristic of 20th century mass production (for often homogeneous markets), while more fragmented markets and multiple co-existing technologies could become more prevalent in the 21st century. Transition research could pay more attention to such transitions toward more diverse regimes.

16.3.4. The Stabilizing and Reinforcing Influence from Landscape Developments

With regard to landscape influences, applications of transition theory often focus on destabilizing or disruptive pressures on the regime (which then open up and create space for the breakthrough of niche-innovations). More attention should, however, be given to landscape influences that stabilize regimes and hinder transitions (as we did in Section 16.1.2.). Additionally, future research could give more attention to landscape trends that override or counteract sustainability developments, for example, geo-political tensions or rival societal aspirations (Cohen, 2010).

16.4. CONCLUDING REMARKS

The various chapters, and this concluding analysis, demonstrate the usefulness of socio-technical transition theory, and the multi-level perspective, for analyzing the possibilities, barriers and drivers of shifts toward sustainable transport. This analysis differs from the economic models, engineering approaches and psychological studies that pervade transport studies. Rather than focusing on a technology fix or behavior change, this book has proposed and applied a co-evolutionary perspective that addresses stability and change, subjective and material dimensions, agency and structures and complex dynamics such as hype cycles, tipping points, fit-stretch patterns and transition pathways.

Nevertheless, transitions remain messy processes. Whereas most books seek to provide a story line about what will happen or what is needed for dealing with car-based transport problems, we have offered an analysis that brings out the vastness of factors involved in various processes of change and looks at various interaction effects. The book, we feel, has gone further than other texts in revealing the contradictions of actor strategies and developments in relation to automobility; it has also uncovered actor-related patterns in various processes of change, revealing the 'method in the madness'.

In order to best reveal the messiness of the world, as well as the 'method' therein, the cooperation between transition scholars and transport experts was essential, and we are pleased to have been able to bring the two communities together in one book project. Furthermore, we believe that we have achieved the goal of writing a book where diversity has been productive in helping us better understand the topic of stability and change in automobility. Because every chapter contributed to this topic, the book is more than a collection of disparate chapters. For transition studies this is an important book because it challenges some theoretical assumptions and offers detailed empirical analysis on transport-related issues, including culture and governance, with a level of empirical sophistication that no transition researcher would be capable of achieving on his or her own. For transport experts the book is valuable because it takes an unusually broad perspective, with special attention to (actor-based) dynamics, without succumbing to a deterministic view or wishful thinking.

We hope to have provided original insights, as well as to have generated an interest in transition processes and transition research. We hope also that other people will engage with the transition topic by building on the book's findings and so address some of the various research puzzles.

NOTES

1. The complexity of different logics also works against this. Commuters want few stops, but few stops mean fewer travelers can board a bus or train.
2. In 2003 the European Union announced the *European Hydrogen and Fuel Cells Partnership*, and in the same year America started the FreedomCAR and Vehicle Technologies program, backed up by $1.2 billion investments in hydrogen and fuel cell research. In 2010, the US Department of Energy's budget for fuel cell technology was cut by 60%. This suggests a moving away from funding hydrogen fuel cells to technologies with more immediate promise.

3. Biofuels made up about 1.4% of European Union-wide fuel in 2005 (http://www.bioenergywiki.net/European_Union).
4. Orsato et al. (Chapter 10) present a figure of national sales targets for BEVs and plug-in HEVs. Figure 10.3 predicts that in 2020 cumulative sales for 14 countries will be only 4 million. Likewise, Wells et al. (Chapter 6) comment on German government plans, which in 2009 "promised €500 million for electric vehicle infrastructures and technology development with the intention of making the country the biggest 'electro-mobility' market in the world. Interestingly this would result in an electric vehicle fleet of just 5 million units by 2030, compared with about 50 million petrol and diesel cars on the road in 2009."
5. Zijlstra and Avelino (Chapter 8) argue that the "frame of conventional system builders is influenced by four central principles: (a) efficiency, (b) calculability, (c) predictability and (d) control by non-human techniques."
6. In terms of vehicle kilometers per gross domestic product, the decline appears to be even more substantial (Chapter 7).

REFERENCES

Castells, M. (1996–1998) *The Information Age: Economy, Society and Culture* (3 vol.), Cambridge, MA: Blackwell.
Cohen, M. (2010) 'Destination unknown: Pursuing sustainable mobility in the face of rival societal aspirations', *Research Policy*, 39(4): 459–470.
Geels, F.W. (2007a) 'Analysing the breakthrough of rock 'n' roll (1930–1970): Multi-regime interaction and reconfiguration in the multi-level perspective', *Technological Forecasting and Social Change*, 74(8): 1411–1431.
Geels, F.W. (2007b) 'Transformations of large technical systems: A multi-level analysis of the Dutch highway system (1950–2000)', *Science Technology & Human Values*, 32(2): 123–149.
Geels, F.W. & Kemp, R. (2007) 'Dynamics in socio-technical systems: Typology of change processes and contrasting case studies', *Technology in Society*, 29(4): 441–455.
Giddens, A. (2009) *The Politics of Climate Change*, Cambridge, England: Polity.
Hoogma, R., Kemp, R., Schot, J. & Truffer, B. (2002) *Experimenting for Sustainable Transport Futures. The Approach of Strategic Niche Management*, London: Spon Press.
Kagan, R. (2008) *The Return of History and the End of Dreams*, New York: Knopf.
Kemp, R., Rotmans, J. & Loorbach, D. (2007) 'Assessing the Dutch energy transition policy: How does it deal with dilemmas of managing transitions?' *Journal of Environmental Policy and Planning*, 9(3–4): 315–331.
King, J. (2007) *The King Review of Low-Carbon Cars*, London: H.M. Treasury.
Konrad, K., Truffer, B. & Voss, J. (2008) 'Multi-regime dynamics in the analysis of sectoral transformation potentials: Evidence from German utility sectors', *Journal of Cleaner Production*, 16(11): 1190–1202.
Nieuwenhuis, P. & Wells, P. (2003) *The Automotive Industry and the Environment: A Technical, Business and Social Future*, Cambridge, England: Woodhead.
Orsato, R.J. & Wells, P. (2007) 'U-turn: The rise and demise of the automobile industry', *Journal of Cleaner Production*, 15(11–12): 994–1006.
Raven, R.P.J.M. (2007) 'Co-evolution of waste and electricity regimes: Multi-regime dynamics in the Netherlands (1969–2003)', *Energy Policy*, 35(4): 2197–2208.

Raven, R.P.J.M. & Verbong, G.P.J. (2007) 'Multi-regime interactions in the Dutch energy sector. The case of combined heat and power in the Netherlands 1970–2000', *Technology Analysis and Strategic Management*, 19(4): 491–507.

Rifkin, J. (2002) *The Hydrogen Economy: The Creation of the Worldwide Energy Web and the Redistribution of Power on Earth*, New York: Tarcher/Putnam.

Romm, J.J. (2005) *The Hype About Hydrogen: Fact and Fiction in the Race to Save the Climate*, Washington, DC: Island Press.

Rothaermel, F.T. (2001) 'Complementary assets, strategic alliances, and the incumbent's advantage: An empirical study of industry and firm effects in the biopharmaceutical industry', *Research Policy*, 30(8): 1235–1251.

Shove E. (2010) 'Beyond the ABC: Climate change policy and theories of social change', *Environment and Planning A*, 42(6): 1273–1285.

Sperling, D. & Gordon, D. (2009) *Two Billion Cars. Driving Toward Sustainability*, Oxford, England: Oxford University Press.

Teece, D. (1986) 'Profiting from technological innovation: Implications for integration, collaboration, licensing and public policy', *Research Policy*, 15(6): 285–305.

Contributors

Flor Avelino has worked at the Dutch Research Institute for Transitions (DRIFT) since 2005 as an action researcher and advisor regarding the management of sustainability transitions. She is currently finalizing a PhD thesis on the role of (social) power and empowerment in sustainability transitions, focusing on the interface between transport and land-use planning. Meanwhile she is also setting up a new research project on transitions to sustainable communities and regions, focusing on bottom-up innovation, social movements and social entrepreneurs. Flor's background is in political science and international relations (Master Leyden University).

Frank Boons is associate professor at the Department of Public Administration of Erasmus University, Rotterdam and Director of the international off-campus PhD program on cleaner production, cleaner products, industrial ecology and sustainability. His research deals with the governance of material and energy flows, which includes studies of industrial ecology and sustainability of production and consumption systems. His work has been published in journals such as *Ecological Economics*, *Journal of Cleaner Production*, *Technological Forecasting and Social Change*, and the *Journal of Industrial Ecology*. His latest book, entitled *Creating Ecological Value*, presents an evolutionary approach to environmental business strategies."

Kiron Chatterjee is a Senior Lecturer in Transport Planning at the University of the West of England, Bristol, where he is course leader for the MSc in Transport Planning. He has 20 years experience in transport research with this including studies in road safety, traffic management, travel behavior, transport modeling and appraisal and transport policy. A major focus of Kiron's current research is on the study of travel behavior change using longitudinal data. This involves studying the effects on travel behavior of social change (e.g., changing work practices) and transport initiatives (e.g., bus rapid transit). It also involves methodological development of panel data travel choice models. He is currently technical advisor for the design of social research in a 4-year national evaluation study of cycling investment. In 2003, he completed a project for the Department for Transport which explored alternative future scenarios for transport in Great Britain in 2030.

Marloes Dignum is a PhD candidate at the School of Innovation Sciences of Eindhoven University of Technology. Her research concerns the impact of large technological visions by focusing on hydrogen energy. The analyses provide insight in the adoption of a vision into a (policy) discourse and combine this insight with actions taken to realize the vision. Prior Marloes obtained an MSc in Technology and Society at Eindhoven University of Technology, which included a guest position at the University of California, Berkeley. She also conducted research at the Energy Research Centre of The Netherlands (ECN). Her research interests include sustainability transitions, futures studies and governance.

Marc Dijk is a research fellow at Maastricht University (International Center for Integrated Assessment and Sustainable Development, ICIS). He was educated in aerospace engineering at Delft University of Technology for which he wrote his master's thesis at the product development department of Jaguar Cars in Coventry, England where he compared the product development organizations of Jaguar and Volvo Cars (Sweden). As a PhD student he developed a model for analyzing paths of innovation in car mobility, within a micro-macro co-evolutionary framework, which incorporates feedback loops, actor frames and competition between technologies. He has published applications of the framework for the cases of hybrid-electric engines on the automobile market after 1990 and for the case of park and rides in Europe.

Iain Docherty is Professor of Public Policy and Governance at the University of Glasgow. Iain's research and teaching address the interconnecting issues of public management, institutional change and regional competitiveness. He has authored and edited five books on transport issues: *Making Tracks: The Politics of Local Rail Transport* (Ashgate, 1999); *A New Deal for Transport? The UK's Struggle With the Sustainable Transport Agenda* (Blackwell, 2003); *Diverging Mobilities? Devolution, Transport and Policy Innovation* (Elsevier, 2008); *Transport Geographies: Mobilities, Flows and Spaces* (Blackwell, 2008); and *Traffic Jam: Ten Years of 'Sustainable' Transport in the UK* (Policy Press, 2008). Iain has held a number of advisory roles including Non-Executive Director of Transport Scotland and is a member of the UK Commission for Integrated Transport's Expert Academic Panel.

Geoff Dudley is a Research Fellow in the Centre for Transport & Society (CTS) at University of the West of England, Bristol. He is a political scientist by discipline and has worked at the Universities of Oxford, Essex, Warwick, Staffordshire and Strathclyde. He has a particular interest in change process dynamics and has worked on a number of Economic and Social Research Council projects with these themes. This work includes (with Jeremy Richardson) Why Does Policy Change? Lessons From British Transport Policy 1945–99 (Routledge, 2000). He has published widely in leading public policy journals, including Political Studies, Public Administration and the Journal of European Public Policy (of which he is a member of the editorial board). While at CTS, Geoff has worked on a number of projects for the UK Department for Transport, including evidence-based reviews of both public and business attitudes on transport. He has also written the official history of the Government Car Service.

Oliver Ehret is the Programme Manager for Hydrogen Infrastructure at the National Organization Hydrogen and Fuel Cell Technology (NOW), the central organization supporting hydrogen and fuel cell technologies in Germany. His tasks include assessing project proposals, supervising projects, project initiation and coordination and strategic work regarding agenda-setting for technology development. Prior to joining NOW in 2008, he worked as a researcher of innovation in aeronautics with Philip Cooke at Cardiff University, United Kingdom. In 2005, Oliver obtained a PhD from Cardiff University. His thesis developed an interdisciplinary social science analytical framework that explored the emergence and social control of technological innovation. This framework was then applied to the introduction of hydrogen vehicles in Germany. The Clean Energy Partnership discussed in the co-authored chapter in this book represented an important topic of his PhD research.

Frank W. Geels is professor at Science Policy Research Unit (SPRU), one of the founding institutes of innovation studies, at the University of Sussex. He is one of the world-leading scholars on socio-technical transitions, which investigates the relations between social and technical developments in large-scale systemic change. Frank is well known for his conceptual and empirical work on the multi-level perspective. Many of his case studies come from the transport domain, including transitions in aviation, shipping and car transport. His work is inter-disciplinary and mobilizes insights from science and technology studies, evolutionary economics, history of technology, (neo)institutional theory and sociology. He has published four books and more than 30 peer-reviewed articles, some of which are highly cited.

Phil Goodwin is Emeritus Professor of Transport Policy at University College London and the University of West England. He was formerly head of the transport research units at Oxford University and UCL, a member of the UK standing advisory committee on trunk road assessment, a non-executive director of the Port of Dover and a policy advisor to the UK Deputy Prime Minister, the EC Transport Commissioner, the European Union 'Civitas' group of cities implementing sustainable transport policies and various other international, national and local transport agencies. His published work comprises around 200 articles, chapters, reports and books, mostly on dynamic approaches to travel demand analysis and transport strategies aimed at reducing car dependence by means of pricing, 'soft' measures, priorities and investment.

Peter Harman is Technical Director of deltatheta UK Limited where he leads the development of engineering modeling and simulation software products and provides consultancy in the application of simulation in engineering development. Peter is a member of the not-for-profit Modelica Association, and the core philosophy of deltatheta is the support and promotion of the open-standard Modelica language for modeling dynamic engineering systems. From 2006 to 2009 Peter was responsible for the development of mathematical models of mechanical, electrical and fluid systems at Honda Racing F1. In this time he has revolutionized the methods used in development and application of such models, in particular quality control and integration with company software. This work was heavily used in the development of the 2009 car, which went on to dominate the 2009 F1 season. Prior to this Peter led a team of dynamics and simulation engineers at Ricardo UK Ltd providing engineering consultancy in the automotive and motorsport industries.

Reg Harman has since 2000 been an independent consultant in transport policy and practice and in urban and regional planning. His previous career included full time positions with various commercial, public sector and research organizations, latterly as Policies Director of the UK Chartered Institute of Transport. More recently he has also been a part time senior research fellow at University College London and contributed to research and teaching work at other universities. Reg's principal expertise and interest lie in public transport planning, especially for railways and light rail; developing urban and regional transport strategies; improving accessibility at city and local level; comparing transport and spatial planning practice across Europe; and professional education and training. In 1996 Reg served as an adviser on transport policy to the Hungarian government. He has been on the experts group for two European Union transport projects (LIBERTIN and TRANSFORUM) and on a number of British professional committees and working groups.

Juliet Jain is a Senior Research Fellow at the Centre for Transport and Society, University of the West of England (UWE), and her academic background is in geography and sociology. Her research interest lies primarily with the intersection of information and communication technologies and mobility, with a focus on how travel time is used and experienced by public transport passengers. She has also worked on the EPSRC/DoT/TSB funded 'Ideas in Transit' research and considered the social context in which *user innovation* might occur or be applied.

René Kemp is Professor of Innovation and Sustainable Development at International Centre for Integrated Assessment and Sustainable Development (ICIS) and professorial fellow at UNU-MERIT in Maastricht, the Netherlands. He is an expert on eco-innovation and sustainability transitions and has published in innovation journals, environmental economics journals and policy journals. He was research director at STEP in Oslo and visiting researcher at IPTS in Seville (Spain), Harvard University in Boston (United States), Foscari University in Venice (Italy), SPRU in Sussex (United Kingdom) and CIRUS in Duebendorf (Switzerland). He is member of the editorial board of the journal *Environmental Innovation and Societal Transitions*, editor of *Sustainability Sciences*, member of the scientific board of ARTEC, University of Bremen and advisory editor of Research Policy. He co-authored the book *Experimenting for Sustainable Transport* (SPON Press, 2002), and together with Jan Rotmans he developed the steering model of transition management, which is used by the Dutch government as a model for sustainable innovation policy.

Glenn Lyons is Professor of Transport and Society and founder of the Centre for Transport & Society (CTS) at the University of the West of England, Bristol (UWE). Glenn's aim, and that of CTS, has been to improve and promote understanding of the inherent links between lifestyles and personal travel in the context of continuing social and technological change. Glenn was formerly Director of the international Transport Visions Network and Chairman of the Transport Planning Society; he was Chairman of the UK's Universities Transport Study Group until 2010. In the field of traveler information Glenn acted for several years as an external advisor to the UK Department for Transport. He is a member of the international specialists network 'ICT—Mobilising Persons, Places and Spaces' and is a member of the US Transportation Research Board's Committee on Telecommunications and Travel Behavior.

Andrew May is a Research Fellow within the Design School at Loughborough University. His research interests include the user centred design and evaluation of information and communication technologies, with a particular focus on mobile technologies and workplace and transport sectors. He is currently working on the 'Ideas in Transit' project. This is funded by various UK funding bodies (EPSRC/DoT/TSB) and is investigating how sustainable travel and transport services can be informed by user driven innovation.

Val Mitchell is a Research Fellow within the Loughborough Design School at Loughborough University. She specializes in the development of User Centred Design methodologies for eliciting user requirements for future technologies and services, in particular understanding user needs and requirements for mobile communication and other interactive products. Her research interests focus on the use of information and communication technologies to promote sustainability both within the transport and domestic sectors. She is currently a researcher on the EPSRC/DoT/TSB funded Ideas in Transit project which is investigating how user driven innovation can be used to inform the design of sustainable travel and transport services.

Paul Nieuwenhuis joined the Centre for Automotive Industry Research (CAIR) at Cardiff University in 1991 and became one of its two directors in 2006. CAIR studies economic and strategic aspects of the world motor industry. He is a founder member of the ESRC Centre for Business Relationships, Accountability, Sustainability and Society (BRASS). His main interests are historic and environmental and publications have been in these areas, e.g. *The Green Car Guide* (1992), *The Death of Motoring?* (1997), *The Automotive Industry and the Environment* (2003), *The Business of Sustainable Mobility* (ed. 2006). He also contributed to the *Beaulieu Encyclopaedia of the Automobile* which won a Cugnot Award from the Society of Automotive Historians.

Renato J. Orsato is Professor at the São Paulo Business School of Getulio Vargas Foundation (EAESP-FGV) São Paulo, Brazil. During 2006–2010 he worked as a Senior Research Fellow at the INSEAD Social Innovation Centre in Fontainebleau (France). As a researcher and educator in the past 15 years Prof. Orsato taught at Masters, MBA and Executive Programs at Lund University (Sweden), INSEAD and HEC Paris (France), the University of Technology Sydney (Australia), and worked with universities, public organizations and private businesses in more than 20 countries. Dr. Orsato is the author of *Sustainability Strategies* (Palgrave Macmillan, INSEAD Business Press (2009), a runner up for the 2010 Best Book Award of the Organizations and the Natural Environment division of the Academy of Mangement. He has also written several book chapters and teaching cases, and published in academic journals such as *California Management Review, Organization Studies, Journal of Cleaner Production* and *Journal of Industrial Ecology*.

Graham Parkhurst is Professor of Sustainable Mobility and Director of the Centre for Transport & Society, University of the West of England, Bristol, UK. Graham has two decades of experience researching and teaching in relation to transport policy and practice, previously at the University of Oxford and University College London. The central narrative to Graham's research has been the interaction between individuals' evolutionary, social and psychological contexts on the one hand, and technological and policy innovation on the other. Within this theme Graham has examined the behavioural, network, socioeconomic and environmental outcomes of new transport systems, services, infrastructure and regulations, through evaluation studies, literature reviews and empirical investigations. These studies have drawn on theoretical perspectives from psychology, sociology, economics, geography, political science and planning.

Bonno Pel is a PhD candidate at the Governance of Complex Systems research group at the Department of Public Administration, Erasmus University of Rotterdam. He has a background in transportation planning and socio-political philosophy. His PhD thesis research concerns innovation attempts, (intersecting) translation sequences and system synchronization in the Dutch traffic management field. This research into 'system innovation in the making' is part of the research program of the 'Knowledge Network on System Innovations and Transitions' (KSI).

Jon Shaw is Professor of Transport Geography at the University of Plymouth. He has a particular interest in issues of transport governance and policy, and has authored and/or edited five books in this field. He is also co-editor of the major textbook *Transport Geographies: Mobilities, Flows and Spaces* (Blackwell, 2008). Jon is Associate Editor of the *Journal of Transport Geography* and a former Chair of the Royal Geographical Society's Transport Geography Research Group. He has held a number of advisory positions, including Specialist Adviser to the House of Commons' Transport Committee, and he is currently a member of the First Great Western Advisory Board. Jon is currently working on three new book projects, including an ambitious venture that brings together transport geographers and mobilities geographers in an effort to develop further linkages between their different approaches to travel and mobility research.

Mimi Sheller is Professor of Sociology and Director of the Center for Mobilities Research and Policy at Drexel University. She is also Senior Research Fellow and former co-Director of the Centre for Mobilities Research at Lancaster University (United Kingdom) and founding co-editor of the journal *Mobilities*. She is the author of the books *Consuming the Caribbean* (2003), *Democracy After Slavery* (2000) and forthcoming *Citizenship From Below* (Duke University Press). She is co-editor with John Urry of *Mobile Technologies of the City* (Routledge, 2006), *Tourism Mobilities* (Routledge, 2004) and a special issue of *Environment and Planning A* on 'Materialities and Mobilities'. She has held recent Visiting Fellowships in the Davis Center for Historical Studies at Princeton University (2008–2009); Media@McGill in Montreal (2009);

the Center for Mobility and Urban Studies at Aalborg University, Denmark (2009); and the Penn Humanities Forum at the University of Pennsylvania (2010–2011).

Henrietta Sherwin is a Research Fellow at the Centre for Transport and Society at the University of the West of England, Bristol where she is currently working on a 4-year national evaluation study of cycling. Her main interest is the integration of land use planning and transport but more recently she has been involved in work exploring the potential for travel behavior change toward cycling in combination with rail use. This led to two collaborations using action research: the first implementing a pilot of bike-sharing scheme in Bristol (Hourbike) and the second testing an intervention designed to change the behavior of those who drive to work to rail travel with walking or cycling access.

Geert Teisman is full professor at the department of Public Administration, Erasmus University Rotterdam and Chair of the Science Group Governance of Complex Systems (GOCS). The 15 senior researchers in GOCS have expertise in the fields of water, land-use, environmental affairs, building processes, mobility and infrastructure. The group is known for its research in complexity theory. Teisman published on public private relations, complex decision making and complexity management. A recent publication is *Managing Complex Governance Systems. Dynamics, Self-Organisation and Coevolution in Public Investments* (Routledge, 2009), edited with M.W. van Buuren and L. Gerrits. He was scientific director of a knowledge impulse program in the field of water governance and is co-director of an innovation program on area development sponsored by 40 public, private and knowledge organizations. He is member of the scientific advisory committee of the Netherlands Environmental Assessment Agency.

Wijnand Veeneman is senior advisor at inno-V Amsterdam and associate professor at Delft University of Technology. In addition, he a member of the Advisory Board of the Netherlands Institute of Government and program leader in the Next Generation Infrastructures research program. In 2009 he was chairing the local organizing committee of the Thredbo International conference on Competition and Ownership in Land Passenger Transport. His main research work is in two fields: public transport governance and managing large engineering projects. He has carried out a wide variety of projects on those topics as diverse as advising on light rail projects in Tel Aviv, knowledge transfer between projects at Shell and tendering and governance strategies at many different transport authorities.

Peter Wells is a Reader of Logistics and Operations Management, Cardiff Business School and a Director of the Centre for Automotive Industry

Research, which he joined in 1990. He has a wide knowledge of economics, strategy and business models in the automotive industry, particularly in an environmental context. Other research interests include the distribution, retail and marketing of cars and the history of car design. Dr Wells has also published papers on wealth and sustainability, celebrity, totalitarianism and local eco-industrialism among other diverse interests. In 2002 Dr Wells became a founder member of the ESRC-funded Centre for Business Relationships, Accountability, Sustainability and Society (BRASS) working on sustainable automobility. He is a Director at evfutures.com. His latest book is *The Automotive Industry in an Era of Eco-Austerity* (Edward Elgar, 2010).

Masaru Yarime is Associate Professor of the Graduate Program in Sustainability Science (GPSS) of the Graduate School of Frontier Sciences of the University of Tokyo. His research interests include corporate strategy, public policy and institutional design for sustainability innovation, university-industry collaboration and structural analysis of knowledge creation, diffusion and utilization. He serves as the Editor-in-Chief for the *Journal of Science Policy and Research Management* as well as an editor for the journals *Sustainability Science* and *Environmental Innovation and Societal Transitions*. He received a BS in Chemical Engineering from the University of Tokyo, MS in Chemical Engineering from the California Institute of Technology and PhD in the Economics and Policy Studies of Technological Change from the University of Maastricht, the Netherlands. Previously he worked as Senior Research Fellow at the National Institute of Science and Technology Policy (NISTEP) of Japan.

Toon Zijlstra studied Urban Design & Planning at the Technical University of Eindhoven. He finished his study in 2009. His master's thesis was about car-central thinking among planners and designers, resulting in a spatially defined form of automobile dependency. He currently works for the Dutch Ministry of Transport and Water management, serving as a policy advisor on bicycling policies. He also participates in several study groups in the field of mobility management.

Index

A
Abrantes, P. 108, 120
Acerete, B. 107, 119
actor network 23, 24, 182, 194, 244
Adams, J. 120, 252, 265
add-on 59, 61–62, 327, 329, 338–339
air quality 8, 10, 101, 231, 258, 331
Akrich, M. 251, 253, 265
Aldcroft, D. 87, 103
Aleklett, K. 127, 137
Allen, S. 116, 120
Allenby, B. 211, 227
Allsop, R. 33, 47
An, F. 127, 137
Anable, J. 158, 159
Andera, J. 131, 137
Andrews, D. 127, 137
Ansell, C. 184, 200
Arthur, W. 13, 27
automobile landscape 162–163, 168, 172–174
automobile regime 3, 19, 29–30, 42, 46, 55, 60, 84–102, 107–117, 125, 156, 160–175, 283, 308, 311–313, 335–336, 340, 346–357, 360, 364–365
Avelino, F. 22, 73, 76, 77, 160, 337, 340–341, 347, 366, 372, 375
Axhausen, K. 38, 47

B
Bach, B. 168–169, 175–176
Baeten, G. 160, 165–166, 175–176
Bailey, D. 130, 137
Baillie, B. 162, 169, 177
Bakker, P. 163, 176
Bakker, S. 209, 226
Balcombe, N. 307
Ball, M. 6, 7, 229, 247

Banister, D. 10, 27, 41, 47, 161, 166, 168–169, 175–176
Barbier, E. 110, 121
Barker, T. 87, 102
Bartolomeo, M. 133, 139
Bassett, D. 8, 27
Batchelor, R. 125, 137
battery electric vehicles 10, 12, 23, 24, 59, 65, 67, 69, 71, 124, 131, 205, 229, 231–233, 239, 241, 245, 246, 247, 249, 343–345, 352–353, 357–358, 368–370, 372
Bauknecht, D. 217, 226, 267
Baumol, W. 108, 110, 119
Benford, R. 182, 185, 200
Berggren, C. 218, 227
Berkhout, F. 49, 78
bicycle infrastructure 15, 191–192, 330, 359
Bijker, W. 251, 253, 266
Binsbergen, A. van 333
Bissell, D. 184, 200
Blauwhof, G. 253, 266
Blow, C. 313, 332
Böhm, S. 161, 162, 164, 165, 176
Bolden, T. 306
Bondi, L. 186, 200
Bonhoff, K. 230, 242, 248–249
Boons, F. 24, 250, 265, 267, 338, 342, 375
Borraz, O. 113, 119
Borroni-Bird, C. 11, 28, 136, 137, 138, 227
Bovy, P. 333
Bowler, P. 50, 77
Boyle, G. 228–229, 249
Brand, A. 160, 170, 176, 354
Breakthrough 67, 231, 247, 259, 261, 264–265, 338, 346, 370

386 Index

Bressers, N. 76–77
Britton, E. 69, 76
Broadbent, J. 184, 200
Brown, J. 162, 176, 283
Brugnach, M. 56, 77
Bruijn, H. 296, 306
Brundtland, G. 91, 102, 110
Buchan, K. 111, 119
Buchanan, C. 90, 102, 143
Buehler, R. 8, 27
Burnham, J. 317, 333
Burns, L. 11, 28, 136, 138, 227
bus network 114, 119, 301
buses 237–240, 244, 292, 298–306, 318–322
Butler, E. 108, 119
Buuren, A. van 254, 267, 381

C

Cairns, S. 151, 158–159
Callon, M. 251, 253–254, 265–266
Camera, C. 129, 139
Carrington, P. 201
Castells, M. 163–164, 176, 348, 372
Causality 16, 58, 365
Cavill, N. 99, 103
Chandler, A. 123, 137
Chapman, R. 231, 248
Charlesworth, G. 88, 102, 107, 119
Chase, P. 44, 47
Chatterjee, K. 8, 20, 83, 95, 103, 337, 348, 355, 356, 364, 366, 375
Christensen, C. 69, 76
Ciferri, L. 134, 137
Clark, A. 130, 137
Clearly, J. 108, 119, 121–122
Cochrane, A. 107, 119
co-evolution 16, 18–20, 45–46, 49–50, 64, 75, 90, 123, 250, 364–365, 371
Cohen, M. 8, 349, 354, 370, 372
Coles, A. 52, 56, 77
Collantes, G. 231, 248
Competition 16, 39, 56, 72, 105–108, 126, 164, 171, 208, 210, 214–216, 224, 229, 246, 290, 299, 317–318, 321, 344, 347, 352, 367
Congestion 9–14, 98, 102, 144–157, 282, 332
congestion charge 6, 11–13, 19, 21, 85, 95–98, 113–114, 116, 118–119, 180, 359
Conley, J. 28, 177

Connell, D. 136, 137
Conybaere, J. 176
Cope, A. 99, 103
Corn, J. 70, 76
Courtenay, V. 125, 137
Coutard, O. 167, 176
Cowan, R. 205, 226
Crouter, S. 8, 27
cultural landscape 185–186, 191, 193–194, 350
Currah, A. 272, 285
Currie, G. 167, 176

D

Dant, T. 197, 200
David, P. 10, 126, 137, 307
Davidson, J. 186, 200
Debande, O. 107, 119
DeCicco, J. 127, 137
Dennis, K. 11, 27, 160–161, 164, 167, 176, 182, 199–200
Destabilization 58
Diamond, J. 124, 137
Diffusion 19, 59, 64, 67–72, 181, 184, 211, 214, 216, 221, 251, 262, 279, 324, 341, 346, 350, 358, 361, 364, 368–369
Dignum, M. 23, 229, 338, 344, 353, 369–370, 376
Dijk, M. 23, 26, 205, 209, 226, 308, 326, 332, 338–339, 376
DiMaggio, P. 210, 227
Discontinuity 221
Docherty, I. 21, 103, 104, 115, 118–122, 148, 158, 337, 347, 352, 362–364, 376
Dodge, M. 264, 266
Dudley, G. 3, 8, 20, 83, 87–92, 96, 98, 102, 103, 335, 337, 348, 355–356, 364, 366, 376
Duncan, J. 125, 137
Dunford, M. 130, 138
Dupuy, G. 160–161, 163–164, 167, 175, 176–177
Dyerson, R. 54, 76, 123, 139, 218, 228
Dynamics 5, 10, 16–17, 20, 46, 56–57, 74, 83, 85, 101, 153, 158, 173, 182, 184–185, 197, 203, 240, 243, 251, 262, 265, 357, 364–365, 371
Dyos, H. 87, 103

E

Echols, A. 133, 139

Index 387

Eddington, R. 98, 103
Edensor, T. 186, 200
Egan, D. 71, 77
Ehret, O. 23, 229–230, 233, 234, 235, 237, 242, 248–249, 338, 344, 353, 369, 370, 377
Eiser, J. 38, 47
Electricity 11, 18, 49, 58, 61, 62–63, 65, 76, 193, 205, 217–218, 226, 236, 239, 357–358, 368
Eliasoph, N. 184, 200
Elzen, B. 49, 51, 76
emergence 16, 17, 72, 85, 90, 95, 101, 106, 110, 129, 164, 180, 183–184, 199, 231, 250, 270, 273, 299, 308, 312, 315, 329, 340, 361
Emirbayer, M. 182, 200
Enoch, M. 321–322, 332
environmental problems 11, 49–50, 134, 234, 345, 353
established regime 63, 87, 98, 224–225, 331, 355, 367
evolutionary change 4, 5, 255
evolutionary economics 253, 377
Ewing, P. 127, 137

F
Fabri, J. 234, 248
Farag, S. 281, 285
Farrington, F. 113, 120
Faulks, R. 287, 306–307
Filarski, R. 165–166, 168, 178
Finer, S. 88, 103
Fletcher, R. 61, 76
Flink, J. 125, 138
Flohr, S. 272, 285
Flotow, P. 229, 249
Fol, S. 167, 176
form and function 64, 221
Foster, M. 106, 108, 120
Frame 15, 19, 44, 49, 85, 90–91, 161, 172, 181–182, 184–187, 189, 230, 331, 361, 364, 368, 376
Franey, J. 134, 137
French, S. 3, 95, 194, 200, 202, 208, 214, 216, 217
Freudendal-Pedersen, M. 183, 185, 201
Friedman, L. 107, 200–201
fuel efficiency 14, 18, 106
fuel price 84, 95, 218, 346, 349, 360, 366
functional transition 152

functionality 56, 63–64, 65, 153, 209, 221–222, 328
Fung, F. 127, 137
Furness, Z. 183, 188, 201

G
Gakenheimer, R. 288, 294, 307
Garche, J. 230, 241, 249
Gärling, T. 179
Garud, R. 138, 253, 267
Gaunt, M. 116, 120
Geels, F. 3, 11, 16–17, 20, 27–28, 44–63, 67–72, 76–79, 85–86, 89, 103, 172, 177, 182, 185, 201, 221–222, 224, 227, 230, 249–250, 253, 265–267, 335, 342, 365, 368, 372, 377
Geerlings, H. 76
Gehl, J. 166–167, 177
Geitmann, S. 229, 249
Genus, A. 52, 56, 77
Gerrits, L. 254, 267, 381
Geurts, K. 170, 173, 177
Giddens, A. 54, 77, 111, 120, 366, 372
Gilbert, C. 327, 333
Gill, N. 111, 121
Gilroy, P. 192, 201
Glaeser, E. 106, 120
Glaister, S. 109, 120, 317, 333
Glanville, W. 143, 159
Gleeson, B. 165, 175, 177
Gleick, J. 163, 177
Glynn, M. 70, 78
Goeverden, C. 295, 307
Gomez-Ibanez, J. 107, 121
Goodwin, P. 21–22, 31, 41, 47–48, 92–93, 103, 107–108, 110, 120, 140, 144, 146–147, 152, 158–159, 280, 285, 337, 342, 353, 355–356, 366–367, 377
Gordon, D. 8, 28, 124, 349, 362, 373
Gorz, A. 166, 177
Grabher, G. 272, 285
Graedel, T. 211, 227
Graham, B. 108, 120
Graham, S. 194, 198, 201–202
Gray, D. 113, 115, 120
Green, C. 323, 333
Green, K. 49, 76
Greene, D. 110, 120
Greenfield, A. 201
Greer, S. 112, 120
Grin, J. 51–52, 77, 79
Guest, P. 161, 178

H
Haas, T. 170, 177
Hajer, M. 170, 177
Hall, D. 115, 119
Hall, P. 323, 333
Hallett, S. 92, 103, 146, 159
Hambleton, R. 116, 120
Hamer, M. 87, 103
Hamilton, K. 108, 119, 121–122, 162, 169, 177
Han, N. 129, 138
Hanley, N. 110, 120, 201
Hanna, J. 108, 119, 121, 122
Harman, R. 25, 269, 285–286, 291, 306–307, 338–340, 378
Harms, L. 160, 162–163, 177
Hartig, T. 50, 77
Harvey, D. 107, 120, 164, 177
Hazell, R. 112, 120
Hedeker, D. 201
Heinberg, R. 127, 138
Hekkert, M. 46–47
Heldmann, H. 106, 120
Hemmelskamp, J. 133, 139
Henderson, J. 11, 28, 163–164, 170, 177
Hendriks, C. 52, 77
Heye, D. 270, 285
Hibbs, J. 288, 307
Hickman, R. 41, 47
Hinterhuber, H. 129, 139
Hippel, E. von 270, 285
Hitchens, D. 133, 139
Hoed, R. van den 129, 138, 231, 249
Holley, D. 281, 285
Holz, G. 56, 77
Honnery, D. 133, 138
Hoogma, R. 17, 24, 28, 49, 53, 72, 77, 137–138, 207–208, 222, 227, 230–231, 242, 249, 272, 285, 365, 372
Horton, D. 183, 201
Huber, J. 50, 77, 206, 227
Hulten, S. 205, 226
Humphrey, J. 129, 138
hybrid-electric 12, 18, 62, 205, 208, 213, 221, 223, 345, 352
hybridization 59, 61–62, 198, 221
hydrogen 11–12, 14, 18, 23–25, 61, 131, 136, 208–209, 225, 229–247, 344–346, 349, 358, 362, 371
hype 5, 16, 20, 51, 60, 65–67, 150, 165, 205, 210, 212, 220, 246, 329, 344–345, 371

I
Ibert, O. 272, 285
Ichijo, K. 129, 138
Ilbery, B. 171, 179
Illich, I. 160, 163, 167, 171, 177
incremental innovation 224, 250–254, 262–264, 353
instability 20, 41, 83
institutions 13, 50, 55, 84, 104–105, 111–112, 118, 146, 206, 268
intermodal 5–7, 10, 14, 18, 20, 65, 255, 257, 260, 308–331, 336–339, 356–357, 359, 365, 369
internal combustion engine 3, 8, 12, 18, 21, 23, 59, 60, 125–126, 134, 208, 229, 287, 306, 343
invention 63, 125, 270–273, 279–280
Ison, S. 110, 120

J
Jacobs, J. 162, 166, 171, 177
Jacobsson, S. 46–47
Jain, S. 24, 184, 192, 201, 268, 281, 285, 338, 341, 378
Jalilian, H. 327, 333
Jeffery, C. 120
John, P. 52, 74, 91, 93–94, 113, 119, 199, 276, 381
Jones, C. 161, 176
Jones, D. 129, 139
Jones, P. 112, 121
Jordan, A. 10, 28
Judge, D. 121

K
Kable, G. 134, 138
Kagan, R. 349, 372
Kaiser, F. 50, 77
Karnoe, P. 138
Kaufman, V. 184, 201
Keating, M. 112, 121
Kee, B. van der 166, 178
Kemp, R 3, 10, 17, 20, 23, 26, 28, 49, 51–57, 63, 72–73, 76–78, 85, 123, 133, 137–139, 205, 207, 224–225, 227, 230, 249, 267, 272, 285, 308, 335, 338–339, 365, 372, 379
Kennedy, A. 99, 103
Kenny, F. 92, 103, 146, 159
Kenworthy, J. 11, 28, 160, 167, 169–170, 178
Kenyon, S. 37, 48
Kern, F. 52, 78

Index 389

King, J. 159, 353–354, 372
Kingdon, J. 86, 88, 96, 103, 264, 266
Kirkbride, A. 158
Kitchin, R. 264, 266
Kleiner, B. 129, 138
Klijn, E. 253, 266
Knowles, R. 108, 120
Koelemeijer, J. 295, 307
Kohlbacher, F. 129, 138
Köhler, J. 41, 48
Konrad, K. 368, 372
Koppenjan, J. 253, 266
Kornai, J. 75, 77
Koyabashi, S. 130, 137
Kramer, G. 51, 78
Kunstler, J. 166, 170, 177
Kuroki, Y. 208, 228

L
l'Hostis, A. 291, 307
Lanberg, B. 11, 28
Land, C. 161, 176, 315, 382
Landau, R. 78, 103
landscape development 17, 27, 67, 74, 250, 346, 348, 361, 370
landscape pressure 27, 60, 86, 174, 243, 245, 250, 346, 348, 351, 356
Latour, B. 251, 265
Laube, F. 160, 177
Law, J. 182, 201, 251, 253, 266, 289
Le Grand, J. 111, 120
Leadbeater, L. 271, 285
learning curves 51, 346
learning processes 18, 53, 54, 67, 72, 230, 245, 364
Lefevre, C. 107, 108, 121
Lente, H. van 226
Levy, D. 71, 77
Leydesdorff, L. 264, 266
Liberalization 163, 172, 258, 347, 364
Lichterman, P. 184, 200
Lie, M. 71, 77
Linkages 11, 55, 58, 63, 253, 357, 368
Litman, T. 160, 164–165, 177
Little, S. 38, 48
Lloyd, A. 234, 249
Loomes, G. 280, 285
Loorbach, D. 51–52, 72, 77, 133, 139, 365, 372
Louçã, F. 123, 138
Lounsbury, M. 70, 78
Luhmann, N. 251, 266
Lüthje, C. 271, 285

Lynas, M. 127, 138
Lyons, G. 20, 24, 29–30, 34, 37–38, 48, 95, 103, 167, 178, 255, 266, 268–269, 280–281, 285, 331, 338, 341, 379

M
Mackett, R. 168, 178, 307
Macki, P. 120
Mackinnon, D. 118, 120
MacNeill, S. 130, 137
Magnusson, T. 218, 227
Malone, L. 28, 78
Markandya, A. 110, 121
Markard, J. 52, 56, 78
market demand 15, 53, 150, 217, 338
market development 260, 263, 342
Marsden, G. 114, 121
Marsh, D. 102
Martens, P. 10, 28
Martin, S. 113, 120, 161, 166, 177
Marx, K. 164, 178
Matthews, M. 161, 178
May, A. 24, 77, 114, 121, 210, 233, 240, 248, 268, 275, 338, 341, 379
Maye, D. 171, 179
McCulloch, R. 144, 159
McGinity, B. 133, 139
Meadowcroft, J. 52, 78
Mechanisms 4–6, 11, 26, 40, 50–54, 58–59, 68, 73–74, 107, 158, 181, 184, 335, 351, 359
Melaina, M. 123, 138
Mendonça, S. 123, 138
Menerault, P. 291, 307
Merton, R. 44, 48
Metz, D. 161, 168, 178
Meyer, J. 107, 121
Mikler, J. 125, 138
Miller, D. 183, 201
Mingardo, G. 326, 333
Mische, A. 182, 184, 201
Mitchell, D. 178
Mitchell, M. 11, 28, 136, 138, 164, 166, 227
Mitchell, V. 24, 268, 338, 341, 380
mobility pattern 14, 152, 175, 205, 210, 288, 294, 303, 335, 365
Mol, A. 206, 227, 228
Mom, G. 165–166, 168, 178, 206, 227
Momentum 53, 57, 58, 143, 165, 205, 210–211, 216, 220, 265, 268, 339, 341–344, 356–358

Morgan, K. 112, 121
Moriarty, P. 133, 138
Morton, H. 32, 48
Motavalli, J. 132, 138
Moulton, S. 111, 121
Muller, L. 99, 103
Murkens, J. 112, 121
Mytelka, L. 228, 229, 249

N
Næss, A. 50, 78
natural gas 25, 216, 235, 236, 301, 340
negative externalities 49, 133, 351, 353–354, 361, 363, 367
nested hierarchy 52, 58, 254
Newman, P. 11, 28, 160, 167, 169–170, 178
Newson, C. 158–159
niche development 25, 52, 72, 85, 184, 192, 198, 206, 210, 230–231, 242–246, 268, 270, 276, 283, 336, 343
Nieuwenhuis, P. 21, 123, 125, 127, 136–139, 207, 227, 352, 358, 372
Nill, J. 18, 28
Noise 8, 10, 62, 70, 136, 167, 169–170, 331
Nooteboom, S. 73, 78

O
Oates, W. 110, 119
oil price 4–5, 23, 73, 106, 206, 247, 284, 341, 349–350, 360–361
Olleros, F. 70, 78
Oltra, V. 208, 227
Ong, P. 167, 178
Orsato, R. 21, 23, 123, 131, 139, 205–206, 210, 213, 218, 226–228, 338, 343, 352–353, 358, 363, 369, 372, 380
Owens, S. 140, 159

P
Paaswell, R. 189, 193–194, 201
Packer, J. 194–195, 201
Pahl-Wostl, C. 56, 77
Painter, J. 111, 121
Palmer, C. 125, 139
Papacostas, C. 287, 307
Paradigm 10, 33, 74, 93, 100, 125–129, 132, 136, 163, 168–169, 172, 174, 253, 341, 350, 353, 356

Parker, S. 133, 139
Parkhurst, G. 26, 98, 103, 308, 310, 321, 325, 329, 332–333, 338–339, 365, 369
Paterson, M. 161, 163, 164–167, 176, 178
path dependence 13, 50, 54, 126, 136, 200, 309, 351
Paulley, N. 295, 307
Peake, S. 110, 120
Pearce, D. 110, 121
Peck, J. 107, 121
Peeters, P. 295, 307
Pel, B. 24, 250–251, 256, 265, 267, 338, 342, 380
Peters, P. 160, 162, 164, 165, 166–167, 178
Petrick, I. 133, 139
Pickett, K. 165–166, 179
Pilkington, A. 54, 76, 123, 139, 218, 228
Pirie, M. 108, 119
policy innovation 112
Polley, D. 253, 267
Potter, S. 321, 332
Powell, W. 210, 227
Preston, J. 108, 121, 307
Prevedouros, P. 287, 307
Proctor, T. 128, 139
Profitability 131
public health 18, 83, 92, 187–188, 281, 284, 312, 341
public transport regime 291, 312, 313, 367
Pucher, J. 8, 27, 107–108, 121
Putnam, R. 28, 166, 178, 373

R
radical innovation 3, 16–17, 52–53, 67, 70, 175, 245, 250, 262, 264, 344, 353, 356
Rafferty, J. 48
rail infrastructure 113, 289, 296, 318
rail network 114, 116–117, 297, 301, 320
Ranney, M. 50, 77
Raven, R. 51, 67, 76, 79, 230, 249, 368, 372, 373
Rayner, S. 28, 78
regime change 127, 189, 190, 245, 270, 308
Reihe, B. 122
Rennings, K. 133, 139
research and development 52, 134, 199, 207, 209, 212–213, 216,

218, 232, 237, 240–241, 243, 270, 302, 343, 344, 346, 350, 365
Richardson, J. 85, 88, 90–91, 103, 377
Richmond, J. 297, 307
Richter, E. 201
Rietveld, P. 294–295, 307, 320, 332–333
Rifkin, J 11, 28, 344, 373
Riley, R. 126, 139
Rip, A. 17, 28, 52, 54, 56–57, 63, 78, 123, 138, 264, 267
Ritzer, G. 161–162, 178
road infrastructure 53, 58, 74, 106, 109, 112–113, 124, 146, 156, 364
road network 31, 93, 116–117, 140, 143, 145, 152, 161, 257–258, 323
Roberts, D. 113, 120
Roberts, J. 108, 119, 121–122
Robertson, A. 168, 178
Robinson, P. 120
Rodaway, P. 186, 201
Rodriguez-Pose, A. 111, 121
Rogers, E. 279, 285
Rogers, R. 163, 166, 178
Roland, G. 75, 78
Romm, J. 11, 28, 246, 249, 345, 373
Rosenberg, N. 63, 76, 78, 85, 103
Rothaermel, F. 369, 373
Rotmans, J. 51–52, 72, 77, 78, 250, 252, 267, 365, 372, 379
Routine 71, 183, 254, 270, 280, 310, 331, 335, 357, 360–361
Rye, T 116, 120
Ryley, T. 121

S
safety 8, 21, 22, 37, 58, 106, 126, 136, 162, 167, 176, 195, 215, 313, 351–354, 375
sailing ship effect 69, 221
Saint Jean, M. 208, 227
Sartorius, C. 28, 227
Sautter, A. 294, 307
Savage, C. 87, 102
Schaeffer, K. 110, 121
Schelling, T. 265, 267
Schot, J. 17, 28, 49, 50–51, 53, 60, 71–72, 76–78, 86, 103, 123, 137–138, 207, 227, 230, 249–250, 265–267, 272, 285, 365, 372

Schulz, J. 188, 201
Schumacher, E. 171, 178
Schumpeter, J. 105–106, 109, 121, 254
Sclar, E. 110, 121
Scott, J. 201
Seifert, T. 167, 178
selection environment 370
Senbergs, Z. 167, 176
Service, R. 232, 249
Shaoul, J. 107, 119, 121
Shaw, J. 21, 103, 104, 113, 115, 118, 120–122, 148, 158, 337, 347, 352, 362–364, 380
Sheller, M. 12–13, 22–23, 28, 33, 48, 71, 163, 178, 180–185, 193–194, 197–198, 201–202, 338, 340–341, 343, 347, 350–351, 358, 366, 368, 381
Sherwin, H. 26, 320, 329, 333, 339, 381
Shiftan, Y. 76
Shiroyama, H. 208, 228
Shove, E. 52, 78, 183, 202, 364, 373
Sigfússon, A. 231, 249
Slinn, M. 161–162, 178
Sloman, L. 99, 103, 151–152, 158–159
Smeed, R. 143, 149, 159
Smit, W. 11, 28, 44–45, 48
Smith, A. 49, 51–52, 75, 78–79, 175, 178, 253, 262, 267
Smith, M. 186, 190, 200, 321, 332
Snow, D. 182, 185, 200
social change 17, 42, 126, 156, 184, 280
social learning 364–365
social network 37–38, 42, 44, 53, 72, 230–231, 244–245, 277, 341
societal landscape 5, 96, 100, 102
socio-technical change 51, 186
socio-technical regime 17, 52, 54–56, 58, 126, 132–133, 351
Sørensen, K. 71, 77
Spaargaren, G. 206, 227–228
Spash, C. 110, 120
Sperling, D. 8, 28, 231, 248–249, 349, 362, 373
Stabilization 184, 253–254
Stadler, C. 129, 139
Stafford, A. 107, 119, 121
Stapleton, P. 107, 121
Stead, D. 76
Steg, L. 160, 178, 179
Steger, U. 229, 249
Sterman, J. 123, 139

Stern, N. 50, 78
Stevens, H. 317, 333
Stirling, A. 49, 78
Stoep, J. van der 166, 178
Stoker, G. 112, 121
Stokes, G. 92, 103, 146, 159, 310, 333
Stradling, S. 38, 48
strategic niche management 24, 72, 230, 265, 272
Struben, J. 123, 139
sub-regime 308–309, 312–314, 318, 339
subsidies 14, 52, 72, 207, 208, 212, 220, 261, 288, 310, 313, 317–318, 321, 327, 330, 339, 346, 348, 356, 364, 366
substitution 45, 60, 62, 86, 206, 250, 323
Suchman, L. 182, 202
Sugden, R. 280, 285
Sussman, J. 255, 267
Sutton, J. 107, 122
Sweeting, D. 116, 120
symbiotic relation 60
system innovation 12, 19, 174–175, 216, 250, 252–254, 260, 262–265, 358, 380

T
Taylor, B. 54, 167, 178
technological change 11, 14, 16, 18, 83, 206, 268, 339, 344, 360, 369
technological learning 235, 236, 272, 365
technological niche 89, 230, 245
technological regime 54–55
Teece, D. 70, 78, 369, 373
Teisman, G. 24, 250, 251–252, 254, 267, 338, 342, 381
Terry, F. 307
Thompson, D. 8, 27
Thomson, J. 109, 122
Thrift, N. 185, 194, 196, 197, 202
Tickell, A. 107, 121
Tiessen, J. 18, 28
Tigar McLaren, A. 28, 177
Tilly, C. 182, 202
Torrance, H. 111, 122
Trajectory 21, 23, 45, 50, 54, 67–68, 86, 125, 133, 136, 193, 206, 220, 223, 225–226, 231, 244, 253, 255–258, 262, 343
tram infrastructure 297, 300

trams 58–59, 157, 296–297, 305, 312, 327
transformation path 14, 60, 86, 270, 281, 357
transition path 14, 59–61, 86, 119, 197, 250–251, 270, 353, 356, 371
transition pattern 20, 51, 59, 329
Travers, T. 317, 333
Trench, A. 120
Tripp, A. 142, 143, 159
Tripsas, M. 70, 78
Truffer, B. 17, 28, 52–53, 56, 77–78, 137–138, 227, 230, 249, 285, 365, 368, 372
Tyme, J. 91, 103

U
Unruh, G. 13, 28, 51, 79
urban planning 22, 39, 101, 109, 165, 168, 170–171, 190, 294, 340
urban transport 23, 58, 190, 193, 288, 314, 324, 356, 359
Urry, J. 11, 27, 33, 48, 74, 126, 139, 160–165, 167, 176, 178–179, 182, 184, 193, 197, 199, 200–202, 252, 264, 267, 381

V
Van Bree, B. 51, 78
Van Oosten, F. 61, 79
Vannini, P. 183, 184, 200–202
Veeneman, W. 25, 286, 288–289, 296, 306–307, 331, 338–339, 382
Vellema, S. 133, 139
Ven, A. van de 15, 251, 253, 267
Venkataraman, S. 253, 267
Verbeek, F. 106, 122
Verbong, G. 51, 67, 78–79, 368, 373
Vergragt, P. 137
Verhees, B. 70, 77, 182, 185, 201
Ververs, R. 168, 179
Vilhelmson, B. 175, 179
Voß, J. 52, 79, 267, 368, 372

W
Wajcman, J. 166–167, 179
Walker, G.5 2, 78
Walker, W. 13, 28, 51, 79
Wall, S. 110, 120
Ward, W. 221, 228
Wardman, M. 307, 310, 334
Wassonove, L. 131, 139, 226, 228
Watts, D. 171, 179
Webb, S. or B. 142, 159

Weber, R. 129, 139
Wee, B. van 170, 173, 177, 294, 307
Wegener, M. 110, 120
Weinstein, A. 162, 179
Wells, P. 21, 123, 125, 129, 131, 136–139, 207, 213, 226–228, 337, 352–353, 356, 358, 362–363, 372, 382
Werner, K. 167, 178
Westwood, S.119
White, P. 118, 122, 307, 327, 334
Wietschel, M. 229, 247
Wilkinson, R. 165–166, 179
Willeke, R. 106, 122
Williams, J. 119
Wise, C. 111, 121

Wolman, H. 121
Wolmar, C. 287, 307
Womack, J. 129, 139
Wood, D. 194, 202

Y
Yarime, M. 23, 205, 208–209, 216, 218, 226, 228, 338, 382

Z
Ziegelaar, A. 168, 179
Zijlstra, T. 22, 160, 176, 179, 337, 340–341, 347, 366, 372, 383
Zonneveld, W.170, 177
Zundel, S. 28, 227
Zwaneveld, P.1 63, 176